T0310300

IMMUNOASSAYS IN AGRICULTURAL BIOTECHNOLOGY

IMMUNOASSAYS IN AGRICULTURAL BIOTECHNOLOGY

Edited by

GUOMIN SHAN

A JOHN WILEY & SONS, INC., PUBLICATION

Published by John Wiley & Sons, Inc., Hoboken, New Jersey
Published simultaneously in Canada

For general information on our other products and services or for technical support, please contact our Customer Care Department within the United States at (800) 762-2974, outside the United States at (317) 572-3993 or fax (317) 572-4002.

Wiley also publishes its books in a variety of electronic formats. Some content that appears in print may not be available in electronic formats. For more information about Wiley products, visit our web site at www.wiley.com.

Library of Congress Cataloging-in-Publication Data is available.

ISBN: 978-0-470-28952-5
Printed in Singapore

oBook: 978-0-470-90993-5
eBook: 978-0-470-90992-8
ePub: 978-0-470-92268-2

10 9 8 7 6 5 4 3 2 1

CONTENTS

FOREWORD

The powerful analytical technique of immunoassay has experienced ever-expanding use in a variety of applications and settings since the pioneering work of Yallow and Berson a half century ago. The application of immunoassays to agricultural biotechnology is expanding and thus creates a great need for a wide variety of users to understand and implement immunoassays and related technologies.

This book describes the application of immunoassay technology in agricultural biotechnology product discovery, research and development, manufacturing, quality control, and national and international regulatory compliance. One unique feature of this book is its contributors, who are experienced practitioners from major agricultural biotech companies, immunoassay reagent manufacturers, academic research institutions, and government agencies. Covering a wide variety of topics, the book offers practical guidance for assay/kit development, assay validation, data analysis and implementation as well as discussions of new technologies and perspectives. It is an annotated window into previous literature as well as a text providing both theory and practical instructions. This book will find widespread utility among all those involved in this exciting new field.

BRUCE HAMMOCK

Distinguished Professor of Entomology &
Cancer Research Center
University of California, Davis

PREFACE

Since the first genetically engineered (GE or GM) crops were commercialized in 1996, adoption of GM crops has become one of the most exciting movements in agricultural history. In the past 14 years, the annual growth rate of GM-crop acreage worldwide has been steady at ~20%, and this trend is expected to continue in the foreseeable future. In 2009, more than 134 million hectares of GM crops were planted in 25 countries worldwide (Source: ISAAA Briefs 2010, http://www.isaaa.org/). The novel genes and proteins in GM plants need to be monitored and tracked in every phase of product development and in the supply chain. From early discovery to product development, farmer cropping, food/feed processing, grain import and export, environmental monitoring, and risk assessment, a rapid and reliable qualitative or quantitative detection method is required. Immunochemistry technology, which serves as a unique detecting tool, has played a very important role in this revolutionary movement.

Since the introduction of GM crops, immunoassay has become so popular in agriculture that even farmers are performing assays in the field to monitor specific traits. Each year hundreds of new assays are developed, and millions of test kits are consumed by university and industrial research laboratories, breeders, farmers, grain handlers, food processors, contract service laboratories, and government agencies. A very broad range of professionals are using immunoassay technology daily in GM related areas, and many are completely new to this technology. The purpose of this book is to help assay users by providing technical and practical guidance, and describing limitations and pitfalls of applying immunoassay in agricultural biotechnology.

I take this opportunity to express my gratitude and sincerity to all the authors for working with me over the past two years. I also would like to thank Laura Tagliani, Kathryn Clayton, Yelena Dudin, and Brian Skoczenski for taking the time and effort to review chapters and provide valuable comments. Finally, I thank the staff at John Wiley & Sons, especially Jonathan Rose and Lisa Van Horn, and the project manager at Thomson Digital, in particular Sanchari Sil, for their efficient coordination during planning, review and production phases of the book publication.

GUOMIN SHAN

Indianapolis, Indiana
October 2010

CONTRIBUTORS

Clara Alarcon, Pioneer Hi-Bred International, Inc., Johnston, Iowa

Mehdi Arbabi-Ghahroudi, Institute for Biological Sciences, National Research Council of Canada, Ottawa, Ontario, Canada

Murali Bandla, APHIS, United States Department of Agriculture, Washington, District of Columbia

Michael C. Brown, Strategic Diagnostics Inc., Newark, Delaware

Gina M. Clapper, American Oil Chemists' Society (AOCS), Urbana, Illinois

Thomas Currier, Bayer CropScience LP, Research Triangle Park, North Carolina

Giorgio De Guzman, School of Biological Sciences, Monash University, Melbourne, Victoria, Australia

Patrick Doyle, Department of Environmental Biology, University of Guelph, Guelph, Ontario, Canada

Ai-Guo Gao, Monsanto Company, St. Louis, Missouri

Gregory Gilles, Dow AgroSciences LLC, Indianapolis, Indiana

G. David Grothaus, Monsanto Company, St. Louis, Missouri

J. Christopher Hall, Department of Environmental Biology, University of Guelph, Guelph, Ontario, Canada

Leslie A. Harrison, Monsanto Company, St. Louis, Missouri

Rod A. Herman, Dow AgroSciences LLC, Indianapolis, Indiana

G. Ronald Jenkins, GIPSA, United States Department of Agriculture, Kansas City, Missouri

Lulu Kurman, Solae LLC, St. Louis, Missouri

John Lawry, Dow AgroSciences LLC, Indianapolis, Indiana

Zi Lucy Liu, Monsanto Company, St. Louis, Missouri

Beryl Packer, Monsanto Company, St. Louis, Missouri

Thomas Patterson, Dow AgroSciences LLC, Indianapolis, Indiana

Jean Schmidt, Pioneer Hi-Bred International, Inc., Johnston, Iowa

Tandace A. Scholdberg, GIPSA, United States Department of Agriculture, Kansas City, Missouri

Claudia Sheedy, Agriculture and Agri-Food Canada, Lethbridge, Alberta, Canada

Guomin Shan, Dow AgroSciences LLC, Indianapolis, Indiana

Robert P. Shepherd, School of Biological Sciences, Monash University, Melbourne, Victoria, Australia

Ray Shillito, Bayer CropScience LP, Research Triangle Park, North Carolina

Andre Silvanovich, Monsanto Company, St. Louis, Missouri

Rick Thompson, American Bionostica, Inc., Swedesboro, New Jersey

Amanda M. Walmsley, School of Biological Sciences, Monash University, Melbourne, Victoria, Australia

Michele Yarnall, Syngenta Biotechnology, Inc., Research Triangle Park, North Carolina

Kerrm Y. Yau, Dow AgroSciences LLC, Indianapolis, Indiana

CHAPTER 1

INTRODUCTION

Guomin Shan

The concept of immunoassay was first revealed in 1945 when Landsteiner found that an antibody could selectively bind to a small molecule when conjugated to a larger carrier molecule (Landsteiner, 1945). However, the first immunoassay was not reported until the late 1950s when Yalow and Berson developed the technique of radioimmunoassay while studying insulin metabolism. They used ^{131}I-labeled insulin to monitor insulin levels in humans (Yalow and Berson, 1959, 1960). This pioneering work set the stage for the rapid advancement and wide application of immuno-chemical methods in medicine, agriculture, and the environment. To recognize their contribution, Rosalyn Yalow received the Nobel Prize in Medicine and Physiology in 1977.

For the first decade after Yalow's discovery, immunoassay was primarily confined to the clinical area. The first application of this technology in agriculture was reported in 1970 when Centeno and Johnson developed antibodies that specifically bound to the insecticides DDT and malathion (Centeno and Johnson, 1970). A few years later, radioimmunoassays were developed for the pesticides aldrin, dieldrin, and parathion (Langone and Van Vunakis, 1975; Ercegovich et al., 1981). In 1972, Engvall and Perlman published on the use of enzymes as labels for immunoassay and introduced the term enzyme-linked immunosorbent assay or ELISA (Engvall and Perlman, 1972). In 1980, Hammock and Mumma 1980 described the potential use of enzyme immunoassay (EIA) for agrochemicals and organic environmental pollutants. Since then, the application of immunoassay to agriculture has increased exponentially (Vanderlan et al., 1988; Kaufman and Clower, 1991; Van Emon et al., 1989; Van Emon and Lopez-Avila, 1992). In addition to applications for pesticide analysis, this technology has been widely used for food toxins such as mycotoxins in grain (Casale et al., 1988) and feed (Yu et al., 1999), gibberellin hormones in plant (Yang et al., 1993), plant growth regulators (Weiler, 1984), and pathogens (Webster et al., 2004), as well as for the determination of the species of origin of milk (Hurley et al., 2004) and meats (Mandokhot and Kotwal, 1997).

Since the introduction of genetically engineered (GE) or genetically modified (GM) crops in mid 1990s, immunoassay has become the leading choice as an analytical method for detection and quantification of GM proteins in every phase of trait discovery research, product development, seed production, and commercialization. After GM crop commercialization, this technology plays an important role in

Immunoassays in Agricultural Biotechnology, edited by Guomin Shan
Copyright © 2011 John Wiley & Sons, Inc.

post-launch activities such as product stewardship, regulatory compliance, food chain, risk assessment, environmental monitoring, and international trade (Grothaus et al., 2006). Scientists and researchers previously focused on the application of conventional analytical procedures are now involved in immunoassay development and applications. In particular, many professionals, including government agencies, university researchers, and farmers, who are now using immunoassays, are new to this technology and the advantages and limitations of its application. To successfully develop a quality immunoassay and validate it for plant matrices requires a wide range of skills and knowledge. These include an understanding of the biochemical properties of the analyte in the plant matrix, protein chemistry, purification techniques for immunogen production, and enzyme conjugation for the production of labels. In addition, knowledge of general biochemical techniques, statistical principles for the development and optimization of the assay system, data interpretation, and basic skills of troubleshooting is required. To enable professionals in immunoassay development, and to assist them in mastering the use of this technology in the agricultural biotechnology field, it is important to possess these skills in the necessary depth.

Several immunoassay books or handbooks have been published in the past two decades (Tijssen, 1985; Paraf and Peltre, 1991; Law, 1996; Van Emon, 2006); however, these books either are specifically focused on clinical diagnostics or are targeted to immunoassays for small molecules and applications in environmental pollution, animal husbandry, and food contaminants. Numerous journal articles are published each year specific to immunoassays in the agricultural biotechnology field, including assay development, validation, application, and review papers. The need for a discipline-specific, comprehensive review with educational purposes, and offering practical guidance and broad technical coverage, has not been met. The purpose of this book is to provide assistance to a wide range of readers with varying technical backgrounds. It is aimed to help those who are new to the technology to find their way through the mass of publications and literature. It should also help those who are already working in the field to build a greater sense of confidence.

In this book, we focus on three key areas of immunoassay in agricultural biotechnology. The first five chapters thoroughly cover the development and validation of immunoassays. Antibodies are the critical reagent of any immunoassay and mainly govern assay sensitivity and selectivity. Polyclonal and monoclonal antibodies are the most commonly used antibodies, and there are numerous publications available that describe the principles and practical processes for antibody production and characterization. Due to the recent advances in biotechnology, the growing need of new approaches to rapidly produce immunoreactive reagents for academic and industrial research, and the increasing awareness of animal use, recombinant antibody or antibody engineering has emerged as one of the most studied fields in the past decade, and recombinant antibodies may become the dominant source of future immunochemical reagents. Chapter 3 provides a thorough overview of established and emerging technologies available for the *de novo* generation of antigen-specific recombinant antibody fragments. Microtiter plate-based ELISA and lateral flow device (LFD) are the two most common assay formats used in the agricultural biotechnology field. The basic principles and detailed procedures regarding ELISA kit and LFD development are discussed in Chapters 4 and 5, providing readers with practical guidance in all aspects of assay development including assay design, reagent selection and screening, assay optimization, and troubleshooting. To demonstrate the quality and reliability of a resulting assay in target matrices, a thorough validation is required (Grothaus et al., 2006). Chapter 6 describes method validation criteria and process steps for both qualitative and quantitative immunoassays in plant matrices.

The second focus of this book is the application of validated immunoassay in GM crop research, product development, and commercialization. Antibody-based immunochemical methods are critical tools for gene discovery, plant transformation, event selection, introgression, trait characterization, and seed production. They enable researchers to characterize the protein of interest in the target plant and to qualitatively and quantitatively detect the expression of GM protein in plant tissues (Chapter 10). During product development, a large number of samples are usually generated and require timely analysis. High-throughput assay systems have been widely adopted in many agricultural biotechnology research organizations. Chapter 8 describes immunoassay automation and its application to plant sample analysis. After GM crop commercialization, immunoassay continues to play an important role in grain production and in the food industry. Due to the loss of immunoreactivity of GM proteins during processing, antibody-based immunochemical methods normally are not suitable for processed food. Therefore, they are primarily used for nonprocessing food materials. One common application is in the identity preservation of GM or non-GM products in the supply chain and quality management systems (Chapter 11). In international trade, lateral flow immunoassay is actively used as a detection method for meeting required grain threshold testing requirements, as well as to detect the low-level presence of seeds or plants in a sample. In addition, Chapter 15 contains a review of the latest developments in global harmonization of the use of immunoassay along seed, grain, feed, and food chains.

Since the commercial release of GM crops, transgenic proteins (especially *Bacillus thuringiensis* (Bt) proteins) have entered the environment through continuous cropping, which has drawn attention for risk assessment and monitoring. Due to the low extractability of protein from soil matrix, it has been extremely difficult to quantitatively monitor protein residue in soil. In Chapter 12, we review the recent development of soil extraction systems and provide practical guidance for the use of immunoassays for protein detection in soil. For a quantitative immunoassay, data analysis and result interpretation are always the key, but often overlooked. Various assay characteristics can influence the interpretation and potential sources of error for a particular assay. Plant tissues and soil are tough matrices to deal with in sample analysis, where the analyst constantly faces challenges including matrix effect, extraction efficiency, and dilution parallelism. Moreover, properly modeled standard curves are important for accurately estimating the concentration of proteins in samples evaluated in an immunoassay (Herman et al., 2008). In Chapter 9, we analyze the factors that need to be considered in data interpretation and provide detail practical guidance for assay troubleshooting.

Finally, other than for crops, GM plant systems have been extended to animal vaccine development and biopharma research. Again, immunoassay is a primary analytical tool for the detection of recombinant proteins or antibodies in these areas. Chapters 13 and 14 provide an overview of immunoassay applications in both animal health and biopharma.

REFERENCES

Casale, W. L., Pestka, J. J., and Hart, L. P. Enzyme-linked immunosorbent-assay employing monoclonal-antibody specific for deoxynivalenol (vomitoxin) and several analogs. *J. Agric. Food Chem.* **1988**, *36*, 663–668.

Centeno, E. R. and Johnson, W. J. Antibodies to two common pesticides, DDT and malathion. *Int. Arch. Allergy Appl. Immunol.*, **1970**, *37*, 1–13.

Engvall, E. and Perlman, P. Enzyme-linked immunosorbent assay, ELISA. *J. Immunol.*, **1972**, *109*, 129–135.

Ercegovich, C. D., Vallejo, R. P., Gettig, R. R., Woods, L., Bogus, E. R., and Mumma, R. O. Development of a radioimmunoassay for parathion. *J. Agric. Food Chem.*, **1981**, *29*, 559.

Grothaus, G. D., Bandla, M., Currier, T., Giroux, R., Jenkins, G. R., Lipp, M., Shan, G., Stave, J. W., and Pantella, V. Immunoassay as an analytical tool in agricultural biotechnology. *J. AOAC Int.*, **2006**, *89*, 913–928.

Hammock, B. D. and Mumma, R. O. Potential of immunochemical technology for pesticide analysis. In Harvey, J. J. and Zweig, G. (Eds.), *Pesticide Analytical Methodology*, American Chemical Society, Washington, DC, 1980, pp. 321–352.

Herman, R. A., Scherer, P. N., and Shan, G. Evaluation of logistic and polynomial models for fitting sandwich-ELISA calibration curves. *J. Immunol. Methods*, **2008**, *339*, 245–258.

Hurley, I. P., Coleman, R. C., Ireland, H. E., and Williams, J. H. H. Measurement of bovine IgG by indirect competitive ELISA as a means of detecting milk adulteration. *J. Dairy Science*, **2004**, *87*, 215–221.

Kaufman, B. M. and Clower, M. Immunoassay of pesticides. *J. Assoc. Off. Anal. Chem.*, **1991**, *74*, 239–247.

Landsteiner, K. *The Specificity of Serological Reactions*, Harvard University Press, Cambridge, MA, 1945.

Langone, J. J. and Van Vunakis, H. Radioimmunoassay for dieldrin and aldrin. *Res. Commun. Chem. Pathol. Pharmacol.*, **1975**, *10*, 163.

Law, B. *Immunoassay: A Practical Guide*, Taylor & Francis, London, UK, 1996.

Mandokhot, U. V. and Kotwal, S. K. Enzyme-linked immunosorbent assays in detection of species origin of meats: a critical appraisal. *J. Food Sci. Technol.*, **1997**, *34*, 369–380.

Paraf, A. and Peltre, G. *Immunoassays in Food and Agriculture*, Kluwer Academic Publishers, Dordrecht, 1991.

Tijssen, P. *Practice and Theory of Enzyme Immunoassays*, Elsevier, Amsterdam, 1985.

Vanderlan, M., Watkins, B. E., and Stanker, L. Environmental monitoring by immunoassay. *Environ. Sci. Technol.*, **1988**, *22*, 247–254.

Van Emon, J. M. *Immunoassay and Other Bioanalytical Techniques*, CRC Press, Boca Raton, FL, 2006.

Van Emon, J. M. and Lopez-Avila, V. Immunochemical methods for environmental analysis. *Anal. Chem.*, **1992**, *64*, 79A–88A.

Van Emon, J. M., Seiber, J. N., and Hammock, B. D. Immunoassay techniques for pesticide analysis. In Sherma, J. (Ed.), *Analytical Methods for Pesticides and Plant Growth Regulators. Volume XVII. Advanced Analytical Techniques*, Academic Press, San Diego, CA, 1989.

Voller, A., Bidwell, D. E., and Bartlett, A. Enzyme immunoassays in diagnostic medicine: theory and practice. *Bull. World Health Organ.*, **1976**, *53*, 55.

Webster, C. G., Wylie, S. J., and Jones, M. G. K. Diagnosis of plant viral pathogens. *Curr. Sci.*, **2004**, *86*, 1604–1607.

Weiler, E. W. Immunoassay of plant growth regulators. *Annu. Rev. Plant Physiol.*, **1984**, *35*, 85–95.

Yalow, R. S. and Berson, S. A. Assay of plasma insulin in human subjects by immunological methods. *Nature*, **1959**, *184*, 1648–1649.

Yalow, R. S. and Berson, S. A. Immunoassay of endogenous plasma insulin in man. *J. Clin. Invest.*, **1960**, *39*, 1157–1175.

Yang, Y. Y., Yamaguchi, I., Murofushi, N., and Takahashi, N. Anti-GA$_3$-Me Antiserum with High Specificity toward the 13-Hydroxyl Group of C$_{19}$-Gibberellins. *Biosci. Biotechnol. Biochem.*, **1993**, *57*, 1016–1017.

Yu, W., Yu, F.-Y., Undersander, D. J., and Chu, F. S. Immunoassays of selected mycotoxins in hay, silage and mixed feed. *Food Agric. Immunol.*, **1999**, *11*, 307–319.

CHAPTER 2

PRINCIPLES OF IMMUNOASSAYS

Claudia Sheedy
Kerrm Y. Yau

2.1 INTRODUCTION

Immunological assays are based on the use of antibodies. Antibodies can be produced against several types of antigens, depending on the needs of the study. Several reviews and books have been written on the principles of immunoassays (IA), and several excellent review articles on the use of immunological and molecular techniques for the detection of genetically engineered organisms have already been published (Ahmed, 2002; Grothaus et al., 2006; Stave, 2002; Van Duijn et al., 2002). This chapter provides a brief overview of the immunological and chemical principles involved in the development of such immunoassays.

Initially, antibodies used in immunoassays were polyclonal in nature. Hybridoma technology was a major improvement in immunochemistry, as it allowed the production of antibody-producing immortalized cell lines expressing a single type of antibody, that is, monoclonal antibodies (Köhler and Milstein, 1975). The advent of

Immunoassays in Agricultural Biotechnology, edited by Guomin Shan
Copyright © 2011 John Wiley & Sons, Inc.

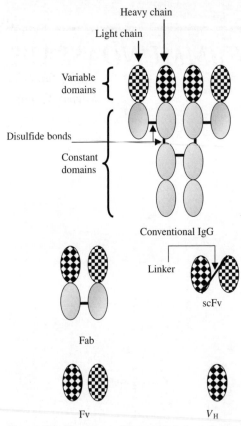

Figure 2.1 Diagrammatic representation of a complete immunoglobulin, formed by two heavy chains and two light chains connected by disulfide bonds, and antibody fragments thereof: antibody fragment (Fab), single-chain variable fragment (scFv), variable fragment (Fv), and variable fragment of the heavy chain (V_H).

recombinant antibody and phage display technologies (Smith, 1985) on the other hand, where antibody fragments could be displayed on the surface of filamentous phage, led to the production of antibodies and fragments more suitable for biomedical research and therapy (Little et al., 2000). The development of phage display as a selection strategy for antibody fragments has, therefore, been a major breakthrough in antibody engineering and molecular biology (Ghahroudi et al., 1997). Instead of relying solely on intact immunoglobulins, their antigen binding fragments (Fabs), variable fragments (Fvs), single-chain Fvs (scFvs), and single-domain antibodies (sdAbs) (Figure 2.1) could be cloned and selected with little or no loss in antigen binding affinity (Davies and Riechmann, 1996; Little et al., 2000) and efficiently expressed in *Escherichia coli* (Kipriyanov et al., 1997).

Immunoglobulins and fragments have tremendous potential for use in a variety of research, diagnostic, and therapeutic applications (Hayden et al., 1997). Recombinant antibodies for most biotechnological applications can be easily obtained from several library formats (phage display, ribosome display, and yeast and bacterial display). Moreover, several different strategies to improve antibody properties such as affinity and specificity are available. Techniques for the production of antibodies and fragments, and their engineering, have constantly evolved over the past several decades (Churchill et al., 2002). More detailed information on antibody engineering and its application in agricultural biotechnology is provided in Chapter 3.

For most applications, antibodies with high affinity are desirable (Mian et al., 1991) since high affinity often corresponds to an increase in binding properties, resulting in increased chances of successful treatment or diagnosis. Although antibodies isolated from immune libraries often possess high affinity, this is not always the case for those isolated from naïve and synthetic libraries, especially with hapten-specific antibodies (Charlton et al., 2001). In addition, some antibody fragments have higher cross-reactivity than those of their parental immunoglobulins (Charlton et al., 2001). This has been observed not only with protein antigens, but also with small molecules. Several approaches have therefore been proposed to further increase the affinity of recombinant antibody fragments. These approaches consist of optimizing panning procedures, along with further engineering of the antibodies through mutagenesis and gene shuffling.

Library panning, the process through which antibodies with required properties are isolated from libraries, can be optimized by including stringent selection conditions, antibody elution with the free hapten or free hapten analogues, and subtractive panning. Several strategies can be used to "affinity mature" the antibodies derived from the original libraries, namely, random mutagenesis, site-directed mutagenesis, and antibody shuffling. Random mutagenesis consists of randomly mutating the antibody gene, whereas site-directed mutagenesis generally "directs" or assigns mutations to certain positions along the antibody gene sequence. Alternatively, when multiple clones of recombinant antibodies specific to the same target antigen are available, through homologous recombination, different antigen binding site sequences from different antibodies can be "shuffled" on the DNA level to create novel sequences that may have improved binding affinity and specificity. Following maturation, desired antibodies can be selected using biopanning followed by characterization using surface plasmon resonance (SPR), titration calorimetry, immunoassays, or other means. In this chapter, we will discuss the development of immunoassays using conventional antibody molecule, that is, immunoglobulins. Antibody engineering will be covered in more detail in Chapter 3.

2.2 ANTIGENS

Antigens are molecules that can provoke the onset of a specific immune response during animal immunization. We can therefore refer to these molecules as "antigenic." Following immunization with an antigen, antibodies produced by B cells can specifically recognize this antigen. The antibodies can therefore be used to develop a specific immunoassay for detection and quantification of the antigen. Several different types of antigens (proteins, peptides, haptens, etc.) have been targeted for immunoassay development. However, genetically engineered organisms (GE or GM)-specific immunoassays have been developed mainly against protein antigens.

Proteins are probably some of the easiest antigens to work with for immunoassay development, as they are large molecules easily recognized by the animal's immune system. Haptens and peptides being much smaller in size then tend to elicit a lower or less specific immune response, if at all recognized by the immune system. Peptides and haptens are therefore often conjugated to larger carrier proteins to allow a stronger immune response. Antigens used for animal immunization and GM traits detection and quantification are reviewed below.

2.2.1 Plant-Made Proteins

Currently, a few proteins that are being expressed as a result of the introduction of transgenes in plants for herbicide tolerance or insect resistance from proteins of *Bacillus thuringiensis* (Bt) have been extracted from plant materials and used for immunoassay development. Plant-expressed proteins tend to be extracted from plant tissues at low concentrations and low purity. It is therefore easier to express the same protein or a close relative into microorganisms such as *E. coli*, after which standardized and highly efficient purification protocols can be used to purify the target protein in sufficient amounts to conduct protein characterization and provide materials for subsequent immunizations. As plant- and bacteria-expressed proteins may have different posttranslation modifications such as methylation and glyco-sylation, there may be lower specificity and/or sensitivity when immunoassays are developed using the bacterially expressed protein for plant material quantification and detection. These aspects of GM-specific immunoassays will be discussed in a later section.

2.2.2 Bacteria-Expressed Proteins

The vast majority of immunoassays developed against GM protein antigens have used bacterially produced and purified proteins for animal immunization and assay development. The bacterially expressed proteins chosen for assay development reflect the current market shares of the various GM crops: both Bt proteins and the CP4 EPSP synthase are the proteins most often targeted for immunoassay detection and quantification in Ag Biotech. Bt proteins for which assays have been developed include Cry1Ab, Cry1Ac (Paul et al., 2008; Fantozzi et al., 2007; Roda et al., 2006; Petit et al., 2005), and Cry9C (Quirasco et al., 2004) from *E. coli*, and Cry1Ac and Cry1F (Shan et al., 2007, 2008) from *Pseudomonas fluorescens*. Other proteins include PAT (phosphinothricin-*N*-acetyltransferase) (Xu et al., 2005) and CP4 EPSPS (Lipp et al., 2000) from *E. coli*.

2.2.3 Peptides

There is one published example of peptides being used for antibody production for a GM protein immunoassay (Van Duijn et al., 2002). Due to the limited availability of the targeted enzyme (EPSP synthase), the antibodies were produced against three peptides, each conjugated to the carrier protein bovine serum albumin. Following immunizations, monoclonal antibodies were obtained via hybridoma technology and purified by immunoaffinity chromatography prior to being assessed for immunoassay development and antigen quantification (Van Duijn et al., 2002).

2.2.4 DNA

It is possible to immunize animals with DNA. This option has been used to produce antibodies against various genes and gene products; however, DNA detection is mainly performed by real-time PCR in agricultural biotechnology and no work has been published yet on the use of this option for detecting transgenic traits. Interestingly, upon animal immunization with DNA and/or plasmids coding for a given protein, the immune system may recognize the protein coded for and the corresponding antiserum can be used to detect the protein via ELISA (Choudhury et al., 2008).

2.3 ANTIBODIES

Antibodies are Y-shaped proteins produced by the vertebrate immune system. They consist of four polypeptide (two heavy and two light) chains, stabilized through disulfide bonds. The antigen binding domains form the tips of the Y structure, whereas the effector (or constant) domains form the stem of the molecule. In animals, a first encounter with an antigen results in the production of low-affinity antibodies. Upon this primary exposure, cells producing higher affinity antibodies are selected and undergo a process called "affinity maturation," where the antibody genes from the initial repertoire are randomly mutated, thereby generating antibodies with varying and heightened affinity for the given antigen. The diversity thereby obtained is astronomical. Upon secondary exposure, cells displaying the highest affinity antibodies are rapidly induced and generate a highly specific response. Antibodies are a crucial component of the immunoassay. They determine to a large extent the sensitivity and specificity of the immunoassay for the targeted antigen.

2.3.1 Polyclonal Antibodies

Naturally, an animal's immune system produces several thousands of different antibodies at one given time. This diversity in antibody specificity allows the animal to respond rapidly to most threats such as disease causing organisms. Upon immunization with an antigen such as a recombinant protein, the animal's immune system detects the antigen and undergoes a selection process by which the animal's serum contains a heightened concentration of a subset of antigen-specific antibodies within days. Since several clones of cells produce these antibodies, the serum is a source of polyclonal antibodies, or antibodies issued or produced by several different clones of antibody-producing cells (B cells). Polyclonal antibodies are easy to produce and can lead to very sensitive immunoassays since various antibodies recognize and bind the antigen by different epitopes or structures present on the antigen's surface. However, since polyclonal antiserum consists of numerous antibodies with differing specificity (as opposed to monoclonal antibodies), it can bind to other antigens with the same or similar epitopes on their surfaces and may lead to cross-reactivities that may not be desirable for GM protein quantification. Despite this, many immunoassays developed for GM detection and quantification are polyclonal antibody based as the low cost and short time requirement outweigh the shortcomings.

2.3.2 Monoclonal Antibodies

Monoclonal antibodies, as opposed to polyclonal antibodies, are issued from one single clone of cells. To obtain monoclonal antibodies, individual B cells are fused to myeloma cells and isolated by serial dilution, resulting in a fusion product, or hybridoma cell line, each of which can produce one specific antibody for extended periods of time via tissue culture. Similarly, any recombinant antibody produced in bacteria, yeast, or plant can be thought of as monoclonal, since only one antibody originating from a single clone is expressed. Several monoclonal antibodies have been used for immunoassay development against GM proteins such as Cry1Ab (Ermolli et al., 2006; Fantozzi et al., 2007). Monoclonal antibodies are the most common antibody type used for detection of transgenic proteins from plants.

2.3.3 Recombinant Antibodies

Recombinant antibodies refer to any antibody expressed in an organism other than its original host. Antibodies expressed in bacteria, yeast, plants, and on phage are all recombinant antibodies: the genetic sequence coding for the antibody has been recombined with a genome other than its original one. Recombinant antibodies have several advantages compared to polyclonal and monoclonal antibodies: following cloning, they can be easily produced in bacteria in large quantities, which is ideal when large amounts of antibodies are required, such as for the development of a commercial assay. However, recombinant antibodies may not necessarily possess the same affinity as their parental, full-size counterparts. Recombinant antibodies often necessitate fine-tuning via affinity maturation and mutagenesis and may also require adjustments for high expression in the host. To date, a few recombinant antibodies have been used for GM protein detection. However, the versatility and unique characteristics of recombinant antibodies provide novel tools for research. For example, an scFv has been used to characterize the functional domain of the insecticidal protein Cry1Ab (Gomez et al., 2001).

2.4 ANTIBODY DEVELOPMENT AND PRODUCTION

Antibodies can be produced in several different ways, depending on the desired antibody format. Polyclonal antibodies can be produced only via animal immunization and subsequent harvest of serum. Monoclonal antibodies can be obtained either *in vitro* by tissue culture or *in vivo* by intraperitoneal injection into mice and collection of ascitic fluid produced. Recombinant antibodies can be produced in a variety of organisms, such as bacteria, yeast, and plants.

For polyclonal antibody production, typically a purified protein (e.g., Cry1Ab) is mixed with an adjuvant and saline buffer and injected into animals (mice, rabbits, etc.). Following several immunizations and regular monitoring of the immune response against the antigen, blood from the animal is retrieved and the antiserum containing the antibodies is collected for further analysis and assay development. As described previously, this antiserum contains antibodies issued from many different clones of antibody-producing cells and, therefore, is referred to as polyclonal antibodies.

Lipp et al. (2000) prepared their immunogen by mixing the purified protein (Cry1A) with complete Freund's adjuvant (CFA) for the initial injection and with Freund's incomplete adjuvant (FIA) for subsequent booster injections. Intramuscular or subcutaneous injections of 250–500 µg of purified protein with adjuvant were performed 21–28 days apart, with test bleeds 14 days after every injection. Similarly, Paul et al. (2008) injected 500 µg of purified protein in phosphate buffered saline (PBS) and emulsified with FIA. Injections were performed every 4 weeks, intracutaneously, and test bleeds 14 days after each injection for titer determination (Paul et al., 2008).

For monoclonal antibody, typically mice are immunized with the purified protein following the same immunization protocols as for polyclonal antibody production. However, mouse splenocytes are collected following the immunization process and fused with myeloma cells, thereby forming hybridomas. These hybridomas can then be screened, and the best clones cultured in standard tissue culture facilities. Ascitic fluid used to be the method of choice for monoclonal antibody production. The hybridoma cell line obtained for a given antibody was injected into the peritoneal cavity of mice, where it grew and simultaneously produced the antibodies. After a given length of time, ascitic fluid containing the antibodies was harvested from the peritoneal cavity. However, due to animal care issues associated with this technique, it is not as often used nowadays for standard monoclonal antibody production. Hybridoma technology via tissue culture now tends to be the main source of monoclonal antibodies, along with recombinant antibody technologies.

To develop CP4 EPSPS monoclonal antibodies, Lipp et al. (2000) immunized mice intraperitoneally with 100 µg of purified CP4 EPSPS protein every 14–21 days. The protein was emulsified in CFA for the first injection and in FIA for subsequent booster injections (Lipp et al., 2000). Mouse splenocytes were fused with a myeloma cell line using polyethylene glycol and supernatant from grown hybridomas was screened by immunoassay to determine which hybridomas displayed the greatest affinity for CP4 EPSPS (Lipp et al., 2000). The selected hybridoma was injected into mice intraperitoneally for ascitic fluid production (Lipp et al., 2000). Ascitic fluid was collected for 2 weeks beginning 3 weeks after the cell injection (Lipp et al., 2000). The monoclonal antibodies thereby produced were purified by affinity chromatography on protein A (Lipp et al , 2000).

Antibodies produced by either bacteria or hybridoma cultures can be grown in fermenters, which allow large batch cultures to be obtained and from which antibodies can be purified. With hybridoma cultures, fermenters are usually referred to as bioreactors, and containers of sizes from 500 mL to 10 L in which medium is passed allow hybridoma growth and antibody production. With bacterial cultures, larger fermenters can be used.

Following antibody production by either tissue culture or ascitic fluid, monoclonal and recombinant antibodies must be purified to obtain an effective reagent. Common purification methods include affinity chromatography using immunoaffinity or protein A/G columns. Immobilized metal affinity chromatography (IMAC) and size-exclusion chromatography are also used for recombinant antibody purification from bacterial production.

In immunoaffinity chromatography, antigens or antibodies specific for the desired antibodies are immobilized via conjugation to a solid support such as agarose beads and this mixture is placed in a column or a container. The desired antibodies are incubated with the solid support and bound to the antibodies or antigens on the solid support. Following washing steps to remove unbound materials, the desired antibody can be eluted with a buffer or a solvent. Similarly, antibodies can be purified by protein A and G columns, in which these proteins naturally bind antibodies via their effector end (C-terminal).

IMAC is often used to isolate and purify recombinant antibodies, which are generally labeled with a histidine tag (five or six histidine residues in frame with the antibody molecule) or a c-myc tag. Based on the ionic structure of the tag, antibodies will bind to the nickel-charged column and can be specifically eluted from the column following washing steps through the use of a reagent able to disrupt the nickel–histidine interaction. Finally, in size-exclusion chromatography, which is often used to further purify recombinant antibodies following IMAC, beads with specific pore sizes are used to separate molecules based on their molecular weight. Molecules that exceed the pore size and are thereby excluded from entering the beads elute earlier than smaller molecules that can enter the beads. This is an effective method when impurities are significantly smaller than the desired protein.

2.5 ANTIBODY–ANTIGEN INTERACTIONS

The understanding of antibody characteristics is important for effective antibody selection and assay design. In general, antibodies interact with antigens by several different means. Van der Waals forces, ionic and electrostatic interactions, and flexibility of the antibody binding sites to accommodate various antigens by molding themselves around those may all be involved in antigen binding.

The determination of antibody–antigen affinity constants and binding kinetics is an important part of antibody characterization (Neri et al., 1996; Nieba et al., 1996) and is pivotal in relating the structure of biological macromolecules to their function (Myszka, 1997; Paci et al., 2001). Several methods can be used to measure antibody affinity: equilibrium dialysis, band-shift assay, competitive ELISA, fluorescence quenching, titration calorimetry, and, more recently, real-time interaction analysis, also referred to as surface plasmon resonance. However, the measurement of weak affinity interactions, such as those between haptens and antibodies, has been hampered by the scarcity of analytical methods available (Ohlson et al., 1997).

Real-time interaction analysis of antigen–antibody interactions by optical biosensor technology has established itself as a powerful and general methodology for the determination of affinity constants (Karlsson, 1994). Since the first commercial biosensor was introduced in 1990, several types of optical biosensors have become available (Myszka, 1997). They have been used to characterize a wide variety of molecular interactions, including antibody–antigen, ligand–receptor, and protein–carbohydrate interactions (Myszka, 1997).

Surface plasmon resonance allows the real-time analysis of molecular interactions without labeling and consumes only small amounts of sample (Myszka, 1997). Moreover, reactions occurring simultaneously in different flow cells can be monitored so that several surfaces can be analyzed at the same time (Myszka, 1997). A reference flow cell is normally used to validate data for nonspecific interaction and refractive index changes, improving data quality (Myszka, 1997). However, although generating data from surface plasmon resonance is fairly easy, interpreting the data and the interaction kinetics accurately has proven to be more difficult (Myszka, 1997). The way in which a ligand, especially a hapten, is attached to the sensor surface is critical. Orienting the ligand through various functional groups or capturing the ligand using antibodies and fusion tags will all influence the resulting data (Myszka, 1997).

In a basic surface plasmon resonance experiment, one reactant (the ligand) is attached to the sensor surface. The other reactant (the analyte) flows over this surface in solution (Myszka, 1997). As the analyte binds to the ligand, the refractive index fluctuates and this change is measured (Neri et al., 1996). Figure 2.2 illustrates the binding of an antigen to a sensor chip coated with the antibody. Initially, buffer flows over the chip surface to establish a flat sensorgram baseline (Neri et al., 1996). The antigen solution is then injected and binding is detected as an increase in the resonance units (RUs) as a function of time (Neri et al., 1996). As the antigen injection stops, the complex is washed with buffer and the signal decays as the antigen and the antibody dissociate. Once injections of various concentrations of analyte are performed, their sensorgrams can be overlaid (Figure 2.3a) and data are analyzed (Figure 2.3b and c).

In surface plasmon resonance, the response detected is proportional to the mass of the analyte that binds to the surface (Karlsson, 1994). It therefore means that for haptens, which are very small molecules, surface plasmon resonance is challenging since the interaction of small molecules with immobilized ligands may not be detected (Karlsson, 1994). Biosensor technology keeps improving, however, and low molecular weight-specific biosensors have been developed as well.

2.5.1   ELISA

Immunoassays are ideal for qualitative and quantitative detection of proteins in complex matrices. Both monoclonal and polyclonal antibodies have been successfully used to develop commercial kits for qualitative and quantitative measurements. Depending on the specificity of the detection system, particular application, time allotted, and cost, assays are customarily designed and formatted to meet the

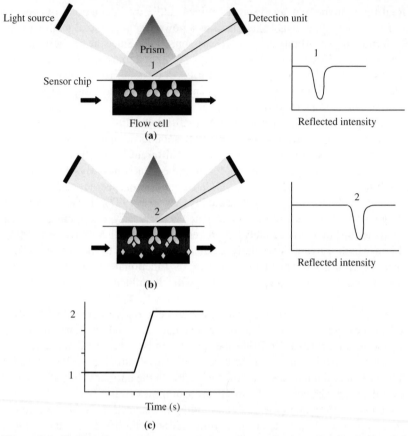

Figure 2.2 Surface plasmon resonance detects changes in the refractive index of the surface layer of a solution in contact with the sensor chip. (a) SPR is observed as a sharp dip in reflected intensity at an angle that depends on the refractive index of the medium on the nonilluminated side of the surface. (b) The SPR angle shifts when biomolecules bind to the surface and change the refractive index of the surface layer. (c) The sensorgram is a plot of the SPR angle against time and displays the progress on the interaction at the sensor surface. (Adapted from *BIA Applications Handbook*, version AB, 1998.)

particular needs (Ahmed, 2002). Microwell, lateral flow strips or devices, microtiter plates, and coated tubes are the most commonly used ELISA formats: they are quantitative, highly sensitive, economical, high throughput, and ideal for high-volume laboratory analysis. The lateral flow device (LFD) and antibody-coated tube format are also suitable for field testing.

Sandwich ELISA is the most commonly used format of IA for the quantification of protein in GM matrices (Stave, 2002; Grothaus et al., 2006). As shown in Figure 2.4, coating antibody specific to the protein of interest is first attached, through nonspecific hydrophobic interactions, to the surface of a support, for example, the well of a microtiter plate. Blocking agent is then used to occupy any open sites to reduce the

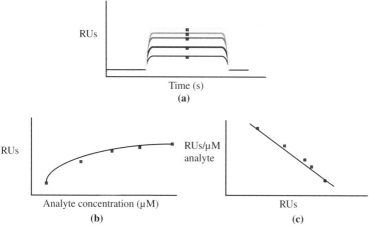

RUs

Time (s)

(a)

RUs

Analyte concentration (µM)

(b)

RUs/µM
analyte

RUs

(c)

Figure 2.3 Data collection and analysis. Sensorgrams obtained for each analyte concentration injected are overlaid (a). The average value in RUs obtained for each concentration is then plotted against the concentration (steady-state affinity plot, (b) or against the average RU value/concentration (Scatchard plot, c).

background noise. Frequently used blockers include bovine serum albumin, skim milk, gelatin, or many other commercially available formulations. Many off-the-shelf ELISA kits have precoated and preblocked plates that allow immediate sample analysis. With the coating antibody, target protein in the samples will be captured and in turn detected by another specific antibody, the detection antibody. Depending on the assay, the detection antibody may be itself conjugated to an enzyme, for example, horseradish peroxidase (HRP) for signal generation, or may be detected by a subsequent secondary antibody, for example, goat anti-rabbit IgG antibody/HRP

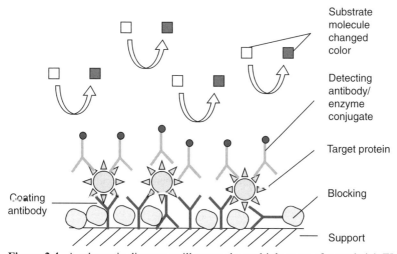

Substrate
molecule
changed
color

Detecting
antibody/
enzyme
conjugate

Target protein

Blocking

Coating
antibody

Support

Figure 2.4 A schematic diagram to illustrate the multiple steps of a sandwich ELISA.

conjugate. The signal is directly proportional to the number of the enzyme conjugated antibodies present; hence, the protein of interest can be quantified using a standard curve to calibrate the signal.

Several limitations are anticipated for the quantitative determination of genetically engineered traits with protein-based analytical methods in general. Since the expression of introduced traits is tissue specific and developmentally regulated, sample tissue types, methodology, and timing have to be well designed and monitored for an accurate report of the expression of the target protein in the crop tissue sample. Storage, handling, and processing of materials should also be taken into account to prevent the exposure of samples to thermal, mechanical, or chemical damage between the field and the laboratory (Miraglia et al., 2004). As an example, ELISA was used to detect the presence of Bt protein at 0.1–0.5% of total soluble protein but because precision of the ELISA in this range was very low, the results were inconsistent (Ma et al., 2005). ELISA is the method of choice to screen for the presence of a protein associated with a particular GM trait in raw materials, semiprocessed foods, and processed ingredients if the expressed protein is not degraded, and hence detectable. However, because ELISA is less sensitive than polymerase chain reaction (PCR) methods, it is less suitable for testing finished or complex food products containing many other ingredients and typically at lower concentrations. Although analytical tests for whole proteins are practical and effective, some GM products do not express a detectable level of protein (Ahmed, 1995). Quantitative determination of GM traits is important when mandatory labeling is required and/or special commodities must be GM free. Therefore, further easy-to-use, cheap, and fast methods for the quantitative detection of GM traits are required and current methodologies have to continue to evolve to better meet the needs of regulatory agencies and product stewardship.

ELISA development, validation, and improvement for different applications will each be discussed subsequently in this book. For example, immunoassay development and applications for trait discovery, grain products, or environmental monitoring will be separately discussed in Chapters 10–12. Here, we give some examples to demonstrate how the versatility of ELISA allows modifications to meet the needs of researchers in the lab or specialists in the field. ELISAs can be simplified to reduce the time required for analysis. For example, a rapid competitive assay for fumonisin B1 (FB1) needs only 20 min (Wang et al., 2006). Although FB1 is a small hapten, the techniques applied to the assay discussed in this particular paper are useful for GM ELISA development and optimization. In the FB1 ELISA, analytical standards or diluted corn leaf sample extracts and HRP–FB1 conjugate are premixed in a glass tube, and this mixture is added to FB1-specific antibody-coated microtiter wells and incubated for 10 min. After washing with buffer to remove any unbound FB1–HRP, a colorimetric substrate is added. The color development reaction is stopped after 10 min at room temperature and absorbance determined by spectrometry. In the case of this competitive binding ELISA, the absorbance is inversely proportional to the FB1 concentration of the sample. The limits of detection, defined as the FB1 concentration resulting in a 15% inhibition of color development (IC_{15}), for the conventional and rapid assays were 0.2 ± 0.1 and $0.5 \pm 0.2 \, \mu g/L$, respectively. However, the assay time was shortened from over 3 h to half an hour. The throughput of the assay is therefore increased dramatically with little influence on the sensitivity of the assay.

Recently, fluorometric detection systems have come to be regarded as a very sensitive detection system superior to both radioisotopic and enzyme labels in immunoassays (Velappan et al., 2008). However, conventional fluorometric assays are often limited by high background from nonspecific fluorescence in complex mixtures. A successful alternative has recently been realized in time-resolved fluorometry where the desired signal is distinguished from the interfering background by temporal resolution. Time resolution (measuring the fluorescence intensity after a certain delay time has elapsed following the excitation pulse) technology is a strategy to avoid interference caused by short-lived background. Lanthanide chelates europium(III) (Eu^{3+}), samarium(III) (Sm^{3+}), terbium(III) (Tb^{3+}), and dysprosium(III) (Dy^{3+}) are very effective labels in time-resolved fluorometry due to their long emission duration and reasonably high quantum yield (Hemmila and Harju, 1994). A dissociation-enhanced lanthanide fluoroimmunoassay (DELFIA) has recently been developed that is based on time-resolved fluorometry (http://www.wallac.com). DELFIA assays are based on two different lanthanide chelates: a nonfluorescent chelate optimized for labeling of the binding components and a highly fluorescent chelate created after the binding reactions during the fluorescence enhancement step (Hemmila and Harju, 1994). The enhancement step, similar to the conventional ELISA using secondary antibody, is the key to high sensitivity. Following binding of target antigen with its specific antibody, which is also the nonfluorescent chelate, a secondary antibody is introduced to bind to the primary antibody in an enhancement step, with the fluorescent chelate providing a time-resolved signal that avoids nonspecific interference from fluorescence from nontarget sources. In the assay reported here, the primary antibody was allowed to associate, in liquid phase, with purified antigen in cultured cell or tissue lysates. Antibody left unassociated with liquid phase antigen was then bound to purified antigen immobilized on 96-well plates, and the plate-bound antibody was detected using DELFIA technology. The extent of binding to the immobilized antigen was an inverse measure of the amount of the antigen in liquid phase. The assay procedure was comparable to the current ELISA detection method in specificity and sensitivity of such an antigen detection.

The major advantage of the DELFIA over the ELISA is its higher sensitivity and lower background. It has been shown that the DELFIA is more sensitive than conventional ELISA due to the use of europium chelates as labels (Hemmila and Harju, 1994). The fluorescence properties of the europium chelates, long Stokes shift, narrow-band emission, and exceptionally long fluorescence lifetimes make them ideal labels for time-resolved fluoroimmunoassays. Two different chelate systems are employed: the labeling chelate optimized for labeling purposes and the fluorescence enhancement system optimized for highly sensitive fluorometry. During the fluorescence enhancement step, the label ion, originally coupled to the binding reagent with labeling chelate, is released by lowering the pH (<3.5). Thereafter, the components of the enhancement solution form highly fluorescent chelates and simultaneously create an optimized micellar environment for fluorescence.

In addition to conventional solid-phase ELISAs, assays where paramagnetic particles are coated with a capture antibody and the reaction carried out in a test tube can be utilized to detect GM traits. The particles with antibody-bound reactants are separated from solution by a magnet and the unbound reactants are washed away. This semihomogeneous assay can provide superior sensitivity because magnetic particles provide a much larger surface area than that of a microtiter well for antibody coating. In addition, because the reactive surface is distributed throughout the reaction volume, kinetics are improved over the diffusion-dependent microwell format. Using this assay format, a collaborative validation study in Europe involving 38 laboratories in 13 EU member states and Switzerland was conducted to determine the accuracy and precision of ELISAs against CP4 EPSPS (protein providing glyphosate tolerance) in dried soybean flour (Lipp et al., 2000). Results showed consistent detection on the basis of dry weight for samples containing as low as 0.35% GM materials.

Luminescence oxygen channeling immunoassay (LOCI) is a homogeneous proximity-based assay (Ullman et al., 1994). When two biological molecules are drawn together and interact, that is, binding together, their small distance allows their conjugates to interact. Using this phenomenon, two types of neutral density beads, a donor and an acceptor, conjugated to the target molecules can produce a signal only when the target molecules are bound to each other providing a homogeneous assay. When interaction is established, a photosensitizer in the donor bead converts adjacent ambient oxygen to an excited form of singlet oxygen, which diffuses across to react with thioxene derivatives in the acceptor bead that ultimately emit light at 520–620 nm. The singlet oxygen has a half-life of approximately 4 μs and can diffuse approximately 200 nm from the donor bead during this period. If an acceptor bead is within this range, then light will be emitted. The sensitivity of the technology is enhanced by the anti-Stokes shift effect where the emitted light is of a lower wavelength than the absorbed light. This technology has recently been commercialized as amplified luminescent proximity homogeneous assay technology, hence coined the term Alpha technology, and it is ELISA AlphaLISA™.

2.5.2 Lateral Flow Strip Assays

In recent years, the development and application of LFDs or "dipstick" assays for the determination of GM proteins has gained importance because of the low cost and rapidity of these assays (Stave 2002; Grothaus et al., 2006). LFDs are based on immunochromatographic principles and combine several benefits, including a user-friendly format, short assay times, long-term stability over a wide range of temperatures, and cost-effectiveness. These characteristics render LFDs ideal for the screening of large numbers of samples without requiring highly trained personnel or specialized equipment and therefore are well suited for on-site use. More importantly, these LFDs have shown sensitivity and specificity comparable to those of conventional assays. In many cases, LFDs have been directly compared to other analytical methods to confirm their accuracy and sensitivity. For example, Garber et al. (2005) compared ELISA and LFD for the qualitative and quantitative detection of three different toxins (ricin, amanitin, and T-2 toxins) in beverages, produce, dairy, and baked goods. In all cases, the dipsticks successfully detected all three toxins at concentrations below those at which there could be a health concern. More importantly, no significant background due to the food matrices was observed. In a different study, 27 egg samples and 28 chicken muscle samples were analyzed for the presence of sulfadiazine, a sulfonamide antibiotic. The results showed that the differences in accuracy between test strips and high-performance liquid chromatography (HPLC) were from 0.8% to 11.2% for egg samples and from 2.2% to 34% for chicken muscle samples for quantitative detection (Wang et al., 2007). An AOAC Performance Tested Method study (Thompson and Lindhardt, 2006) evaluated Singlepath Salmonella™, an LFD assay for the presumptive qualitative detection of *Salmonella* spp. in food. Testing in dried skimmed milk, black pepper, dried pet food, desiccated coconut, cooked peeled frozen prawns, raw ground beef, and raw ground turkey showed no significant difference between Singlepath Salmonella™ and the ISO 6579:2002 reference method, a direct culturing method using various type of growth media. An LFD assay for the detection of human metapneumovirus (hMPV) was developed (Kikuta et al., 2008) using two mouse monoclonal antibodies to the nucleocapsid protein of hMPV. A study was done to compare the virus detection rate in nasopharyngeal secretions by the LFD assay to that by real-time reverse transcription PCR (RT-PCR). Nasopharyngeal swab samples collected from infected patients and nonsymptomatic volunteers were analyzed by both the LFD strips and real-time RT-PCR. Although the sensitivity of the LFD assay is less than that of real-time RT-PCR, the LFD assay is a rapid and useful test for the diagnosis of hMPV infections in children. Consensus is that LFD can function rapidly and accurately in complex matrices.

In Chapter 5, LFD technology and its development, validation, manufacturing, and troubleshooting will be discussed in detail; hence, here only basic principles of this technology are briefly reviewed. Depending on the nature of the analyte, for example, small hapten where sandwich assays are not possible (Kaur et al., 2008) or protein molecule (Ma et al., 2005), the LFD design will differ. Hapten-specific LFDs are based on the principle of competitive inhibition immunoassay, using an anti-analyte antibody immobilized onto a membrane and a gold-labeled hapten conjugate as the reporting agent. In a null sample, the hapten conjugate moves along the assay strip and binds to the immobilized antibody, resulting in a detectable band. In a positive sample, the free hapten in the sample competes with the hapten conjugate for the available yet limited antibody binding sites, therefore decreasing or completely inhibiting the signal. A test strip with no or low-intensity band indicates the presence of the target molecule in the sample.

For GM protein analysis, the LFD design is based on the sandwich ELISA format. Protein-specific antibodies were coupled to a signal label and incorporated into the testing strip (Ma et al., 2005). When the test strip is inserted into a sample containing the GM protein, the latter is bound by the incorporated antibodies. The testing strip contains two capture zones: one capturing the antibody-bound protein and the other capturing the labeled antibodies. These zones exhibit a visible color when the sandwich and/or unreacted chromophores are captured on the membrane. These LFDs are simple, rapid (1–2 min), and easy to use, but provide only qualitative results (presence or absence of GM trait).

2.6 CONCLUSION

Antibody-based detection technologies are well developed for diagnostic applications testing complex matrices with sufficient sensitivity to meet the requirements of researchers and regulatory agencies. Antibodies exhibit properties that provide specific and high-affinity recognition of target molecules. These are the attributes that detection technologies are built upon. From SPR that can be used to monitor and evaluate the kinetics of binding in real time, to ELISA that can be adapted to be high-throughput analysis, to LFD that can be used to quantify an analyte in 10 min, each technology has its own strengths and weaknesses and has found a niche in the diagnostic arena. Immunoassays provide a tool that can benefit researchers and field specialists, or when costs permit, be introduced into the developing countries. Currently, polyclonal and monoclonal antibodies dominate the reagents used for immunoassay development. In the future, *in vitro* production of synthetic binding partners might serve an important role by providing an economic and homogeneous source of analytical reagents. In Chapter 3, the technology for recombinant antibody production will be discussed. Tailor-made antibodies can be produced to serve purposes of the researchers and may provide better options and utilities that are not available to date.

REFERENCES

Ahmed F. E. Application of molecular biology to biomedicine and toxicology. *J. Environ. Sci. Health C*, **1995**, *11*, 1–51.

Ahmed F. E. Detection of genetically modified organisms in foods. *Trends Biotechnol.*, **2002**, *20*, 215–223.

Allen, R. C., Rogelj, S., Cordova, S. E., and Kieft, T. L. An immuno-PCR method for detecting *Bacillus thuringiensis* Cry1Ac toxin. *J. Immunol. Methods*, **2006**, *308*, 109–115.

Campbell, K., Fodey, T., Flint, J., Danks, C., Danaher, M., O'Keeffe, M., Kennedy, D. G., and Elliott, C. Development and validation of a lateral flow device for the detection of nicarbazin contamination in poultry feeds. *J. Agric. Food Chem.*, **2007**, *55*, 2497–2503.

Charlton, K., Harris, W. J., and Porter, A. J. The isolation of super-sensitive anti-hapten antibodies from combinatorial antibody libraries derived from sheep. *Biosens. Bioelectron.*, **2001**, *16*, 639–646.

Choudhury, S., Ganguly, A., Chakrabarti, K., Sharma, R. K., and Gupta, S. K. DNA vaccine encoding chimeric protein encompassing epitopes of human ZP3 and ZP4: immunogenicity and characterization of antibodies. *J. Reprod Immunol.*, **2008**, *79*(2):137–147.

Churchill, R. L. T., Sheedy, C., Yau, K. Y. F., and Hall, J. C. Evolution of antibodies for environmental monitoring: from mice to plants. *Anal. Chim. Acta*, **2002**, *468*, 185–197.

Davies, E. L., Smith, J. S., Birkett, C. R., Manser, J. M., Anderson-Dear, D. V., and Young, J. R. Selection of specific phage-display antibodies using libraries derived from chicken immunoglobulin genes. *J. Immunol. Methods*, **1995**, *186*, 125–135.

Davies, J., and Riechmann, L. Single antibody domains as small recognition units: design and in vitro antigen selection of camelized, human VH domains with improved protein stability. *Protein Eng.*, **1996**, *9*(6), 531–537.

Emslie, K. R., Whaites, L., Griffiths, K. R., and Murby, J. Sampling plan and test protocol for the semiquantitative detection of genetically modified canola (*Brassica napus*) seed in bulk canola sees. *J. Agric. Food Chem.*, **2007**, *55*, 4414–4421.

Ermolli, M., Prospero, A., Balla, B., Querci, M., Mazzeo, A., and Vend en Eede, G. Development of an innovative immunoassay for CP4EPSPS and Cry1AB genetically modified protein detection and quantification. *Food Addit. Contam.*, **2006**, *23*, 876–882.

Fantozzi, A., Ermolli, M., Marini, M., Scotti, D., Balla, B., Querci, M., Langrell, S. R. H., and Van den Eede, G. First application of a microsphere-based immunoassay to the detection of genetically modified organisms (GMOs): quantification of Cry1Ab protein in genetically modified maize. *J. Agric. Food Chem.*, **2007**, *55*(4), 1071–1076.

Garber, E. A., Eppley, R. M., Stack, M. E., McLaughlin, M. A., and Park, D. L. Feasibility of immunodiagnostic devices for the detection of ricin, amanitin, and T-2 toxin in food. *J. Food Prot.*, **2005**, *68*, 1294–1301.

Ghahroudi, M. A., Desmyter, A., Wyns, L., Hamers, R., and Muyldermans, S. Selection and identification of single domain antibody fragments from camel heavy-chain antibodies. *FEBS Lett.*, **1997**, *414*, 521–526.

Gomez, I., Oltean, D. I., Gills, S. S., Bravo, A., and Soberon, M. Mapping the epitope in cadherin-like receptors involved in *Bacillus thuringiensis* Cry1A toxin interaction using phage display. *J. Biol. Chem.*, **2001**, *276*, 28906–28912.

Grothaus, G. D., Bandla, M., Currier, T., Giroux, R., Jenkins, G. R., Lipp, M., Shan, G., Stave, J. W., and Pantella, V. Immunoassay as an analytical tool in agricultural biotechnology. *J. AOAC Int.*, **2006**, *89*, 913–928.

Hayden, M. S., Gilliland, L. K., and Ledbetter, J. A. Antibody engineering. *Curr. Opin. Immunol.*, **1997**, *9*, 201–212.

Hemmila, I. and Harju, R. Time-resolved fluorometry. In: Hemmila, I., Stahlberg, T.,and Mottram, P. (Eds.), *Bioanalytical Applications of Labelling Technologies*, Wallac Oy, Turku, 1994, pp. 83–120.

Karlsson, R. Real-time competitive kinetic analysis of interactions between low-molecular-weight ligands in solution and surface-immobilized receptors. *Anal. Biochem.*, **1994**, *221*, 142–151.

Kaur, J., Boro, R. C., Wangoo, N., Singh, K. R., and Suri, C. R. Direct hapten coated immunoassay format for the detection of atrazine and 2,4-dichlorophenoxyacetic acid herbicides. *Anal. Chim. Acta*, **2008**, *607*(1), 92–99.

Kikuta, H., Sakata, C., Gamo, R., Ishizaka, A., Koga, Y., Konno, M., Ogasawara, Y., Sawada, H., Taguchi, Y., Takahashi, Y., Yasuda, K., Ishiguro, N., Hayashi, A., Ishiko, H., and Kobayashi, K. Comparison of a lateral-flow immunochromatography assay with real-time reverse transcription-PCR for detection of human metapneumovirus. *J. Clin. Microbiol.*, **2008**, *46*, 928–932.

Kipriyanov, S. M., Moldenhauer, G., and Little, M. High level production of soluble single chain antibodies in small-scale *Escherichia coli* cultures. *J. Immunol. Methods*, **1997**, *200*, 69–77.

Köhler, G. and Milstein, C. Continuous cultures of fused cells secreting antibody of predefined specificity. *Biotechnology*, **1975**, *24*, 524–526.

Laffont, J. L., Remund, K. M., Wright, D., Simpson, R. D., and Gregoire, S. Testing for adventitious presence of transgenic material in conventional seed or grain lots using quantitative laboratory methods: statistical procedures and their implementation. *Seed Sci. Res.*, **2005**, *15*, 197–204.

Lipp, M., Anklam, E., Stave, J. W., Bahrs-Windsberger, C. I., Crespo, M. T. B., et al. Validation of an immunoassay for detection and quantitation of a genetically modified soybean in food and food fractions using reference materials: interlaboratory study. *J. AOAC Int.*, **2000**, *83*, 919–927.

Little, M., Kipriyanov, S. M., Le Gall, F., and Moldenhauser, G. Of mice and men: hybridoma and recombinant antibodies. *Immunol. Today*, **2000**, *21*, 364–370.

Ma, B. L., Subedi, K., Evenson, L., and Stewart, G. Evaluation of detection methods for genetically modified traits in genotypes resistant to European corn borer and herbicides. *J. Environ. Sci. Health B*, **2005**, *40*, 633–644.

Mian, I. S., Bradwell, A. R., and Olson, A. J. Structure, function and properties of antibody binding sites. *J. Mol. Biol.*, **1991**, *217*, 133–151.

Miraglia, M., Berdal, K. G., Brera, C., Corbisier, P., Holst-Jensen, A., Kok, E. J., Marvin, H. J., Schimmel, H., Rentsch, J., van Rie, J. P., and Zagon, J. Detection and traceability of genetically modified organisms in the food production chain. *Food Chem. Toxicol.*, **2004**, *42*, 1157–1180.

Myszka, D. G. Kinetic analysis of macromolecular interactions using surface plasmon resonance biosensors. *Curr. Opin. Biotechnol.*, **1997**, *8*, 50–57.

Neri, D., Montigiani, S., and Kirkham, P. M. Biophysical methods for the determination of antibody–antigen affinities. *Trends Biotechnol.*, **1996**, *14*, 465–470.

Nieba, L., Krebber, A., and Plückthun, A. Competition BIAcore for measuring true affinities: large differences from values determined from binding kinetics. *Anal. Biochem.*, **1996**, *234*, 155–165.

Ohlson, S., Strandh, M., and Nilshaus, H. Detection and characterization of weak affinity antibody antigen recognition with biomolecular interaction analysis. *J. Mol. Recognit.*, **1997**, *10*, 135–138.

O'Keeffe, M., Crabbe, P., Salden, M., Wichers, J., Van Peteghem, C., Kohen, F., Pieraccini, G., and Moneti, G. Preliminary evaluation of a lateral flow immunoassay device for screening urine samples for the presence of sulphamethazine. *J. Immunol. Methods*, **2003**, *278*, 117–126.

Paci, E., Caflisch, A., Plückthun, A., and Karplus, M. Forces and energetics of hapten–antibody dissociation: a biased molecular dynamics simulation study. *J. Mol. Biol.*, **2001**, *314*, 589–605.

Paul, V., Steinke, K., and Meyer, H. H. D. Development and validation of a sensitive enzyme immunoassay for surveillance of Cry1Ab toxin in bovine blood plasma of cows fed Bt-maize (MON810). *Anal. Chim. Acta*, **2008**, *607*, 106–113.

Petit, L., Baraige, F., Bertheau, Y., Brunschwig, P., Diolez, A., Duhem, K., Duplan, M.-N., Fach, P., Kobilinsky, A., Lamart, S., Schattner, A., and Martin, P. Detection of genetically modified corn (Bt176) in spiked cow blood samples by polymerase chain reaction and immunoassay method. *J. AOAC Int.*, **2005**, *88*, 654–664.

Quirasco, M., Schoel, B., Plasencia, J., Fagan, J., and Galvez, A. Suitability of real-time quantitative polymerase chain reaction and enzyme-linked immunosorbent assay for cry9C detection in Mexican corn tortillas: fate of DNA and protein after alkaline cooking. *J. AOAC Int.*, **2004**, *87*, 639–646.

Remund, K. M., Dixon, D. A., Wright, D. L., and Holden, L. R. Statistical considerations in seed purity testing for transgenic traits. *Seed Sci. Res.*, **2001**, *11*, 101–119.

Roda, A., Mirasoli, M., Guardigli, M., Michelini, E., Simoni, P., and Magliulo, M. Development and validation of a sensitive and fast chemiluminescent enzyme immunoassay for the detection of genetically modified maize. *Anal. Bioanal. Chem.*, **2006**, *384*, 1269–1275.

Shan, G., Embrey, S., and Schafer, B. W. A highly specific enzyme-linked immunosorbent assay for the detection of CryAc insecticidal crystal protein in transgenic WideStrike cotton. *J. Agric. Food Chem.*, **2007**, *55*, 5974–5979.

Shan, G., Embrey, S. K., Herman, R. A., and McCormick, R. W. Cry1F protein not detected in soil after three years of transgenic Bt corn (1507 corn) use. *Environ. Entomol.*, **2008**, *37*, 255–262.

Shim, W.-B., Yang, Z.-Y., Kim, J.-S., Kim, J.-Y., Kang, S.-J., Woo, G.-J., Chung, Y.-C., Eremin, S. A., and Chung, D.-H. Development of immunochromatography strip-test using nanocolloidal gold–antibody probe for the rapid detection of aflatoxin B1 in grain and feed samples. *J. Microbiol. Biotechnol.*, **2007**, *17*, 1629–1637.

Smith, G. P. Filamentous fusion phage: novel expression vectors that display cloned antigens on the virion surface. *Science*, **1985**, *228*, 1315–1317.

Stave, J. W. Protein immunoassay methods for detection of biotech crops: applications, limitations, and practical considerations. *J. AOAC Int.*, **2002**, *85*, 780–786.

Thompson, L. and Lindhardt, C. Singlepath Salmonella. Performance Tested Method 060401. *J. AOAC Int.*, **2006**, *89*, 417–432.

Trucksess, M. W. Determination of Cry9C protein in corn-based foods by enzyme-linked immunosorbent assay: interlaboratory study. *J. AOAC Int.*, **2001**, *84*, 1891–1902.

Ullman, E. F., Kirakossian, H., Singh, S., Wu, Z. P., Irvin, B. R., Pease, J. S., Switchenko, A. C., Irvine, J. D., Dafforn, A., Skold, C. N., and Wagner, D. B. Luminescent oxygen channeling immunoassay: measurement of particle binding kinetics by chemiluminescence. *Proc. Natl. Acad. Sci. USA*, **1994**, *91*, 5426–5430.

Van Duijn, G., Van Biert, R., Bleeker-Marcelis, H., Van Boeijen, I., Adan, A. J., Jhakrie, S., and Hessing, M. Detection of genetically modified organisms in foods by protein and DNA-based techniques: bridging the methods. *J. AOAC Int.*, **2002**, *85*, 787–791.

Velappan, N., Clements, J., Kiss, C., Valero-Aracama, R., Pavlik, P., and Bradbury, A. R. Fluorescence linked immunosorbent assays using microtiter plates. *J. Immunol. Methods*, **2008**, *336*, 135–141.

Wang, S., Quan, Y., Lee, N., and Kennedy, I. R. Rapid determination of fumonisin B1 in food samples by enzyme-linked immunosorbent assay and colloidal gold immunoassay. *J. Agric. Food Chem.*, **2006**, *54*, 2491–2495.

Wang, X., Li, K., Shi, D., Xiong, N., Jin, X., Yi, J., and Bi, D. Development of an immunochromatographic lateral-flow test strip for rapid detection of sulfonamides in eggs and chicken muscles. *J. Agric. Food Chem.*, **2007**, *55*, 2072–2078.

Xu, W., Huang, K., Zhao, H., and Luo, Y. Application of immunoaffinity column as cleanup tool for an enzyme linked immunosorbent assay of phosphinothricin-*N*-acetyltransferase detection in genetically modified maize and rape. *J. Agric. Food Chem.*, **2005**, *53*, 4315–4321.

CHAPTER 3

ANTIBODY ENGINEERING IN AGRICULTURAL BIOTECHNOLOGY

Patrick Doyle
Mehdi Arbabi-Ghahroudi
Claudia Sheedy
Kerrm Y. Yau
J. Christopher Hall

Immunoassays in Agricultural Biotechnology, edited by Guomin Shan
Copyright © 2011 John Wiley & Sons, Inc.

3.1 INTRODUCTION

Antibodies (Abs) are gamma globulin proteins produced by the immune system of vertebrates to recognize and bind to a virtually limitless repertoire of antigens (Ags). Abs are inherently unique as they possess high affinity and binding specificity for a target Ag. These properties have made Abs invaluable reagents within various biotechnological and biomedical applications ranging from standard diagnostic immunoassays (e.g., ELISA) to integrated immunoaffinity-based systems designed to isolate, enrich, and purify specific target reagents (proteins, immunoconjugates, etc.).

Initial efforts to produce *in vitro* Ag-specific Abs led to the introduction of hybridoma technology in 1975 in which Ag-stimulated B cells were fused with their myloma counterparts to generate stable hybrid or "hybridoma" cell lines to enable large-scale production of Ag-specific monoclonal Abs (mAbs) (Köhler and Milstein, 1975). Subsequent advances in the fields of genetics, molecular biology, and immunology have greatly enhanced applications of novel technologies required for the development, isolation, and application of recombinant Ab (rAb) fragments for an expanded range of applications. For example, the routine use of phage display technology to screen combinatorial rAb libraries (Winter et al., 1994; Hoogenboom, 2005) coupled with new software tools to support an expanding structure–function database (Harding et al., 2004) has opened unlimited opportunities to develop tailor-made rAb fragments. Moreover, ongoing improvements associated with rAb optimization and subsequent expression in recombinant hosts offer the possibility of large-scale and low-cost production platforms as an alternative to more expensive, laborious, and often cumbersome methods associated with the mAb development using mammalian cell lines.

Antibody engineering has emerged as a new discipline focused on the identification, cloning, optimization, and expression of Ag-specific rAb genes within various prokaryotic and eukaryotic host systems (Filpula, 2007; Holliger and Hudson, 2005; Jain et al., 2007; Lo, 2004; Maynard and Georgiou, 2000). This chapter presents an overview of established and emerging technologies available for the *de novo* generation of Ag-specific rAb fragments as an alternative to traditional immunoassays based on monoclonal antibodies or polyclonal serum. Applications and future research needs specific to agricultural biotechnology and genetically modified plants are presented as concluding remarks.

3.2 RECOMBINANT ANTIBODY FRAGMENTS

The development and production of Ag-specific rAb fragments derived from full-size conventional Abs (described in Chapter 2) was first reported in the late 1980s following the discovery of the genetic mechanism associated with Ab gene rearrangements. Concurrent advances in subcloning and expression systems coupled with the introduction of polymerase chain reaction (PCR) technology further enabled the recombinant expression of rAb fragments within bacterial hosts (Plückthun and Skerra, 1989; Winter and Milstein, 1991). In general terms, the production of rAb fragments through these new processes involved the isolation of mRNA for specific Ab genes derived from hybridoma, spleen cells, or lymph node leukocytes followed by reverse transcription PCR (RT-PCR) to form complementary DNA (cDNA) and eventual PCR amplification by using gene-specific primer sets to generate a complete Ab gene sequence (Maynard and Georgiou, 2000).

Given the inherent reliance on PCR amplification, the most significant initial advancements were related to the application of optimized oligonucleotides or the so-called primers to anneal and amplify target rAb genes. Within 10 years of adoption of PCR, numerous primer sets were published based on N-terminal sequences of isolated rAb genes (Benhar and Pastan, 1994), Ab leader sequences (Larrick et al., 1989), and known variable domain framework sequences.

A wide variety of rAb fragments have been produced, characterized, and applied in various biological systems (Hudson, 1998, 1999; Little et al., 2000; Holliger and Hudson, 2005; Yau et al., 2003). The following section provides a general description of the three most common and well-documented rAb fragments with potential applications in agricultural biotechnology.

3.2.1 Fab Fragment

The term "Fab," or fragment for antigen binding (Better et al., 1988), refers to a relatively large (MW $\approx$ 50–55 kDa) heterodimer molecule comprised of two polypeptide chains, one containing the constant and variable light chain domains ($C_L + V_L$) of a parent Ab paired with its corresponding variable and first constant heavy chain domains ($C_H1 + V_H$) (Jeffrey et al., 1993) (Figure 3.1). Fabs have the advantage of enhanced durability within *in situ* applications as they are inherently more stable and less prone to dissociation and proteolytic degradation relative to other conventional rAb fragment formats (e.g., scFv). Enhanced stability is attributed to the presence of a covalent disulfide bond that serves to retain tertiary folding and to facilitate more extensive interface and noncovalent interactions between the two paired immunoglobulin chains or four Ab domains (Maynard and Georgiou, 2000). Despite their relatively large size compared to other rAb fragment types, Fab proteins are amendable to standard Ab engineering techniques to improve attributes associated with biological function, recombinant expression, affinity for target Ag, and so on (Filpula, 2007; Holliger and Hudson, 2005; Jain et al., 2007; Maynard and Georgiou, 2000).

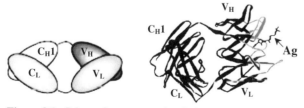

Figure 3.1 Schematic representation (left) and ribbon diagram (right) of recombinant antibody fragment (Fab). Constant and variable light chain domains ($C_L + V_L$), corresponding first constant and variable heavy chain domains ($C_H1 + V_H$), and target Ag (i.e., digoxin) are labeled accordingly (adapted from Joosten et al., 2003 and Jeffrey et al., 1993).

3.2.2 scFv Fragment

The term single-chain variable fragment (scFv) refers to a rAb fragment in which the variable heavy and light domains (V_H and V_L, respectively) of a parent immunoglobulin (reviewed in Chapter 2) are joined by a hydrophilic and flexible peptide linker (Figure 3.2) (Bird et al., 1989; Huston et al., 1988). The order of the domains can be designed for expression as either V_H-linker-V_L or V_L-linker-V_H (Huston et al., 1991, 1995). In most cases, the length of the peptide linker is 15 amino acids based on a $(Gly_4Ser)_3$ configuration (Huston et al., 1995).

Specific length and composition of the peptide linker can have a dramatic impact on posttranslational folding and valency of scFv fragments (Robinson and Sauer, 1998; Turner et al., 1997). For example, Atwell et al. (1999) demonstrated that three amino acid linkers often form scFv dimers whereas linkers with two residues or less induce the formation of trimers. In general, increasing the length of linker will lead to monomeric variants. However, monomeric and multimeric scFv variants have been reported with linkers of up to 30 residues (Desplancq et al., 1994; Nieba et al., 1997).

Once valency is accounted for, one cannot assume that all scFv fragments possess comparable Ag binding affinity to larger Ab formats (e.g., mAb and Fab) because they all have the same V_H/V_L domain pairing. Numerous publications, based on various Ags, have demonstrated that relative scFv affinity is best described on a case by case basis. For example, Choi et al. (2004) noted that a scFv fragment raised against deoxynivalenol (haptenic Ag, MW $\approx$ 296 Da) had a binding affinity that was approximately two orders of magnitude less than the parent mAb. However, based on a similar comparison for the same Ag, Wang et al. (2007) noted no differences between mAb and scFv binding affinity.

Similar variability has been noted for scFv stability and efficiency of expression. In general terms, posttranslational scFv folding and function are greatly influenced by specific expression systems and growth conditions (i.e., choice of

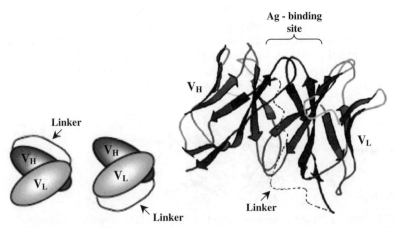

Figure 3.2 Schematic representation (left) and ribbon diagram (right) of recombinant scFv antibody. Variable light and heavy chain domains (V_L and V_H, respectively) and synthetic peptide linker are labeled accordingly (adapted from Joosten et al., 2003 and Filpula, 2007).

expression vector and recombinant host, etc.), as well as the method of purification. In addition, factors such as proteolytic degradation of the linker sequence (Whitlow et al., 1993) and formation of insoluble inclusion bodies (Choi et al., 2004) have great influence on scFv stability and expression.

Various techniques have been used to overcome such limitations. For example, Reiter et al. 1994 reported on the formation of disulfide-stabilized Fvs (i.e., dsFvs) through the introduction of cysteine residues between the V_L and V_H domains that resulted in the formation of an interchain disulfide bond, thereby conferring greater rAb stability. As well, molecular chaperones have been used as part of scFv expression strategy to assist in posttranslational scFv folding and to increase efficiency of expression (Arbabi-Ghahroudi et al., 2005).

3.2.3 V_HH Fragment

The term "V_HH" refers to the variable domain of camelidae family heavy chain Abs (HCAbs). V_HH fragments are among the smallest intact Ag binding fragments known (average 14–15 kDa, 118–136 amino acid residues) (Figure 3.3). The hallmark of V_HH domain resides in key hydrophobic to hydrophilic amino acid substitutions at former V_L interface that imparts increased solubility and stability (e.g., resistance to aggregation). Despite the absence of V_L and V_H combinatorial diversity, single-domain V_HH fragments have demonstrated high-affinity binding to a wide range of Ag types (Ghahroudi et al., 1997; Muyldermans, 2001). HCAbs (reviewed in Chapter 1) are believed to compensate for the loss of V_L domain through extended CDR3 loop regions and a higher rate of somatic hypermutations to form a V_HH paratope that can form a large antigen binding repertoire and additionally can penetrate into and conform to clefts and cavities of Ag epitopes (Muyldermans, 2001). An example of this unique ability is V_HH binding to active sites of enzymes that are not typically recognized by conventional rAb fragments (Desmyter et al., 1996, 2002). Numerous reviews are available that document Ag binding characteristics and potential applications of V_HH fragments (Holliger and Hudson, 2005; Joosten et al., 2003; Muyldermans, 2001).

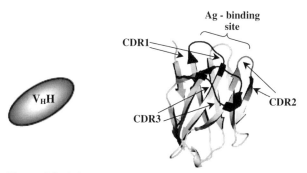

Figure 3.3 Schematic representation (left) and ribbon diagram (right) of recombinant HCAb variable domain (V_HH) of antibody "AMD9." CDRs 1, 2, and 3 in ribbon diagram are labeled accordingly (adapted from Joosten et al., 2003 and Desmyter et al., 2002).

Single-domain $V_H H$ fragments offer a number of advantages relative to larger and more complex rAb fragments such as scFv or Fab. DNA encoding the variable domain of HCAb can be readily isolated from peripheral leukocytes and subcloned into various phage display vectors for highly efficient $V_H H$ library generation since no synthetic linker is required for the pairing of $V_H - V_L$ domains (Ghahroudi et al., 1997). Another beneficial characteristic of $V_H H$ is its hydrophilic amino acid substitutions within FRs (Muyldermans, 2001). Consequently, expression in hosts such as *Escherichia coli* tends to produce high yields of soluble, stable $V_H H$ protein that is inherently free of problems associated with protein aggregation and proteolytic degradation common to scFv expression (Arbabi-Ghahroudi et al., 2005). Finally, recent experiments have demonstrated that $V_H H$ fragments possess enhanced functionality and stability over a range of adverse temperature and solvent regimes where current immunoassays do not typically function (Goldman et al., 2006; Ladenson et al., 2006).

The understanding of amino acid sequences of $V_H H$ fragments provides interesting and useful techniques to improve expression of non-dromedary Abs. For example, it is possible to replace hydrophobic residues on the V_H side of a conventional rAb fragment with "$V_H H$-like" residues to minimize limitations associated with insolubility and nonspecific binding. This process is known as "camelization" and can be used with V_H fragments derived from various animal sources (Davies and Riechmann, 1994; Davies and Riechmann, 1996; Tanha et al., 2001).

3.3   DEVELOPMENT AND TYPES OF ANTIBODY LIBRARIES

The concept of Ab libraries was established following the elucidation of the genetic mechanisms associated with the creation of immunoglobulin gene diversity (Tonegawa, 1983) and the advancement of technologies associated with PCR for amplification of the variable domains of immunoglobulin genes (Orlandi et al., 1989; Songsivilai et al., 1990). In principle, rAb libraries may be classified as immunized, naïve, semisynthetic, or synthetic according to the genetic material used for library construction.

3.3.1   Immunized Libraries

Immunized rAb libraries are constructed from DNA isolated from peripheral leukocytes of a host that has been repeatedly exposed to a single target immunogen or a group of immunogens. Library construction occurs through a stepwise process whereby the B cell genes encoding Ab variable domain regions are amplified by PCR and subcloned into vectors to enable expression of different rAb fragment formats including $V_H H$ (Ghahroudi et al., 1997; Goldman et al., 2006; Ladenson et al., 2006), scFv (McCafferty et al., 1990; Clackson et al., 1991), and Fab (Huse et al., 1989; Mullinax et al., 1990; Orum et al., 1993; Persson et al., 1991).

The inherent advantage of an immunized library format resides in the fact that peripheral leukocytes, and therefore high concentration of the corresponding mRNA used for library construction, are the product of the process of affinity maturation (Hoogenboom et al., 1998). Although the development of a rAb library based on an immunogen-specific response improves the probability of isolating Ag-specific rAbs, overall library size and diversity are often limited relative to other nonbiased library formats (Clackson et al., 1991; Burton et al., 1991).

Depending on the nature of the Ag and overall test system, the development of hyperimmunized libraries can present a number of technical challenges and potential drawbacks associated with isolating rAbs against self-Ags or against Ags that are extremely toxic or chemically unstable. Immunological tolerance is another factor that can reduce the diversity of natural *in vivo* repertoire. In addition, one must account for the various logistical, financial, and ethical considerations associated with the immunization protocol to ensure a robust and predictable response to the Ag of interest (Willats, 2002; Hoogenboom, 1997).

3.3.2   Naïve Libraries

Naïve libraries are prepared according to the same methodology as used for construction of immune libraries except that the animal has not been hyperimmunized with the Ag of interest. The major source of immunoglobulin genes are peripheral blood lymphocytes (PBLs), bone marrow, tonsils, and spleen cells. IgM is the preferred immunoglobulin isotype as it is more diverse and has not been subjected to bias or previous Ag selection (Marks et al., 1991; Griffiths et al., 1993; Little et al., 1993; Schier et al., 1995; Dörsam et al., 1997).

Library size plays a vital role in successfully isolating Ag-specific rAb fragments. This concept has been proven in several reports in which naïve libraries of medium size have only resulted in the isolation of Abs with micromolar affinities (Marks et al., 1991; Griffiths et al., 1994). Accordingly, a library size of approximately 10^{10} is likely to yield Abs with affinities in the low nanomolar range (Vaughan et al., 1996). These results demonstrate that size and heterogeneity of such libraries have critical roles in isolating high-affinity Abs without prior immunization. An advantage of naïve libraries, in particular larger sized ones, is that they are often not antigenically predisposed to self-antigens. As such, large, diverse naïve libraries are a fit as a universal source of Ab binders (Vaughan et al., 1996).

3.3.3   Synthetic/Semisynthetic Libraries

Methodologies associated with synthetic libraries combine the advantages of naïve libraries (e.g., structural diversity) with ongoing advancements associated with improved knowledge of rAb structure and function. Accordingly, these technologies enable *in vitro* production of high-affinity rAbs with application against a range of target Ags that are not affected by inherent controls that may limit expression or bias an immunogenic response to a target Ag (Knappik et al., 2000; Fellouse et al., 2007; Winter and Milstein, 1991).

One of the first semisynthetic libraries was generated by using a repertoire of V_H genes, combined with a synthetic CDR3 construct, and displayed on phage surface as either scFv (Hoogenboom et al., 1992) or Fab (Barbas et al., 1992) fragments. The complexity and diversity were further expanded using a repertoire of V_L genes in conjunction with initial library formats (Nissim et al., 1994; de Kruif et al., 1995). Second-generation libraries evolved from the development of synthetic Ab scaffolds that express well in *E. coli* to help overcome limitations associated with posttranslational rAb folding (Knappik and Plückthun, 1995). It is also possible to construct fully synthetic rAb libraries using available DNA sequences of human immunoglobulins (Tomlinson et al., 1992; Cook and Tomlinson 1995; Andris et al., 1995; Winter, 1998) followed by total gene synthesis for expression in *E. coli* based on optimized rAb scaffolds (Griffiths et al., 1994; Knappik et al., 2000; Krebs et al., 2001).

3.4 DISPLAY SYSTEMS

High-throughput screening of rAb libraries is made possible through techniques that establish a direct physicochemical link between a gene (i.e., genotype) and the resultant rAb fragment (i.e., phenotype). This linkage allows isolation of an Ab fragment with desired Ag binding properties from a large, pooled rAb repertoire, while the genetic material is available for further isolation and engineering. rAb fragments are most commonly displayed on phage, bacteria, yeast, and ribosomes (Li, 2000; Yau et al., 2003).

3.4.1 Phage Display

Phage display was first demonstrated by Smith (1985) who reported expression of foreign peptides as a fusion with pIII viral coat protein on the surface of nonlytic filamentous phage. Within 5 years of Smith's demonstration of a direct link between phenotype, that is, as protein expressed on the phage surface, and genotype, that is, as foreign DNA within phage genome, McCafferty et al. (1990) used phage display techniques to isolate Ag-specific scFv fragments from a diverse filamentous phage-displayed scFv repertoire. Since the time of these reports, phage display technology has matured into a widely adopted technique for selection of rAb fragments from very large phage display libraries.

Phage display techniques are now commonly used to produce Ab fragments with optimized binding affinity, stability, and activity (Forrer et al., 1999) and to isolate Abs that were previously considered to be very difficult to obtain through conventional methods (Griffiths et al., 1993; Hoogenboom and Chames, 2000). Numerous reviews highlight the diverse applications of phage display technologies (Hoogenboom et al., 1998; Paschke, 2006; Smith and Petrenko, 1997; Willats, 2002).

3.4.1.1 Filamentous Bacteriophage Filamentous bacteriophages (Ff) are viruses (e.g., F1, M13, and fd) that can infect and replicate within Gram-negative bacteria (e.g., *E. coli*) without killing the host cell (Russel et al., 2004). These viruses are ideal for applications within rAb engineering since the Ff genome is inherently small and tolerates insertions into nonessential regions without loss of bacterial infectivity. During phage assembly, fusion proteins are expressed on a particular viral coat protein while genetic information encoding the displayed protein is packaged within the single-stranded DNA (ssDNA) of nascent phage particles (Smith and Petrenko, 1997). Direct genotype to phenotype coupling ensures that phages produced within the same infected bacterial cell are comprised of identical clones (Paschke, 2006).

The defining characteristic of Ff structure is a circular 6.4 kbp ssDNA genome encased by a long, flexible tube comprised of ≈2700 copies of the major coat protein (pVIII), with a minor coat proteins (pIII, pVI, pVII, and pIX) expressed at phage tips (Figure 3.4) (Yau et al., 2003). Of these coat proteins, pIII is the most extensively used and well characterized phage display format since it is more amendable than the other coat proteins to large protein insertions without loss of phage infectivity (Hoogenboom et al., 1998; Hoogenboom and Chames, 2000; Russel et al., 2004; Smith and Petrenko, 1997).

3.4.1.2 *Phage Versus Phagemid Vectors*

Phage display rAb libraries are constructed by cloning a diverse PCR-amplified Ab repertoire into either a phage or a phagemid vector (Hoogenboom et al., 1992; Winter et al., 1994). Regardless of library type (i.e., phage or phagemid), the salient goal is to construct a heterogeneous and robust mixture of phage clones that collectively displays rAb–phage fusions capable of target or "Ag-specific" recognition and replication to produce identical progeny phage in large scale (Smith and Petrenko, 1997).

Both phage and phagemid vectors are designed so that the target rAb gene is inserted between a signal sequence and the coat protein gene (e.g., pIII). Likewise, both types of vectors carry antibiotic resistance selectable markers, as well as a phage-derived origin of replication to enable packaging of foreign

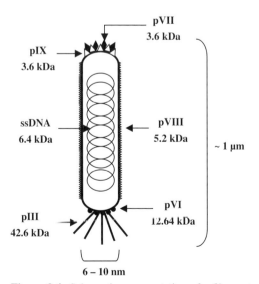

Figure 3.4 Schematic representation of a filamentous bacteriophage and its protein coats (adapted from Willats, 2002 and Yau et al., 2003).

DNA into viral ssDNA and production of nascent phage progeny that displays rAb fused to surface phage coat protein (Figure 3.5) (Russel et al., 2004; Smith and Petrenko, 1997).

Phage vectors are directly derived from the Ff genome whereby the recombinant protein is cloned as a fusion to phage coat protein (e.g., pIII) for production of multivalent rAb phage protein progeny. Conversely, a phagemid vector is a plasmid-based sequence harboring gene III and intergenic region of Fd phage fusion gene under control of a weak to moderate bacterial promoter such as LacZ. Phagemid vectors also have a second origin of replication to allow propagation as a plasmid within host *E. coli* (Figure 3.5). Many phagemid vectors also have an amber stop codon inserted between end of desired protein and pIII gene to enable soluble rAb expression within nonsuppressor strains of *E. coli* (e.g., HB2151) (Russel et al., 2004; Smith and Petrenko, 1997).

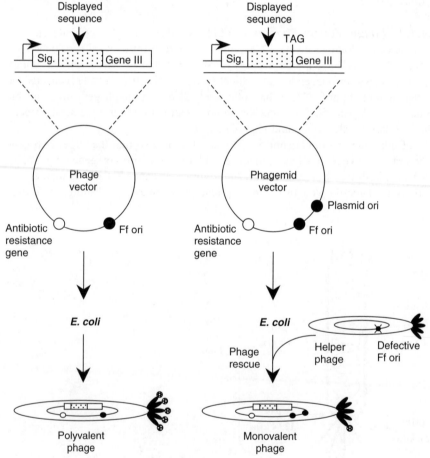

Figure 3.5 Phage (left) and phagemid (right) vectors to construct rAb phage display libraries through the minor coat protein pIII. Ff ori, filamentous phage origin of replication; plasmid ori, plasmid origin of replication; TAG, amber stop codon; Sig., filamentous phage leader sequence (adapted from Russel et al., 2004).

Phage production using phagemid vectors is possible only when additional phage proteins are provided by helper phages through a process known as "phage rescue" (Paschke, 2006) (Figure 3.5). Use of helper phage is, therefore, very critical as it improves the relative proportion of univalent hybrid phage progeny carrying rAb-phagemid genome to wild-type or "bald" phage based only on phage genome (Russel et al., 2004). Helper phage can also impact resultant phage valency within pIII phagemid systems to promote production of monovalent phages carrying only one rAb fragment pIII fusion protein per phage particle (Paschke, 2006; Russel et al., 2004).

Finally, regardless of vector type, during library construction, it is essential to maintain stringent control of fusion protein expression using controllable bacterial promoter such as LacZ (e.g., by using glucose as repressor) through selectable markers and consistent techniques during all propagation steps as continuous expression of rAb-phage coat fusion proteins increases the metabolic burden to host *E. coli* cells. Such techniques are important to ensure consistent selection pressure to limit any ecological advantage for nonfunctional clones that limit library efficiency and function (Paschke, 2006).

3.4.1.3 Panning The selection of Ag-specific Abs from a phage display library occurs through a procedure known as "biopanning" or simply "panning." Although this process can involve many variations and modifications, it relies mainly on the stepwise isolation and amplification of phage displaying Ag binding Ab fragments (Maynard and Georgiou, 2000; Paschke, 2006; Smith and Petrenko, 1997).

After exposure of the full phage display rAb repertoire to the target Ag (e.g., coated onto a solid surface or present in solution or even on the cell surface), washing steps are used to remove phage carrying nonspecific Abs from those displaying Ag-specific Abs. The latter are typically desorbed through low or high pH elution prior to amplification in phage host (typically *E. coli*) (Figure 3.6). After each round of panning, amplified phage are subject to increased selection pressure and washing to continue phage enrichment and amplification based on Ag binding properties of phage-displayed rAbs (Paschke, 2006).

High-affinity phage display Abs are commonly selected within three to four rounds of panning (de Bruin et al., 1999). After panning, rAb fragments are typically expressed in *E. coli* and purified for subsequent characterization and application.

3.4.2 Bacteria Display

Peptide display on the surface of bacteria was first described well over a decade ago (Georgiou et al., 1993). Since the initial reports of recombinant peptide expression in prokaryotes, fully functional Ab fragments have been displayed on surface of *E. coli* (Fuchs et al., 1991; Francisco et al., 1993). As noted for phage display, bacterial display libraries can also be used to select rAbs with improved affinity for target Ag (Daugherty et al., 1998, 1999).

A limitation associated with the expression of foreign peptides on the surface of Gram-negative bacteria (e.g., *E. coli*) is the physiological barrier associated with an extensive network of macromolecules within the bilayer cell envelope. However, this limitation can be circumvented: peptidoglycan chaperones have been used to facilitate

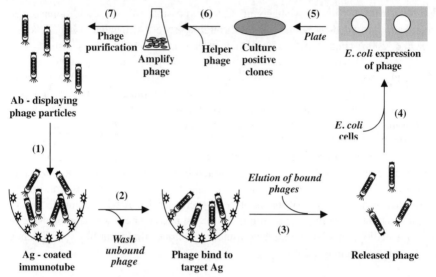

Figure 3.6 Phage display panning cycle: (1) rAb displaying phages are added to immunotubes coated with target Ag, (2) washing is performed to remove nonbinding phages, (3) positive binding phages are eluted, (4) eluted phages are used to infect *E. coli*, (5) positive clones are plated as *E. coli* plaques or colonies, (6) helper phage (if phagemid system) is added for phage rescue to amplify positive clones, and (7) positive phages are purified for reintroduction into next round of panning (de Bruin et al., 1999).

transport and anchoring of recombinant Ab fragments through both the inner and outer membranes without causing an adverse effect on cell growth and viability (Fuchs et al., 1991). Another way to address this problem is to select an appropriate leader sequence and a surface-localized carrier protein (Yau et al., 2003).

It is also possible to express rAbs on the surface of Gram-positive bacteria. For example, scFv expression has been reported on the surface of *Staphylococcus xylosus* and *S. carnosus* (Gunneriusson et al., 1996). Ab display on Gram-positive bacteria has been shown to provide efficient rAb secretion and posttranslational folding.

3.4.3 Yeast Display

As in bacterial surface display, the phenotype and the genotype linkage for yeast surface display models is comprised of the protein on the cell surface and the corresponding DNA within the cell. Given this direct linkage, an additional subcloning step is not required after selection of an Ag binding cell (Yau et al., 2003).

Yeast display of rAbs offers many of the advantages of bacterial display. Yeast has a thick, rigid cell wall that enables stable maintenance of surface-displayed proteins (Georgiou et al., 1997). In contrast to typical prokaryotic expression systems (e.g., *E. coli*), yeast possesses protein folding and secretory mechanisms of higher order eukaryotic (e.g., mammalian) systems. As such, yeast display alleviates some biases associated with prokaryotic protein expression and folding (Boder and Wittrup, 1997).

Despite these advantages, experience has shown that yeast display Ab libraries tend to have limited transformation efficiency, leading to smaller and less diverse repertoires relative to phage or bacterial display (Yau et al., 2003). Nevertheless, high-affinity rAb fragments have been isolated from yeast display libraries. For example, Boder and Wittrup (1997) successfully used yeast surface display to screen a diverse peptide library to select binders to a soluble fluorescein tracer. Likewise, Kieke et al. (1997) reported successful isolation of an anti-T cell receptor scFv using a yeast surface display library.

3.4.4 Ribosome Display

Ribosome display represents an alternative to expression of Ab fragments on the surface of phage, yeast, or bacterial cells. As in other display systems, ribosome display libraries are designed based on a direct link between peptide/Ab phenotype and genotype. Unlike other library formats, ribosome display is entirely *in vitro* or cell free (Hanes and Plückthun, 1997; Schaffitzel et al., 1999).

Ribosome display is unique as the phenotype to genotype construct is designed to include a spacer sequence with no stop codon such that when the nascent Ab fragment is translated from mRNA, both the mRNA and the Ab remain bound to the ribosome molecule to form an "ARM" (antibody–ribosome–mRNA) complex (He and Taussig, 1997) (Figure 3.7).

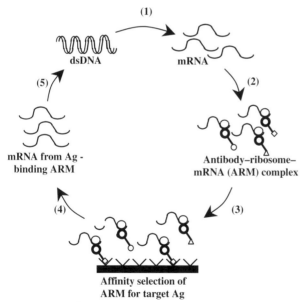

Figure 3.7 Ribosome display panning cycle. (1) Double-stranded DNA (dsDNA) transcription to form messenger RNA (mRNA), (2) *in vitro* translation of mRNA to form ARM complexes, (3) exposure of ARM complexes to immobilized target Ag for affinity selection or Ag-specific panning, (4) elution and isolation of mRNA from Ag-specific ARM complex(es), and (5) reverse transcription PCR amplification of target mRNA to form dsDNA for continued panning or transformation into host organism (e.g., *E. coli*) (adapted from Yau et al., 2003).

A key advantage of cell-free systems is that very large libraries can be handled since transformation efficiency is not a limiting factor as it is for other systems. In addition, *in vitro* systems are not impacted by ongoing random mutation of foreign DNA imparted by the host organism that may limit library diversity. As an example, the typical size limit associated with transformation efficiency for bacterial display libraries ranges from 10^{10} to 10^{11} clones (Yau et al., 2003). With ribosome display, however, libraries with up to 10^{15} members can be created and screened (Roberts, 1999).

Ribosome display also enables the incorporation of continuous diversification during each panning cycle using PCR-based mutagenesis. As a result, ribosome display is a good format to select Ab fragments with improved K_D (dissociation constants) and robust Ag binding affinities (Yau et al., 2003). For example, Hanes et al. (1998) reported the isolation of scFvs from a ribosome display library with affinities to target Ag in the picomolar range.

3.5 OPTIMIZATION OF RECOMBINANT ANTIBODIES

Ab fragments such as scFv, Fab, and V_HH often display lower affinity than their parental, full-size counterparts (Adams and Schier, 1999; Huston et al., 1996). A number of strategies can be used to overcome this limitation, thus enabling isolation of high-affinity Ag-specific rAb fragments. For example, panning procedures versus target Ag can be optimized and *in vitro* mutagenesis (where the natural affinity maturation process is mimicked) can be performed (Adams and Schier, 1999).

3.5.1 Optimized Panning Procedures

Although optimization of the selection or panning process (Figure 3.6) does not inherently improve rAb characteristics, it allows more efficient recovery or selection of high-affinity Ag-specific rAb fragments from a diverse library. Panning strategies used to increase the probability of isolating high-affinity rAbs have generally consisted of gradually decreasing the Ag concentration and increasing the wash stringency during successive rounds of panning (Strachan et al., 2002). In the case of panning against low molecular weight haptenic Ags, panning strategies often use soluble or free hapten to elute the rAb from target Ag. Likewise, structurally similar hapten analogues can be preincubated with which the phage display rAbs prior to exposure to the target hapten conjugate. Finally, subtractive panning, where the rAbs are incubated with the carrier protein prior to incubation with the hapten, can also be used to improve the probability of selecting Ag-specific, high-affinity rAbs (Sheedy et al., 2007).

As more knowledge is gained on Ab structure through X-ray crystallography, specific mutations can be targeted to modify and improve rAb specificity and affinity for target Ag (Lamminmäki et al., 1999; Korpimäki et al., 2003). Ab engineering provides excellent tools to tailor the properties of rAb fragments with respect to affinity, specificity, and performance for different applications (Valjakka et al., 2002).

3.5.2 *In Vitro* Affinity Maturation

Because of the relatively low affinity of rAbs isolated from naïve libraries, the process of somatic hypermutation must be mimicked through antigen-driven affinity maturation (Irving et al., 1996). *In vitro* affinity maturation by site-directed or random mutagenesis is not limited by the restrictions inherent to the natural somatic hypermutation mechanisms (Parhami-Seren et al., 2002). Affinity maturation *in vitro* can, therefore, generate extended and functional variability in rAb molecules (Borrebaeck and Ohlin, 2002) and lead to selection of rAb fragments with increased affinity and specificity for target Ag (Roberts et al., 1987; Wu et al., 1998, 1999).

Strategies to affinity-matured Abs and fragments *in vitro* can be grouped in three broad categories: random mutagenesis, site-directed mutagenesis, and shuffling. Random mutagenesis consists, as its name implies, of the introduction of mutations randomly throughout the gene. Random mutagenesis can be subgrouped into error-prone PCR and bacterial mutator strains. Site-directed mutagenesis includes mutational hot spots and parsimonious mutagenesis, as well as any mutational strategy that assigns the mutations to given positions or residues within an Ab's wild-type gene. Shuffling encompasses chain shuffling, DNA shuffling, and staggered extension process. The most frequently used strategies are those of site-directed mutagenesis, where mutations are directed to specific CDRs or framework regions (Irving et al., 1996), and error-prone PCR, where mutations are randomly introduced across the gene.

3.5.2.1 *Random Mutagenesis*

Error-Prone PCR   Error-prone PCR uses low-fidelity polymerization PCR conditions to introduce point mutations randomly over a gene sequence (Stemmer, 1994). The introduction of random mutations *in vitro* is an efficient method for increasing both specificity and affinity, as long as a strong selection pressure is applied during the panning stage (Miyazaki et al., 1999).

Bacterial Mutator Strains   Affinity maturation of rAbs and increase in their expression levels have been achieved by using *E. coli* mutator cells (Irving et al., 1996, 2002; Coia et al., 2001a, 2001b). This approach consists of the selection of rAbs from a library followed by gene diversification or mutagenesis through amplification in a bacterial mutator strain of *E. coli* such as MutD5, a conditional mutant that produces single base substitutions (transversions or transitions) at high frequency compared to normal *E. coli* cells. Such mutator strains can produce a large number of mutant rAbs that can then be selected by phage display or other methods (Irving et al., 1996). For example, the affinity of a scFv for the glycoprotein glycophorin was increased by passaging the rAb construct through such a bacterial mutator strain. A single point mutation resulted in a 10-fold increase in the scFv binding affinity, and another point mutation located in a framework region near the CDR3 of the V_L resulted in a 1000-fold increase in affinity (Irving et al., 1996). The effects of these point mutations illustrate the power of random mutagenesis in generating high-affinity mutants that could not be obtained by site-directed mutagenesis. However, 4–10 rounds of mutation or passage through mutator cells are required to obtain high-affinity mutants (Coia et al., 2001a).

3.5.2.2   Focused Mutagenesis

Site-Directed Mutagenesis   Improvements in affinity can be achieved *in vitro* by site-directed mutagenesis, where specific or selected residues are mutated at predetermined locations within target rAb genes. Following mutagenesis, a library is constructed and screened for high-affinity mutants. For example, Davies and Riechmann (1996a) have investigated the effect of randomizing CDRs 1 and 2 residues in V_H domains specific for both haptens and protein antigens. This research was successful in terms of generating randomized repertoires that were displayed on phage and affinity selected to improve Ag binding.

Mutational Hot Spots   Chowdhury and Pastan (1999) have developed a strategy that enables the isolation of rAb fragments with increased affinity from small phage display libraries. Their approach is based on the fact that the DNA encoding the variable domains of Abs contains "mutational hot spots," or nucleotide sequences naturally prone to hypermutations during the *in vivo* affinity maturation process. Several types of hot spots have been suggested, such as direct and inverted repeats, palindromes, secondary structures, and certain consensus sequences (Chowdhury and Pastan, 1999). Several consensus hot spot sequences have been studied in great detail; one of these sequences is the tetranucleotide A/GGC/TA/T (Chowdhury and Pastan, 1999). The serine codons AGY, where Y can be either C or T, are other hot spot consensus sequences found in Ab genes that can be used as targets for mutagenesis.

Hot spot mutagenesis techniques have been used to affinity-matured rAb fragments as a precursory step to develop Ag-specific high-affinity rAbs (Chowdhury and Pastan, 1999; Li et al., 2001). For example, hot spot mutagenesis was used to generate scFv fragments with high affinity to mesothelin, a cancer protein Ag. The affinity of the mesothelin-specific scFv PE38 was found to be 11 nM (Chowdhury and Pastan, 1999). Improved affinity was desired to increase its cytotoxicity so that it could serve as a potent immunotoxin. DNA sequences present in the rAb variable domains (hot spots) were selected and random mutations were introduced in the light chain CDR3 (Chowdhury and Pastan, 1999). Thirty-two hot spots were identified in the scFv, 14 being located in the variable heavy domains and 18 in the variable light domains. A library with the affinity-matured clones was constructed and panning of the library yielded several mutants with a 15–50-fold increase in affinity. Mutagenesis of the same original library but outside the hot spots resulted in a mutated library from which the highest affinity clone selected had only a fourfold increased affinity compared to the wild-type Ab (Chowdhury and Pastan, 1999).

Parsimonious Mutagenesis   Parsimonious mutagenesis was developed by Balint and Larrick (1993) as a technique whereby all three CDRs of a variable region gene are simultaneously and thoroughly searched for improved variants in libraries of manageable size (Balint and Larrick, 1993; Schier et al., 1996a, 1996b, 1996c). In parsimonious mutagenesis, synthetic codons are used to mutate about 50% of all targeted amino acids while keeping the other 50% of targeted residues intact (wild type) (Chames et al., 1998). rAb libraries can be constructed with low-redundancy "doping" codons and biased nucleotide mixtures designed to maximize the abundance of combining sites with predetermined proportions of preselected sets of alternative amino acids (Balint and Larrick, 1993).

This technique has several advantages over standard mutagenesis procedures, and one of which is the total number of substitutions per Ab gene is reduced, thereby resulting in a larger number of well-folded and potentially active binders in the library (Chames et al., 1998). Random mutagenesis may target residues essential to Ag binding, leading to a library of mutants that will be mostly nonfunctional, whereas parsimonious mutagenesis will preserve these essential residues, resulting in a library of mainly functional clones (Chames et al., 1998).

3.5.3 Antibody Gene Shuffling

Shuffling of rAb genes can be accomplished in several ways: chain shuffling, DNA shuffling, and staggered extension process (StEP), as well as through variations of these techniques. Chain shuffling consists of shuffling heavy and light chain variable regions of Abs. In DNA shuffling, the Ab DNA is digested with DNase I and randomly reassembled and amplified by PCR. StEP is also a PCR-based method, where template switching, caused by shortened extension time, shuffles various portions of several parental Ab genes.

3.5.3.1 Chain Shuffling Chain shuffling can generate high-affinity rAbs (e.g., scFv) from immunized animals quite rapidly (Kang et al., 1991). Since shuffling approaches mimic somatic hypermutation, they are believed to be more efficient than random or site-directed mutagenesis in producing functional rAbs (Ness et al., 1999, 2002). The heavy and light chains of rAbs isolated from an immune library can be recombined, thereby generating a vast number of functional rAbs from an initially limited genetic repertoire (Kang et al., 1991).

For chain shuffling to be an effective mutagenesis strategy, the preselection of Ag binding rAbs from an immune library is required prior to invoking the shuffling process. In other words, chain shuffling is only feasible to improve rAb affinity within immunized libraries (Park et al., 2000). Accordingly, chain shuffling is not applicable to naïve rAb libraries since heavy and light chains available within the library have not been exposed to target Ag. Moreover, experiments have shown that individual chains might discriminate among partners, rendering the shuffling process difficult between naïve and somatically mutated chains (Kang et al., 1991).

3.5.3.2 DNA Shuffling by Random Fragmentation and Reassembly DNA shuffling (Stemmer, 1994; Stemmer et al., 1995) is based on repeated cycles of point mutagenesis, recombination, and selection that should allow *in vitro* molecular evolution of complex sequences such as proteins. It somewhat mimics the natural mechanisms of molecular evolution (Minshull and Stemmer, 1999), but with a faster rate (Patten et al., 1997; Stemmer, 1994).

The method involves the digestion with DNase I of a large Ab gene to create a pool of random DNA fragments. These fragments can then be reassembled into full-length genes by repeated cycles of annealing in the presence of DNA polymerase (Stemmer, 1994). The fragments prime each other based on homology, and recombination occurs when fragments from one copy of a gene prime another copy, causing a template switch. Because of the template switching events, a certain level of homology is required among the parental genes to be suitable for this method (Stemmer, 1994).

DNA shuffling offers several advantages over more traditional mutagenesis strategies. Compared to methods such as error-prone PCR and site-directed mutagenesis, DNA shuffling can be used with longer DNA sequences and also allows the selection of clones with mutations outside the binding or active site of the protein, whereas site-directed mutagenesis is limited to a given region of the protein due to the limitation in library size that can be efficiently transformed to construct the mutant library (Stemmer, 1994).

DNA shuffling was first performed with 1 kb long interleukin-1β genes. The genes were broken into 10–50 bp fragments and reassembled to their original size and function (Stemmer, 1994). Sequencing of clones following the shuffling process revealed that reassembly introduced point mutations at a rate of 0.7%, a rate similar to that of error-prone PCR. A similar method was used to shuffle β-lactamase genes, where three cycles of shuffling followed by selection yielded a mutant with 32,000-fold increase in enzymatic activity. Error-prone mutagenesis of the same genes resulted in only a 16-fold increase (Stemmer, 1994).

3.5.3.3   Staggered Extension Process   DNA shuffling by staggered extension process was developed in 1998 by Zhao et al. 1998. StEP consists of priming the template sequences followed by several cycles of denaturation and shortened annealing/polymerase-catalyzed extension. During each cycle, the DNA fragments can anneal to different templates based on sequence complementarity and extend further to create recombinant genes. Because of the template switching, the growing polynucleotide chains can contain information from one or various parental clones (Zhao et al., 1998). The whole process can be performed in a single PCR reaction, resulting in a pool of mutants, the majority of which are functional.

3.6   EXPRESSION AND PURIFICATION

There is a growing demand for rAbs for the treatment of human diseases, *in vitro* diagnostic tests, and affinity purification methods, placing pressure on current production capacity that is based largely on bacterial and tissue culture techniques. Alternative systems, such as yeast (Horwitz et al., 1988), baculovirus (zu Putlitz et al., 1990), and plants (Hiatt and Ma, 1992), are very attractive, both for cost-effectiveness and individual advantages such as scalability and safety, for instance, that are absent in bacterial or mammalian systems. Several plant-produced Abs (plantibodies) are undergoing clinical trials and the first commercial approval could be only a few years away. The performance of the first generation of products has been very encouraging so far (Stoger et al., 2005). Ongoing studies are addressing further biochemical constraints, aiming to further improve yields, homogeneity, and authenticity (e.g., its binding characteristics compared to parent Abs). There is no universal expression system that can guarantee high yields of recombinant product, as every Ab will pose unique challenges. The choice of an expression system will depend on many factors: the type of rAb being expressed, the sequence of the individual rAb, and the investigator's preferences.

Given that the focus of this book is on application of immunoassays within agricultural biotechnology, studies and experimental findings that describe mammalian expression and posttranslational modification, and so on, support rAbs as therapeutic treatments will not be addressed.

3.6.1 Prokaryotic Systems

3.6.1.1 E. coli Cytoplasmic Expression The expression of rAb fragments in the reducing environment of the cytoplasm of *E. coli* often leads to the formation of insoluble inclusion bodies that contain unfolded protein (Verma et al., 1998). This necessitates the development of refolding protocols to recover functional Abs. In addition, accumulation of foreign protein in bacterial cells may lead to their poor growth. Inducible promoters can, therefore, reduce the risk of cell toxicity. *lac*, *trp*, and their hybrid *tac* promoters, which are regulated by the *lac* repressor and induced by IPTG, are such examples. Another popular promoter is the λP_L promoter responsible for the transcription of the λ DNA molecule, which is regulated by a temperature-sensitive repressor. The T7 RNA promoter can also be used to obtain tightly controlled, high-level expression. Together with the T7 RNA polymerase (encoded by T7 *gene 1*), it is one of the most powerful systems currently employed for recombinant protein productions. The efficiency of T7 RNA polymerase in processing gene transcripts is superior to that of native *E. coli* translation machinery, which has made it very attractive for high-level production of recombinant proteins (Terpe, 2006).

Even under optimized expression conditions with a suitable promoter system, cytoplasmic expression of functional rAbs and fusion proteins can be very challenging. For example, Peipp et al. (2004) reported that functional rAbs and fusion proteins were obtained only in a minority of the cases for which it has been attempted.

3.6.1.2 E. coli Periplasmic Expression Expression of Ab fragments into the periplasmic space of *E. coli* is the most promising route to produce functional rAb fragments. This methodology was first developed for Fv (Skerra and Plückthun, 1988) and Fab (Better et al., 1988; Plückthun and Skerra, 1989) fragments. The periplasmic route is similar to protein synthesis pathway in eukaryotes, in which the nascent proteins go through the endoplasmic reticulum (ER) to the Golgi apparatus. The transport of secretory proteins to the periplasm of *E. coli* is comparable to that of ER and can lead to proper folding. Signal sequences such as *pelB* (pectate lyase), derived from the *pelB* gene of *Erwinia carotovora*, are added in frame with the genes encoding the H and L chains of Ab fragments, resulting in simultaneous secretion of both chains into the periplasmic space (Plückthun and Skerra, 1989).

The advantage of periplasmic secretion over cytoplasmic expression is that it leads to the production of an assembled, fully functional product without having to refold the protein *in vitro*. The oxidative environment of the periplasm allows disulfide bond formation and the low protease environment permits stable folding of the recombinant proteins (Better et al., 1988). Both intra- and interchain protein folding and heterodimer associations occur in the periplasm, which is necessary for proper folding of a functional Ab (Plückthun and Skerra, 1989). The bacterial environment, however, is not efficient in producing full-length Ab molecules.

3.6.2   Eukaryotic Systems

Production of Ab fragments in eukaryotes has also been widely studied, from unicellular yeast cultures (Freyre et al., 2000; Gach et al., 2007) to higher plants (Hiatt et al., 1989). In this section, different eukaryotic expression systems used for recombinant antibody production will be described and their individual pros and cons will be discussed.

3.6.2.1   Yeast Expression   The main advantages of yeast over bacterial systems are related to the fact that yeast is both a microorganism and a eukaryote. Frenken et al. (1998, 2000) produced both scFv and V_HH in *E. coli* and *Saccharomyces cerevisiae* and found that V_HH production levels in the latter organism was up to 100 mg/L, that is, 1000-fold higher than *E. coli*. Furthermore, *E. coli*-produced V_HH exhibited less specific activity (higher fraction of incorrectly folded soluble species present in the periplasmic preparation) than yeast. In another study (Huang and Shusta, 2006), two scFvs were expressed at 0.25 mg/L in *E. coli*, while yields in *Pichia pastoris* were between 60 and 250 mg/L.

Yeast provides folding pathways for heterologous proteins and when yeast signal sequences are used, correctly folded proteins can be secreted into culture medium. Moreover, yeasts rapidly grow on simple growth media and, therefore, represent an attractive option for industrial-scale production of rAbs. Proteins that may accumulate as insoluble inclusion bodies in *E. coli* are often soluble when expressed in yeast (Freyre et al., 2000). In addition, the degradation of heterologous proteins, often a problem in *E. coli*, is usually reduced in yeast. Another unique feature of yeast is that it has the option of stable transformation without the use of antibiotic resistance, such as ampicillin, or kanamycin resistance commonly used with *E. coli*. *E. coli* uses episomal vectors that propagate extrachromosomally, whereas in yeast the vector carrying the gene of interests is integrated into the host genome by homologous recombination (Joosten et al., 2003).

Secretion of functional Fab and whole immunoglobulin (IgG) Abs in yeast was first demonstrated by Horwitz et al. (1988) in *S. cerevisiae*, although expression levels were rather low (approximately 200 ng Fab/mL). Many efforts have been devoted to improving the expression level of Ab in yeast cells. Yeast secretion yields have been reported to be as high as 1.2 g/L for scFv fragment in fed-batch fermentation cultures (Freyre et al., 2000). Elements that control the expression level of foreign protein in yeast have been well studied. The yeast invertase signal sequence, the PGK promoter, and the polyadenylation signal were used in the study of Fab and IgG expression (Horwitz et al., 1988). The proteins were correctly folded and whole Ab and Fab were purified from the culture supernatants. The antibodies behaved indistinguishably from their lymphoid cell-derived counterparts in both direct and competition binding assays. The chimeric whole Ab from yeast exhibited the same ADCC activity as the chimeric Ab from Sp2/0 cells. Codon optimization and lower repetitive AT and GC content also improve the expression levels of some rAbs by 5–10-fold in standard shake flask cultures (Sinclair and Choy, 2002; Woo et al., 2002), but this is not the case for all rAbs (Outchkourov et al., 2002).

Nonconventional species of yeasts and filamentous fungi have been tested for expression of Abs and rAb fragments. Abdel-Salam et al. (2001) reported the expression of Fab in *Hansenula polymorpha*. Nevertheless, Ab monomers (kappa and gamma chains) did not assemble into a heterodimer and were poorly secreted. *Yarrowia lipolytica* and *Kluyveromyces lactis* have also been reported for scFv production (Joosten et al., 2003). Filamentous fungus *Trichoderma reesei* was also reported to have secreted 1 mg/mL of Fab into the growth medium under shake flask conditions and the yield could be increased to 150 mg/L in bioreactor cultures by fusing the heavy chain Fv to *T. reesei* cellobiohydrolase I (CBHI) enzyme (Nyyssönen et al., 1993; Nyyssönen and Keränen, 1995).

A strong preference has been given to *P. pastoris* for its aerobic growth: a key physiological trait that greatly facilitates culturing at high cell densities compared to the fermentative baker's yeast *S. cerevisiae* (Cregg et al., 1993). An scFv green fluorescent protein was expressed by methylotrophic yeast *P. pastoris* under the methanol-inducible alcohol oxidase 1 (AOX1) promoter and secreted into the growth medium. The soluble fusion protein was properly folded without additional renaturation or solubilization (Petrausch et al., 2007). Rahbarizadeh et al. (2006) found that high inoculum densities limit growth potential but gave rise to a higher level of V_HH production in *P. pastoris*. Medium composition and preinduction osmotic stress were found to have the greatest influence on yield and can be improved when casamino acid or EDTA is included in growth medium. Glycerol supplementation during induction resulted in increased growth rates and biomass accumulation; however, the expression of scFv was repressed (Hellwig et al., 2001). Gasser et al. (2006) first engineered the protein disulfide isomerase (PDI) and the unfolded protein response (UPR) transcription factor HAC1 to be constitutively overexpressed in *P. pastoris* (Mattanovich et al., 2004). While the overexpression of HAC1 led to a moderate increase of Fab secretion, that is, 1.3-fold, PDI enabled an increase of the Fab secretion level by 1.9-fold. Hence, the formation of interchain disulfide bonds can be seen as a major rate-limiting factor to Fab assembly and subsequent secretion.

3.6.2.2 *Insect Cell Expression*   Insect cell expression systems have been shown to be a viable alternative to standard microbial and mammalian systems designed to express rAb fragments and full-size mAbs, particularly for Abs designed for use as therapeutic proteins (Verma et al., 1998; Guttieri et al., 2000; Demangel et al., 2005). Not only the baculovirus-mediated gene expression in insect cells produces large amounts of the foreign protein while allowing it to retain its functional activity, but

also it has a highly restricted host range and is less likely to have contamination during the downstream process and causes harmful effects in end users. Between 1 and 500 mg of recombinant protein per liter of infected cells has been reported (Guttieri et al., 2000; Filpula, 2007). However, the success of a foreign gene expression in insect cells depends on a number of factors. High-quality growth media and careful culturing is required, and for optimal results, the insect cells should be highly viable and in the log phase of growth. Unfortunately, expression of the foreign protein is controlled by a very late viral promoter and peaks while the infection culminates in death of the host cell, thereby allowing only transient expression of the rAb. Previous studies also suggested that during the late phase of baculovirus infection, the host's secretory pathway can become impaired (Reavy et al., 2000).

An alternative to baculovirus-mediated expression is based on stable transformation of the gene, under the control of an appropriate promoter, into insect cells. For example, the *Drosophila* metallothionein promoter has been used to control expression and found to be tightly regulated, directing high levels of transcription when induced by heavy metals, such as cadmium or copper (Guttieri et al., 2000). Moreover, the polyhedrin promoter is considerably stronger than most other eukaryotic promoters, and therefore enables the gene to be transcribed at a high level, causing recombinant proteins to be secreted in the insect cell culture in large amounts (Tan and Lam, 1999). To allow for the continuous expression of Ab genes, Ab DNA can be inserted into the insect cell chromosomes under the control of the baculovirus immediate to the early gene promoter (IEI), which is recognized by the insect cell RNA polymerase (Guttieri et al., 2000). The concentration of IgG recovered from the transformed insect cell culture medium was considerably lower (approximately 0.06 μg of IgG/mL) than the level generated by infection with the baculovirus recombinant (approximately 9 μg of IgG/mL) and with the predicted yield from hybridoma cells (1–10 μg of IgG/mL). Nevertheless, the baculovirus system is still considered a viable alternative to microbial expression systems for whole Ab molecule expression and when posttranslational modification, for example, glycosylation, is required.

3.6.2.3 In Planta Expression   "Plantibodies," a contemporary term to describe full-size Ab or rAb fragments expressed in plant tissues, was first reported almost 20 years ago by Hiatt et al. (1989). Since then many researchers have been trying to express and scale up the production of Ab or fragments in plants (Fischer et al., 1999; Peeters et al., 2001; Stoger et al., 2002). Using plants as bioreactors presents many advantages related to manufacturing facilities, production costs, and biosafety, among others. However, despite the production cost reduction and the biocomparability of plantibodies with their conventional Ab counterparts, contamination risk exists with mycotoxins, alkaloids, and allergenic and immunogenic proteins, especially if the expression host is a tobacco plant. However, the favorable properties of plants are likely to make them a useful alternative for small-, medium-, and large-scale production throughout the development of new Ab-based pharmaceuticals or diagnostic reagents. For example, Khoudi et al. (1999) reported the production of a functional, purified antihuman IgG mAb through expression of its encoding genes in perennial transgenic alfalfa.

The deposition and storage of a scFv and a rAb fragment in pea seeds was reported using a seed-specific USP promoter (Saalbach et al., 2001) in cereals (Stoger et al., 2000), tobacco plants (Ko and Koprowski, 2005), or suspension cells (Yano et al., 2004). In particular, seeds offer special advantages, such as ease of handling and long-term storage stability. Nevertheless, most of the plantibody studies have focused on expression in leaves. Arguably, there may not be a high enough demand for diagnostic Ab that requires the large-scale Ab production within plant expression; nonetheless, the idea of expressing functional rAb molecules in plant has already provided a new way of understanding plant/virus interaction (Sudarshana et al., 2007). The studying and/or altering the function of an antigen *in vivo*, also termed immunomodulation, has also given new insights into plant physiology (Tavlodoraki et al., 1993; Artsaenko et al., 1995, 1998; Conrad and Fiedler, 1998) and plant pathogen resistance (Peschen et al., 2004) or potentially confers herbicide resistance to plants (Almquist et al., 2004; Weiss et al., 2006).

Currently, *Agrobacterium* and particle bombardment are the most commonly used technologies for plant transformation. A single plant binary vector carrying genes encoding heavy and light chains under two different promoters has been used to express both genes and to assemble functional full-size Ab (Düring et al., 1990). The major drawback of the binary vector is low transformation efficiency due to the large size of the Ab DNA, in addition to the enormous binary vector required (Komori et al., 2007). Instead of molecularly stacking the heavy and light Ab genes in the same vector, crossing two transgenic lines separately expressing the heavy and the light chain provides an alternative that is less demanding in terms of molecular biology (Schillberg et al., 2003). The particle bombardment approach can insert genes into the genome of the plant nucleus or the plastid genome, depending on the vector construct. Chloroplasts have been used for stable expression of mAb in the chloroplast genome. Advantages of targeting the chloroplast genome include no position effect, no gene silencing, high expression/accumulation, and minimized environmental concerns (Daniell et al., 2004). Chloroplasts can process proteins with disulfide bridges, which is required for proper folding of proteins. Although stably transformed (transgenic) plants are able to express correctly folded and functional Ab of both IgG and IgA classes (Ma et al., 1995) or its fragments (Owen et al., 1992), yields are generally very low (usually in the range of 1–40 μg/g of fresh biomass). In addition, the time necessary to generate the first grams of research Ab material is very long, requiring over 2 years (Giritch et al., 2006).

To shorten the time required to obtain the functional antibodies from plants, transient expression using agroinfiltration or plant viral vector system has been used (Ko and Koprowski, 2005; Giritch et al., 2006). Using the subgenomic promoter of TMV coat protein and tobacco mild green mosaic virus variant U5 coat protein, two separate virus vectors were used to coinfect *Nicotiana benthamiana* leaf cells. Full-length heavy chain and light chain proteins were produced and assembled into glycosylated functional full-size Ab (Verch et al., 1998). Regeneration time is much shorter than that for stable transformation, and different host plants can be infected by the same viral vector, allowing time-efficient screening for recombinant gene expression. More important, it saves the labor-intensive step of crossing different transgenic plants producing different Ab subunits.

Many different Ab formats have been expressed successfully in plants. These include full-size antibodies (Ma et al., 1995; Giritch et al., 2006), camelid heavy chain antibodies (Ismaili et al., 2007), Fab fragments (Yano et al., 2004; Weiss et al., 2006), and scFvs (Owen et al., 1992; Stoger et al., 2000; Saalbach et al., 2001). Regardless of which form of Ab fragments is being produced by the plant system, ER is an important site of the major biofunctions of synthesis, assembly, and glycosylation of these protein molecules. Antibodies are often targeted to subcellular compartments or the apoplastic space since most proteins are more stable in the subcellular compartment than in the cytosol. The greatest accumulation of full-size antibodies has been obtained by targeting to the apoplastic space and the greatest accumulation of scFv antibodies was obtained when the antibodies were retained in the ER (Daniell et al., 2004).

The average yield of the recombinant protein using plant secretion sequence (SS) is usually 0.1–2% of total soluble protein (Yano et al., 2004). So far, three different leader sequences have been tested: human-derived leader sequence (LS), dicotyledonous calreticulum-derived SS, and monocotyledonous hordothionin-derived SS. The latter did not consistently result in mAb expression, while plants transformed with the dicot SS construct grew more vigorously and expressed the antibodies more consistently than transgenic cells made with human LS. The promoter is another critical element determining the expression level, for example, cauliflower mosaic virus 35S (CaMV35S) promoter for dicotyledonous plants (Olea-Popelka et al., 2005; McLean et al., 2007) and the maize ubiquitin 1 promoter preferable in monocotyledonous plants. Petruccelli et al. 2006 successfully produced a full-length anti-rabies mAb molecule in transgenic tobacco using CaMV35S to drive the heavy chain, while the potato proteinase inhibitor II (pin2) promoter controlling light chain gene produced functional mAb. An effective plant production system for recombinant Ab fragments also requires the appropriate control of posttranslational processing of recombinant products. The ER retention signal sequence (KDEL) has little effect on the accumulation of Ab in the transgenic leaves, but leads to higher Ab yields in seeds. The proteins purified from leaves contain complex N-glycans, including Lewis a epitopes, as typically found in extracellular glycoproteins and consistent with an efficient ER retention and the cis-Golgi retrieval of the Ab. The glycosylated proteins purified from the seed were partially secreted and sorted to protein storage vacuoles (PSVs) in seeds and not found in the ER (Ramirez et al., 2003). More important, full-size antibody (McLean et al., 2007) or scFv fragments (Almquist et al., 2004, 2006; Makvandi-Nejad et al., 2005) produced in tobacco have been proved to be functional. This further supports the fact that "plantibodies" could be an alternate source of both diagnostic and therapeutic antibodies.

3.7 APPLICATION TO IMMUNOASSAYS AND AGRICULTURAL BIOTECHNOLOGY

3.7.1 Immunoassays

Given the widespread adoption of genetically engineered (GE or GM) crops within global agriculture and strict regulations governing their coexistence with non-GM crops (Demont and Devos, 2008), mandatory product labeling (e.g., Codex Alimentarius, EU, 2003), and international trade, there is an increased need to standardize and continuously upgrade the analytical methods used to monitor and verify the presence of GM traits within food and feed (Dong et al., 2008).

Immunoassays for the detection of protein expression within GM crops are currently limited to enzyme-linked immunosorbent assays (ELISA) based on either trait-specific polyclonal serum or mAbs, as well as rapid lateral flow/immunochromatographic diagnostic strip kits (typically based on polyclonal serum) (Lin et al., 2001; Ma et al., 2005; Tripathi, 2005; Van den Bulcke et al., 2007). Both immunoassay formats are widely used in the United States, Canada, and other countries that produce and export GM crops as a primary method for determining minimum thresholds of GM products within commercial food and feed shipments (Stave, 2002).

A wide array of immunoassay products are currently available to assess GM expression within agricultural crops. These testing formats are developed within private or in-house programs or are available as commercial products from companies such as Agdia Inc. (Elkhart, IN, USA), EnviroLogix Inc. (Portland, ME, USA), and Strategic Diagnostics Inc. (Newark, DE, USA). Although these platforms are very useful qualitative and quantitative tools to detect the presence of GM traits within commercial crops, the results generated by current immunoassay products are often limited by detection limit and overall precision, or consistency, of results. For example, Ma et al. (2005) reported that both ELISA and DNA-based (PCR) tests were capable of distinguishing samples with GM concentrations between 0.1% and 0.5%, but the precision at this range was often poor as results were inconsistent. Another consideration is the disparity between the time, costs, and resources required to develop GM-specific immunoassay reagents based on traditional mAb and polyclonal Ab platforms relative to the rapid progression of GM crops toward newer events based on multiple/stacked traits.

Antibody engineering has rapidly evolved over the past 20 years to the point where it is possible to isolate and optimize high-affinity and Ag-specific rAb fragments as a cost-effective alternative to established mAb technologies (Holliger and Hudson, 2005). In this regard, rAb fragments (i.e., Fab, scFv, V_HH) could represent a means to complement, and potentially displace, polyclonal serum and mAbs used within immunoassay formats. In fact, one could argue that the application of rAb technology to detect GM proteins within plant matrices would be a natural extension of published research that demonstrated effective use of rAb technology within immunoassay formats to detect ligands ranging from mycotoxins (Wang et al., 2007, 2008) to caffeine (Ladenson et al., 2006) to large complex protein toxins such as cholera toxin, ricin, and staphylococcal enterotoxin B (Goldman et al., 2006). The application of rAb technology within current ELISA and lateral flow stick GM immunoassay formats would obviously depend on any benefits, or advantages, imparted by this new technology. The following is an initial (and by no means exhaustive) list of "key traits" that rAb fragments could exhibit relative to current mAb and polyclonal technology: enhanced GM trait specificity based on recognition of a single target protein, improved sensitivity to target protein, decreased development costs and timelines, ability to combine rAb reagents to enable simultaneous and customized detection of stacked GM traits, and improved durability and speed of deployment imparting convenient and real-time immunoassay results.

Apart from the advantages associated with improved affinity, cost-effective expression, and Ag specificity (as discussed previously), rAb technology also imparts unique opportunities to optimize the physical characteristics of rAb fragments to suit the intended use pattern. For example, it is possible to express single-domain V_HH fragments in a multimeric pentamer format as a means of increasing antibody avidity (Zhang et al., 2004) with potential use in lateral flow devices and ELISA kits. Depending on the intended application and immunoassay design, it is also possible to couple rAb fragments to larger indicator molecules (e.g., fluorophores) to impart visual GM detection in real time. Finally, given that V_HH fragments have been shown to have high affinity for large, complex enzymes (Desmyter et al., 1996, 2002), such fragments may be particularly suited for detection of GM traits that express enzymes that confer tolerance to the herbicide glyphosate (GAT, EPSPS, etc.).

3.7.2 Research Applications to Agricultural Biotechnology

Apart from standard immunoassay detection procedures, it is also possible to apply rAb technology to study structure–function relationships between GM traits and the target organism. For example, Fernández et al. (2008) reported the use of scFv phage display libraries and various biopanning techniques to characterize epitopes that mediate binding of Cry1Ab and Cry11Aa toxins with target protein receptors within *Manduca sextra* and *Aedes aegypti*, respectively. Such results could provide insights into the mechanism of insect specificity and mode of action of Cry toxins to enable strategies designed to improve target insect toxicity and specificity. Ongoing research in this area could be useful to study, and potentially improve upon, the precise mechanism and toxicity by which target transgenic events work within the host crop and efficacy versus the target organism. Finally, rAbs are also well suited to study and modify metabolic pathways within plants as a means of developing new transgenic traits and GM products (Nölke et al., 2006).

REFERENCES

Abdel-Salam, H. A., El-Khamissy, T., Enan, G. A., and Hollenberg, C. P. Expression of mouse anticreatine kinase (MAK33) monoclonal antibody in the yeast *Hansenula polymorpha*. *Appl. Microbiol. Biotechnol.*, **2001**, *56*, 157–164.

Adams, G. P. and Schier, R. Generating improved single-chain Fv molecules for tumor targeting. *J. Immunol. Methods*, **1999**, *231*, 249–260.

Almquist, K. C., Niu, Y., McLean, M. D., Mena, F. L., Yau, K. Y., Brown, K., Brandle, J. E., and Hall, J. C. Immunomodulation confers herbicide resistance in plants. *Plant Biotechnol. J.*, **2004**, *2*, 189–197.

Almquist, K. C., McLean, M. D., Niu, Y., Byrne, G., Olea-Popelka, F. C., Murrant, C., Barclay, J., and Hall, J. C. Expression of an anti-botulinum toxin A neutralizing single-chain Fv recombinant Ab in transgenic tobacco. *Vaccine*, **2006**, *24*, 2079–2086.

Andris, J. S., Abraham, S. R., Pascual, V., Pistillo, M. P., Mantero, S., Ferrara, G. B., and Capra, J. D. The human antibody repertoire: heavy and light chain variable region gene usage in six alloantibodies specific for human HLA class I and class II alloantigens. *Mol. Immunol.*, **1995**, *32*, 1105–1122.

Arbabi-Ghahroudi, M., Tanha, J., and MacKenzie, R. Prokaryotic expression of antibodies. *Cancer Metastasis Rev.*, **2005**, *24*, 501–519.

Artsaenko, O., Peisker, M., zur Nieden, U., Fiedler, U., Weiler, E. W., Muntz, K., and Conrad, U. Expression of a single-chain Fv antibody against abscisic acid creates a wilty phenotype in transgenic tobacco. *Plant J.*, **1995**, *8*, 745–750.

Artsaenko, O., Kettig, B., Fiedler, U., Conrad, U., and During, K. Potato tubers as a biofactory for recombinant antibodies. *Mol. Breeding*, **1998**, *4*, 313–319.

Atwell, J. L., Breheney, K. A., Lawrence, L. J., McCoy, A. J., Kortt, A. A., and Hudson, P. J. scFv multimers of the anti-neuraminidase antibody NC10: length of the linker between VH and VL domains dictates precisely the transition between diabodies and triabodies. *Protein Eng.*, **1999**, *12*, 597–604.

Balint, R. F. and Larrick, J. W. Antibody engineering by parsimonious mutagenesis. *Gene* **1993**, *137*, 109–118.

Barbas, C. F., 3rd, Bain, J. D., Hoekstra, D. M., and Lerner, R. A. Semisynthetic combinatorial antibody libraries: a chemical solution to the diversity problem. *Proc. Natl. Acad. Sci. USA*, **1992**, *89*, 4457–4461.

Benhar, I. and Pastan, I. Cloning, expression, and characterization of the Fv fragments of the anti-carbohydrate mAbs B1 and B5 as single chain immunotoxins. *Protein Eng.*, **1994**, *7*, 1509–1515.

Better, M., Chang, C. P., Robinson, R. R., and Horwitz, A. H. *Escherchia coli* secretion of an active chimeric antibody fragment. *Science*, **1988**, *240*, 1041–1043.

Bird, R. E., Hardman, K. D., Jacobson, J. W., Johnson, S., Kaufman, B. M., Lee, S. M., Lee, S. H., Pope, G. S., Riordan, G. S., and Whitlow, M. Single-chain antigen-binding proteins. *Science*, **1989**, *242*, 423–426.

Boder, E. T. and Wittrup, K. D. Yeast surface display for screening combinatorial polypeptide libraries. *Nat. Biotechnol.*, **1997**, *15*, 553–557.

Borrebaeck, C. K. and Ohlin, M. Antibody evolution beyond nature. *Nat. Biotechnol.*, **2002**, *20*, 1189–1190.

de Bruin, K., Spelt, J., Mol, R., Koes, R., and Quattrocchio, F. Selection of high-affinity phage antibodies from phage display libraries. *Nat. Biotechnol.*, **1999**, *17*, 397–399.

de Kruif, J., Boel, E., and Logtenberg, T. Selection and application of human single chain Fv antibody fragments from a semi-synthetic phage antibody display library with designed CDR3 regions. *J. Mol. Biol.*, **1995**, *248*, 97–105.

Burton, D. R., Barbas, C. F., Persson, M. A., Koenig, S., Chanock, R. M., and Lerner, R. A. A large array of human monoclonal antibodies to type 1 human immunodeficiency virus from combinatorial libraries of asymptomatic seropositive individuals. *Proc. Natl. Acad. Sci. USA*, **1991**, *88*, 10134–10137.

Chames, P., Coulon, S., and Baty, D. Improving the affinity and the fine specificity of an anti-cortisol antibody by parsimonious mutagenesis and phage display. *J. Immunol.*, **1998**, *161*, 5421–5429.

Choi, G. H., Lee, D. H., Min, W. K., Cho, Y. J., Kweon, D. H., Son, D. H., Park, K., and Seo, J. H. Cloning, expression and characterization of single-chain variable fragment antibody against mycotoxin deoxynivalenol in recombinant *Escherichia coli*. *Protein Expr. Purif.*, **2004**, *35*, 84–92.

Chowdhury, P. S. and Pastan, I. Improving antibody affinity by mimicking somatic hypermutation *in vitro*. *Nat. Biotechnol.*, **1999**, *17* 568–572.

Clackson, T., Hoogenboom, H. R., Griffiths, A. D., and Winter, G. Making antibody fragments using phage display libraries. *Nature*, **1991**, *352*, 624–628.

Coia, G., Hudson, P. J., and Irving, R. A. Protein affinity maturation *in vivo* using *E. coli* mutator cells. *J. Immunol. Methods*, **2001a**, *251*, 187–193.

Coia, G., Pontes-Braz, L., Nuttal, S. D., Hudson, P. J., and Irving, R. A. Panning and selection of proteins using ribosome display. *J. Immunol. Methods*, **2001b**, *254*, 191–197.

Conrad, U. and Fiedler, U. Compartment-specific accumulation of recombinant immunoglobulins in plant cells: an essential tool for antibody production and immunomodulation of physiological functions and pathogen activity. *Plant Mol. Biol.*, **1998**, *38*, 101–109.

Cook, G. P. and Tomlinson, I. M. The human immunoglobulin VH repertoire. *Immunol. Today*, **1995**, *16*, 237–242.

Cregg, J. M., Vedvick, T. S., and Easchke, W. C. Recent advantages in the expression of foreign gene in *Pichia pastoris*. *Bio/Technology*, **1993**, *11*, 905–910.

Daniell, H., Carmona-Sanchez, O., and Burns, B. Chloroplast derived antibodies, biopharmaceuticals and edible vaccines. In Fischer, R.and Schillberg, S. (Eds.), *Molecular Farming*, Wiley-VCH, Weinheim, 2004, pp. 113–133.

Daugherty, P. S., Chen, G., Olsen, M. J., Iverson, B. L., and Georgiou, G. Antibody affinity maturation using bacterial surface display. *Protein Eng.*, **1998**, *11*, 825–832.

Daugherty, P. S., Olsen, M. J., Iverson, B. L., and Georgiou, G. Development of an optimized expression system for the screening of antibody libraries displayed on the *Escherichia coli* surface. *Protein Eng.*, **1999**, *12*, 613–621.

Davies, J. and Riechmann, L. 'Camelising' human antibody fragments: NMR studies on V_H domains. *FEBS Lett.*, **1994**, *339*, 285–290.

Davies, J. and Riechmann, L. Single antibody domains as small recognition units: design and *in vitro* antigen selection of camelized, human V_H domains with improved protein stability. *Protein Eng.*, **1996a**, *9*, 531–537.

Davies, J. and Riechmann, L. Affinity improvement of single antibody VH domains: residues in all three hypervariable regions affect antigen binding. *Immunotechnology*, **1996b**, *2*, 169–179.

Demangel, C., Zhou, J., Choo, A. B. H., Shoebridge, G., Halliday, G. M., and Britton, W. J. Single chain antibody fragments for the selective targeting of antigens to dendritic cells. *Mol. Immunol.*, **2005**, *42*, 979–985.

Demont, M. and Devos, Y. Regulating coexistence of GM and non-GM crops without jeopardizing economic incentives. *Trends Biotechnol.*, **2008**, *26*, 353–358.

Desmyter, A., Transue, T. R., Arbabi Ghahroudi, M., Thi, M. H., Poortmas, F., Hamers, R., Muyldermans, S., and Wyns, L. Crystal structure of a camel single-domain VH antibody fragment in complex with lysozyme. *Nat. Struct. Biol.*, **1996**, *3*, 803–811.

Desmyter, A., Spinelli, S., Payan, F., Lauwereys, M., Wyns, L., Muyldermans, S., and Cambillau, C. Three camelid VHH domains in complex with porcine pancreatic α-amylase. *J. Biol. Chem.*, **2002**, *277*, 23645–23650.

Desplancq, D., King, D. J., Lawson, A. D. G., and Mountain, A. Multimerization behavior of single chain Fv variants for the tumor-binding antibody B72.3. *Protein Eng.*, **1994**, *7*, 1027–1033.

Dong, W., Yang, L., Shen, K., Kim, B., Kleter, G. A., Marvin, H. J. P., Guo, R., Liang, W., and Zhang, D. GMDD: a database of GMO detection methods. *BMC Bioinf.*, **2008**, *9*, Art. 260.

Dörsam, H., Rohrbach, P., Kürschner, T., Kipriyanov, S., Renner, S., Braunagel, M., Welschof, M., and Little, M. Antibodies to steroids from a small human naive IgM library. *FEBS Lett.*, **1997**, *414*, 7–13.

Düring, K., Hippe, S., Kreuzaler, F., and Schell, J. Synthesis and self-assembly of a functional monoclonal antibody in transgenic *Nicotiana tabacum*. *Plant Mol. Biol.*, **1990**, *15*, 281–293.

EU. Regulation (EC) 1829/2003 and 1830/2003 of the European Parliament and of the Council of 22 September 2003 on genetically modified food and feed. *Off. J. Eur. Commun.*, **2003**, *L268*, 1–28, http://eurlex.europa.eu/LexUriServ/site/en/oj/2003/l_268/l_26820031018en00010023.pdf.

Fellouse, F. A., Esaki, K., Birtalan, S., Raptis, D., Cancasci, V. J., Koide, A., Jhurani, P., Vasser, M., Wiesmann, C., Kossiakoff, A. A., Koide, S., and Sidhu, S. S. High-throughput generation of synthetic antibodies from highly functional minimalist phage-displayed libraries. *J. Mol. Biol.*, **2007**, *373*, 924–940.

Fernández, L. E., Gómez, I., Pacheco, S., Arenas, I., Gilla, S. S., Bravo, A., and Soberón, M., Employing phage display to study the mode of action of *Bacillus thuringiensis* Cry toxins. *Peptides*, **2008**, *29*, 324–329.

Filpula, D. Antibody engineering and modification technologies. *Biomol. Eng.*, **2007**, *24*, 201–215.

Fischer, R., Vaquero-Martin, C., Sack, M., Drossard, J., Emans, N., and Commandeur, U. Towards molecular farming in the future: transient protein expression in plants. *Biotechnol. Appl. Biochem.*, **1999**, *30*, 113–116.

Forrer, P., Jung, S, and Plückthun, A. Beyond binding: using phage display to select for structure, folding and enzymatic activity in proteins. *Curr. Opin. Struct. Biol.*, **1999**, *9*, 514–520.

Francisco, J. A., Campbell, R., Iverson, B. L., and Georgiou, G. Production and fluorescence-activated cell sorting of *Escherichia coli* expressing a functional antibody fragment on the external surface. *Proc. Natl. Acad. Sci. USA*, **1993**, *90*, 10444–10448.

Frenken, L. G., Hessing, J. G., van den Hondel, C. A., and Verrips, C. T. Recent advances in the large-scale production of antibody fragments using lower eukaryotic microorganisms. *Res. Immunol.*, **1998**, *149*, 589–599.

Frenken, L. G., van der Linden, R. H., Hermans, P. W., Bos, J. W., Ruuls, R. C., de Geus, B., and Verrips, C. T. Isolation of antigen specific llama V_{HH} antibody fragments and their high level secretion by *Saccharomyces cerevisiae*. *J. Biotechnol.*, **2000**, *78*, 11–21.

Freyre, F. M., Vazquez, J. E., Ayala, M., Canaan-Haden, L., Bell, H., Rodriguez, I., Gonzalez, A., Cintado, A., and Gavilondo, V. Very high expression of an anti-carcinoembryonic antigen single chain Fv antibody fragment in the yeast *Pichia pastoris*. *J. Biotechnol.*, **2000**, *76*, 157–163.

Fuchs, P., Breitling, F., Dübel, S., Seehaus, T., and Little, M. Targeting recombinant antibodies to the surface of *Escherchia coli*: fusion to a peptidoglycan associated lipoprotein. *Biotechnology*, **1991**, *9*, 1369–1372.

Gach, J. S., Maurer, M., Hahn, R., Gasser, B., Mattanovich, D., Katinger, H., and Kunert, R. High level expression of a promising anti-idiotypic antibody fragment vaccine against HIV-1 in *Pichia pastoris*. *J. Biotechnol.*, **2007**, *128*, 735–746.

Gasser, B., Maurer, M., Gach, J., Kunert, R., and Mattanovich, D. Engineering of *Pichia pastoris* for improved production of antibody fragments. *Biotechnol. Bioeng.*, **2006**, *94*, 353–361.

Georgiou, G., Poetschke, H. L., Stathopoulos, C., and Francisco, J. A. Practical applications of engineering gram-negative bacterial cell surfaces. *Trends Biotechnol.*, **1993**, *11*, 6–10.

Georgiou, G., Stathopoulous, C., Daugherty, P. S., Nayak, A. R., Iverson, B. L., and Curtiss, R. Display of heterologous proteins on the surface of microorganisms: from the screening of combinatorial libraries to live recombinant vaccines. *Nat. Biotechnol.*, **1997**, *15*, 29–34.

Ghahroudi, M., Desmyter, A., Wyns, L., Hamers, R., and Muyldermans, S. Selection and identification of single domain antibody fragments from camel heavy-chain antibodies. *FEBS Lett.*, **1997**, *414*, 521–526.

Giritch, A., Marillonnet, S., Engler, C., van Eldik, G., Botterman, J., Klimyuk, V., and Gleba, Y. Rapid high-yield expression of full-size IgG antibodies in plants coinfected with non-competing viral vectors. *Proc. Natl. Acad. Sci. USA*, **2006**, *103*, 14701–14706.

Goldman, E. R., Anderson, G. P., Liu, J. L., Delehanty, J. B., Sherwood, L. J., Olson, L. E., Cummins, L. B., and Hayhurst, A. Facile generation of heat-stable antiviral and antitoxin single domain antibodies from a semisynthetic llama library. *Anal. Chem.*, **2006**, *28*, 8245–8255.

Griffiths, A. D., Malmqvist, M., Marks, J. D., Bye, J. M., Embleton, M. J., McCafferty, J., Baier, M., Holliger, K. P., Gorick, B. D., Hughes-Jones, N. C., Hoogenboom, H. R., and Winter, G. Human anti-self antibodies with high specificity from phage display libraries. *EMBO J.*, **1993**, *12*, 725–734.

Griffiths, A. D., Williams, S. C., Hartley, O., Tomlinson, I. M., Waterhouse, P., Crosby, W. L., Kontermann, R. E., Jones, P. T., Low, N. M., Prospero, T. D., Hoogenboom, H. R., Nissim, A., Cox, J. P. L., Harrison, J. L., Zaccolo, M., Gherardi, E., and Winter, G. Isolation of high affinity human antibodies directly from large synthetic repertoires. *EMBO J.*, **1994**, *13*, 3245–3260.

Gunneriusson, E., Samuelson, P., Uhlen, M., Nygren, P. A., and Stahl, S. Surface display of a functional single-chain Fv antibody on staphylococci. *J. Bacteriol.*, **1996**, *178*, 1341–1346.

Guttieri, M. C., Bookwalter, C., and Schmaljohn, C. Expression of a human, neutralizing monoclonal antibody specific to Puumala virus G2-protein in stably-transformed insect cells. *J. Immunol. Methods*, **2000**, *246*, 97–108.

Hanes, J. and Plückthun, A. *In vitro* selection and evolution of functional proteins by using ribosome display. *Proc. Natl. Acad. Sci. USA*, **1997**, *94*, 4937–4942.

Hanes, J., Jermutus, L., Weber-Bornhauser, S., Bosshard, H. R., and Plückthun, A. Ribosome display efficiently selects and evolves high-affinity antibodies *in vitro* from immune libraries. *Proc. Natl. Acad. Sci. USA*, **1998**, *95*, 14130–14135.

Harding, S. E., Longman, E., Carrasco, B., Ortega, A., and Garcia, J. Studying antibody confirmations by ultracentrifugation and hydrodynamic modeling. In Lo, B. K. C. (Ed.), *Antibody Engineering: Methods and Protocols*, Humana Press, Totowa, NJ, 2004, pp. 93–113.

He, M. and Taussig, M. J. Antibody–ribosome–mRNA (ARM) complexes as efficient selection particles for *in vitro* display and evolution of antibody combining sites. *Nucleic Acids Res.*, **1997**, *25*, 5132–5134.

Hellwig, S., Emde, F., Raven, N. P. G., Henke, M., van der Logt, P., and Fischer, R. Analysis of single-chain antibody production in *Pichia pastoris* using on-line methanol control in fed-batch and mixed-feed fermentations. *Biotechnol. Bioeng.*, **2001**, *74*, 344–352.

Hiatt, A. and Ma, J. K. Monoclonal antibody engineering in plants. *FEBS Lett.*, **1992**, *307*, 71–75.

Hiatt, A., Cafferkey, R., and Bowdish, K. Production of antibodies in transgenic plants. *Nature*, **1989**, *342*, 76–78.

Holliger, P. and Hudson, P. J. Engineered antibody fragments and the rise of the single domains. *Nat. Biotechnol.*, **2005**, *23*, 1126–1136.

Hoogenboom, H. R. Selecting and screening recombinant antibody libraries. *Nat. Biotechnol.*, **2005**, *23*, 1105–1116.

Hoogenboom, H. R. Designing and optimizing library selection strategies for generating high-affinity antibodies. *Trends Biotechnol.*, **1997**, *15*, 62–70.

Hoogenboom, H. R. and Chames, P. Natural and designer binding sites made by phage display technology. *Immunol. Today*, **2000**, *21*, 371–378.

Hoogenboom, H. R., Marks, J. D., Griffiths, A. D., and Winter, G. Building antibodies from their genes. *Immunol. Rev.*, **1992**, *130*, 41–68.

Hoogenboom, H. R., de Bruine, A. P., Hufon, S. E., Hoet, R. M., Arends, J. W., and Roovers, R. C. Antibody phage display technology and its applications. *Immunotechnology*, **1998**, *4*, 1–20.

Horwitz, A. H., Chang, C. P., Better, M., Hellstrom, K. E., and Robinson, R. R. Secretion of functional antibody and Fab fragment from yeast cells. *Proc. Natl. Acad. Sci. USA*, **1988**, *85*, 8678–8682.

Huang, D. and Shusta, E. V. A yeast platform for the production of single-chain antibody-green fluorescent protein fusions. *Appl. Environ. Microbiol.*, **2006**, *72*, 7748–7759.

Hudson, P. J. Recombinant antibody fragments. *Curr. Opin. Biotechnol.*, **1998**, *9*, 395–402.

Hudson, P. J. Recombinant antibody constructs in cancer therapy. *Curr. Opin. Immunol.*, **1999**, *11*, 548–557.

Huse, W. D., Sastry, L., Iverson, S. A., Kang, A. S., Alting-Mees, M., Burton, D. R., Benkovic, S. J., and Lerner, R. A. Generation of a large combinatorial library of the immunoglobulin repertoire in phage lambda. *Science*, **1989**, *246*, 1275–1281.

Huston, J. S., Mudgett-Hunter, M., Tai, M. S., McCartney, J. E., Warren, F., Haber, E., and Oppermann, H. Protein engineering of single-chain Fv analogs and fusion proteins. *Methods Enzymol.*, **1991**, *203*, 46–48.

Huston, J. S., George, A. J. T., Tai, M. S., McCartney, J. E., Jin, D., Segal, M., Keck, P., and Oppermann, H. Single-chain Fv design and production by preparative folding. In Borrebaeck, C. A. (Ed.), *Antibody Engineering*, Oxford University Press, New York, 1995, pp. 185–227.

Huston, J. S., Margolies, M. N., and Haber, E. Antibody binding sites. *Adv. Protein Chem.*, **1996**, *49*, 329–450.

Huston, J. S., Levinson, D., Mudgett-Hunter, M., Tai, M. S., Novotny, J., Margolies, M. N., Ridge, R. J., Bruccoleri, R. E., Haber, E., Crea, R., and Opperman, H. Protein engineering of antibody binding sites: recovery of specific activity in an anti-digoxin single-chain Fv analogue produced in *Escherichia coli*. *Proc. Natl. Acad. Sci. USA*, **1988**, *85*, 5879–5883.

Irving, R. A., Kortt, A. A., and Hudson, P. J. Affinity maturation of recombinant antibodies using *E. coli* mutator cells. *Immunotechnology*, **1996**, *2*, 127–143.

Irving, R. A., Coia, G., Raicevic, A., and Hudson, P. J. Use of *Escherichia coli* mutator cells to mature antibodies. *Methods Mol. Biol.*, **2002**, *178*, 295–302.

Ismaili, A., Jalali-Javaran, M., Rasaee, M. J., Rahbarizadeh, F., Forouzandeh-Moghadam, M., and Memari, H. R. Production and characterization of anti-(mucin MUC1) single-domain antibody in tobacco (*Nicotiana tabacum* cultivar Xanthi). *Biotechnol. Appl. Biochem.* **2007**, *47*, 11–19.

Jain, M., Kamal, N., and Batra, K. Engineering antibodies for clinical applications. *Trends Biotechnol.*, **2007**, *25*, 307–316.

Jeffrey, P. D., Strong, R. K., Sieker, L. C., Chang, C. Y. Y., Campbell, R. L., Petsko, G. A., Haber, E., Margolies, M. N., and Sherfiff, S. 26-10 Fab-digoxin complex: affinity and specificity due to surface complementarity. *Proc. Natl. Acad. Sci. USA*, **1993**, *90*, 10310–10314.

Joosten, V., Lokman, C., van den Hondel, A. M. J. J., and Punt, P. J. The production of antibody fragments and antibody fusion proteins by yeasts and filamentous fungi. *Microb. Cell Fact.*, **2003**, *2*, 1–15.

Kang, A. S., Jones, T. M., and Burton, D. R. Antibody redesign by chain shuffling from random combinatorial immunoglobulin libraries. *Proc. Natl. Acad. Sci. USA*, **1991**, *88*, 11120–11123.

Khoudi, H., Laberge, S., Ferullo, J. M., Bazin, R., Darveau, A., Castonguay, Y., Allard, G., Lemieux, R., and Vézina, L. P. Production of a diagnostic monoclonal antibody in perennial alfalfa plants. *Biotechnol. Bioeng.*, **1999**, *64*, 135–143.

Kieke, M. C., Cho, B. K., Boder, E. T., Kranz, D. M., and Wittrup, K. D. Isolation of anti-T cell receptor scFv mutants by yeast surface display. *Protein Eng.*, **1997**, *10*, 1303–1310.

Knappik, A. and Plückthun, A. Engineered turns of a recombinant antibody improve its *in vivo* folding. *Protein Eng.*, **1995**, *8*, 81–89.

Knappik, A., Ge, L., Honegger, A., Pack, P., Fischer, M., Wellnhofer, G., Hoess, A., Wölle, J., Plückthun, A., and Virnekäs, B. Fully synthetic human combinatorial antibody libraries (HuCAL) based on modular consensus frameworks and CDRs randomized with trinucleotides. *J. Mol. Biol.*, **2000**, *296*, 57–86.

Ko, K. and Koprowski, H. Plant biopharming of monoclonal antibodies. *Virus Res.*, **2005**, *111*, 93–100.

Köhler, G. and Milstein, C. Continuous cultures of fused cells secreting antibody of predefined specificity. *Nature*, **1975**, *256*, 293–299.

Komori, T., Imayama, T., Kato, N., Ishida, Y., Ueki, J., and Komari, T. Current status of binary vectors and superbinary vectors. *Plant Physiol.*, **2007**, *145*, 1155–1160.

Korpimäki, T., Rosenberg, J., Virtanen, P., Lamminmäki, U., Tuomola, M., and Saviranta, P. Further improvement of broad specificity hapten recognition with protein engineering. *Protein Eng.*, **2003**, *16*, 37–46.

Krebs, B., Rauchenberger, R., Reiffert, S., Rothe, C., Tesar, M., Thomassen, E., Cao, M., Dreier, T., Fischer, D., Höss, A., Inge, L., Knappik, A., Marget, M., Pack, P., Meng, X. Q., Schier, R., Söhlemann, P., Winter, J., Wölle, J., and Kretzschmar, T. High-throughput generation and engineering of recombinant human antibodies. *J. Immunol. Methods*, **2001**, *254*, 67–84.

Ladenson, R. C., Crimmins, D. L., Landt, Y., and Ladenson, J. H. Isolation and characterization of a thermally stable recombinant anti-caffeine heavy-chain antibody fragment. *Anal. Chem.*, **2006**, *78*, 4501–4508.

Lamminmäki, U., Paupério, S., Westerlund-Karlsson, A., Karvinen, J., Virtanen, P. L., Lövgren, T., and Saviranta, P. Expanding the conformational diversity by random insertions to CDRH2 results in improved anti-estradiol antibodies. *J. Mol. Biol.*, **1999**, *291*, 589–602.

Larrick, J. W., Danielsson, L., Brenner, C. A., Abrahamson, M., Fry, K. E., and Borrebaeck, C. A. K. Rapid cloning of rearranged immunoglobulin genes from human hybridoma cells using mixed primers and the polymerase chain reaction. *Biochem. Biophys. Res. Commun.*, **1989**, *60*, 1250–1256.

Li, M. Applications of display technology in protein analysis. *Nat. Biotechnol.*, **2000**, *18*, 1251–1256.

Li, Y., Lipschultz, C. A., Mohan, S., and Smith-Gill, S. J. Mutations of an epitope hot-spot residue alter rate limiting steps of antigen–antibody protein–protein associations. *Biochemistry*, **2001**, *40*, 2011–2022.

Lin, H. Y., Chiang, J. W., and Shih, D. Y. C. Detection of genetically modified soybeans by PCR method and immunoassay kits. *J. Food Drug Anal.*, **2001**, *9*, 160–166.

Little, M., Breitling, F., Dübel, S., Fuchs, P., Braunagel, M., Seehaus, T., and Klewinghaus, I. Universal antibody libraries on phage and bacteria. *Year Immunol.*, **1993**, *7*, 50–55.

Little, M., Kipriyanov, S. M., Le Gall, F., and Moldenhauer, G. Of mice and men: hybridoma and recombinant antibodies. *Immunol. Today*, **2000**, *21*, 364–370.

Lo, B. K. C. *Antibody Engineering: Methods and Protocols*, Humana Press, Totowa, NJ, 2004, p. 561.

Ma, J. K.-C., Hiat, A., Hein, M., Vine, N., Wang, F., Stabila, P., van Dollewerd, C., Mostov, K., and Lehner, T. Generation and assembly of secretory antibodies in plants. *Science*, **1995**, *268*, 716–719.

Ma, B. L., Subedi, K., Evenson, L., and Stewart, G. Evaluation of detection methods for genetically modified traits in genotypes resistant to European corn borer and herbicides. *J. Environ. Sci. Health B*, **2005**, *40*, 633–644.

Makvandi-Nejad, S., McLean, M. D., Hirama, T., Almquist, K. C., MacKenzie, C. R., and Hall, J. C. Transgenic tobacco plants expressing a dimeric single-chain variable fragment (scFv) antibody against *Salmonella enterica* serotype paratyphi B. *Transgenic Res.*, **2005**, *14*, 785–792.

Marks, J. D., Hoogenboom, H. R., Bonnert, T. P., McCafferty, J., Griffiths, A. D., and Winter, G. By-passing immunization. Human antibodies from V-gene libraries displayed on phage. *J. Mol. Biol.*, **1991**, *222*, 581–597.

Mattanovich, D., Gasser, B., Hohenblum, H., and Sauer, M. Stress in recombinant protein producing yeasts. *J. Immunol. Methods*, **2004**, *113*, 121–135.

Maynard, J. and Georgiou, G. Antibody engineering. *Annu. Rev. Biomed. Eng.*, **2000**, *2*, 339–376.

McCafferty, J., Griffiths, A. D., Winter, G., and Chiswell, D. J. Phage antibodies: filamentous phage displaying antibody variable domains. *Nature*, **1990**, *348*, 552–554.

McLean, M. D., Almquist, K. C., Niu, Y., Kimmel, R., Lai, Z., Schreiber, J. R., and Hall, J. C. A human anti-*Pseudomonas aeruginosa* serotype 06ad immunoglobulin G1 expressed in transgenic tobacco is capable of recruiting immune system effector function *in vitro. Antimicrob. Agents Chemother.*, **2007**, *51*, 3322–3328.

Minshull, J. and Stemmer, W. P. C. Protein evolution by molecular breeding. *Curr. Opin. Chem. Biol.*, **1999**, *3*, 284–290.

Miyazaki, C., Iba, Y., Yamada, Y., Takahashi, H., Sawada, J. I., and Kurosawa, Y. Changes in the specificity of antibodies by site-specific mutagenesis followed by random mutagenesis. *Protein Eng.*, **1999**, *12*, 407–415.

Mullinax, R. L., Gross, E. A., Amberg, J. R., Hay, B. N., Hogrefe, H. H., Kubitz, M. M., Greener, A., Alting-Mees, M., Ardourel, D., Short, J. M., Sorge, J. A., and Shopes, B. Identification of human antibody fragment clones specific for tetanus toxoid in a bacteriophage lambda immunoexpression library. *Proc. Natl. Acad. Sci. USA*, **1990**, *87*, 8095–8099.

Muyldermans, S. Single domain camel antibodies: current status. *Rev. Mol. Biotechnol.*, **2001**, *74*, 277–302.

Ness, J. E., Welch, M., Giver, L., Bueno, M., Cherry, J. R., Borchert, T. V., Stemmer, W. P., and Minshull, J. DNA shuffling of subgenomic sequences of subtilisin. *Nat. Biotechnol.*, **1999**, *17*, 893–896.

Ness, J. E., Kim, S., Gottman, A., Pak, R., Krebber, A., Borchert, T. V., Govindarajan, S., Mundorff, E. C., and Minshull, J. Synthetic shuffling expands functional protein diversity by allowing amino acids to recombine independently. *Nat. Biotechnol.*, **2002**, *20*, 1251–1255.

Nieba, L., Honegger, A., Krebber, C., and Plckthun, A. Disrupting the hydrophobic paths at the antibody variable/constant domain interface: improved *in vivo* folding and physical characterization of an engineered scFv fragment. *Protein Eng.*, **1997**, *10*, 435–444.

Nissim, A., Hoogenboom, H. R., Jones, P. T., and Winter, G. Antibody fragments from a 'single pot' phage display library as immunochemical reagents. *EMBO J.*, **1994**, *13*, 692–698.

Nölke, G., Fischer, R., and Schillberg, S. Antibody-based metabolic engineering in plants. *J. Biotechnol.*, **2006**, *124*, 271–283.

Nyyssönen, E. and Keränen, S. Multiple roles of the cellulase CBHI in enhancing production of fusion antibodies by the filamentous fungus *Trichoderma reesei. Curr. Genet.*, **1995**, *28*, 71–79.

Nyyssönen, E., Penttilä, M., Harkki, A., Saloheimo, A., Knowles, J. K., and Keränen, S. Efficient production of antibody fragments by the filamentous fungus *Trichoderma reesei. Biotechnology*, **1993**, *11*, 591–595.

Olea-Popelka, F. C., McLean, M. D., Horsman, J., Almquist, K., Brandle, J. E., and Hall, J. C. Increasing expression of an anti-picloram single-chain variable fragment (scFv) Ab and resistance to picloram in transgenic tobacco (*Nicotiana tabacum*). *J. Agric. Food Chem.*, **2005**, *53*, 6683–6690.

Orlandi, R., Güssow, D. H., Jones, P. T., and Winter, G. Cloning immunoglobulin variable domains for expression by the polymerase chain reaction. *Proc. Natl. Acad. Sci. USA*, **1989**, *86*, 3833–3837.

Orum H., Andersen, P. S., Oster, A., Johansen, L. K., Riise, E., Bjørnvad, M., Svendsen, I., and Engberg, J. Efficient method for constructing comprehensive murine Fab antibody libraries displayed on phage. *Nucleic Acids Res.*, **1993**, *21*, 4491–4498.

Outchkourov, N. S., Stiekema, W. J., and Jongsma, M. A. Optimization of the expression of equistatin in *Pichia pastoris. Protein Expr. Purif.*, **2002**, *24*, 18–24.

Owen, M., Gandecha, A., Cockburn, B., and Whitelam, G. Synthesis of a functional anti-phytochrome single-chain Fv protein in transgenic tobacco. *Biotechnology*, **1992**, *10*, 790–794.

Parhami-Seren, B., Viswanathan, M., and Margolies, M. N. Selection of high affinity *p*-azophenylarsonate Fabs from heavy-chain CDR2 insertion libraries. *J. Immunol. Methods*, **2002**, *259*, 43–53.

Park, S.-G., Lee, J.-S., Je, E.-Y., Kim, I.-J., Chung, J.-H., and Choi, I.-H. Affinity maturation of natural antibody using a chain shuffling technique and the expression of recombinant antibodies in *Escherichia coli. Biochem. Biophys. Res. Commun.*, **2000**, *275*, 553–557.

Paschke, M. Phage display systems and their applications. *Appl. Microbiol. Biotechnol.*, **2006**, *70*, 2–11.

Patten, P. A., Howard, R. J., and Stemmer, W. P. C. Applications of DNA shuffling to pharmaceuticals and vaccines. *Curr. Opin. Biotechnol.*, **1997**, *8*, 724–733.

Peeters, K., De Wilde, C., De Jaeger, G., Angenon, G., and Depicker, A. Production of antibodies and antibody fragments in plants. *Vaccine*, **2001**, *19*, 2756–2761.

Peipp, M., Saul, D., Barbin, K., Bruenke, J., Zunino, S. J., Niederweis, M., and Fey, G. H. Efficient eukaryotic expression of fluorescent scFv fusion proteins directed against CD antigens for FACS applications. *J. Immunol. Methods*, **2004**, *285*, 265–280.

Persson, M. A., Caothien, R. H., and Burton, D. R. Generation of diverse high-affinity human monoclonal antibodies by repertoire cloning. *Proc. Natl. Acad. Sci. USA*, **1991**, *88*, 2432–2436.

Peschen, D., Li, H.-P., Fischer, R., Kreuzaler, F., and Liao, Y.-C. Fusion proteins comprising a *Fusarium*-specific antibody linked to antifungal peptides protect plants against a fungal pathogen. *Nat. Biotechnol.*, **2004**, *22*, 732–738.

Petrausch, U., Dernedde, J., Coelho, V., Panjideh, H., Frey, D., Fuchs, H., Thiel, E., and Deckert, P. M. A33scFv-green fluorescent protein, a recombinant single-chain fusion protein for tumor targeting. *Protein Eng. Des. Sel.*, **2007**, *20*, 583–590.

Petruccelli, S., Otegui, M. S., Lareu, F., Tran Dinh, O., Fitchette, A. C., Circosta, A., Rumbo, M., Bardor, M., Carcamo, R., Gomord, V., and Beachy, R. N. A KDEL-tagged monoclonal antibody is efficiently retained in the endoplasmic reticulum in leaves, but is both partially secreted and sorted to protein storage vacuoles in seeds. *Plant Biotechnol. J.*, **2006**, *4*, 511–527.

Plückthun, A. and Skerra, A. Expression of functional antibody Fv and Fab fragments in *Escherichia coli*. *Methods Enzymol.*, **1989**, *178*, 497–515.

Rahbarizadeh, F., Rasaee, M. J., Forouzandeh, M., and Allameh, A. A. Over expression of anti-MUC1 single domain antibody fragments in the yeast *Pichia pastoris*. *Mol. Immunol.*, **2006**, *43*, 426–435.

Ramirez, N., Rodriguez, M., Ayala, M., Cremata, J., Perez, M., Martinez, A., Linares, M., Hevia, Y., Paez, R., Valdes, R., Gavilondo, J. V., and Selman-Housein, G. Expression and characterization of an anti-(hepatitis B surface antigen) glycosylated mouse antibody in transgenic tobacco (*Nicotiana tabacum*) plants and its use in the immunopurification of its target antigen. *Biotechnol. Appl. Biochem.*, **2003**, *38*, 223–230.

Reavy, B., Ziegler, A., Diplexcito, J., MacIntosh, S. R., Torrance, L., and Mayo, M. Expression of functional recombinant antibody molecules in insect cell expression system. *Protein Expr. Purif.*, **2000**, *18*, 221–228.

Reiter, Y., Brinkman, U., Kreitman, R. J., Jung, S. H., Lee, B., and Pastan, I. Stabilization of the F_V fragments in recombinant immunotoxins by disulfide bonds engineered into conserved framework regions. *Biochemistry*, **1994**, *33*, 5451–5459.

Roberts, R. W. Totally *in vitro* protein selection using mRNA–protein fusions and ribosome display. *Curr. Opin. Chem. Biol.*, **1999**, *3*, 268–273.

Roberts, S., Cheetham, J. C., and Rees, A. R. Generation of an antibody with enhanced affinity and specificity for its antigen by protein engineering. *Nature*, **1987**, *328*, 731–734.

Robinson, C. R. and Sauer, S. T. Optimizing the stability of single-chain proteins by linker length and composition mutagenesis. *Proc. Natl. Acad. Sci. USA*, **1998**, *95*, 5929–5934.

Russel, M., Lowman, H. B., and Clackson, T. Introduction to phage biology and phage display. In Clackson, T.and Lowman, H. B. (Eds.), *Phage Display: A Practical Approach*. Oxford University Press, Oxford, UK, 2004, pp. 1–26.

Saalbach, I., Giersberg, M., and Conrad, U. High-level expression of a single-chain Fv fragment (scFv) antibody in transgenic pea seeds. *J. Plant Physiol.*, **2001**, *158*, 529–533.

Schaffitzel, C., Hanes, J., Jermutus, L., and Plückthun, A. Ribosome display: an *in vitro* method for selection and evolution of antibodies from libraries. *J. Immunol. Methods*, **1999**, *231*, 119–135.

Schier, R., Marks, J. D., Wolf, E. J., Apell, G., Wong, C., McCartney, J. E., Bookman, M. A., Huston, J. S., Houston, L. L., Weiner, L. M., and Adams, G. P. *In vitro* and *in vivo* characterization of a human anti-c-erbB-2 single-chain Fv isolated from a filamentous phage antibody library. *Immunotechnology*, **1995**, *1*, 73–81.

Schier, R., Balint, R. F., McCall, A., Apell, G., Larrick, J. W., and Marks, J. D. Identification of functional and structural amino-acid residues by parsimonious mutagenesis. *Gene* **1996a**, *169*, 147–155.

Schier, R., Bye, J., Apell, G., McCall, A., Adams, G. P., Malmqvist, M., Weiner, L. M., and Marks, J. D. Isolation of high-affinity monomeric human anti-c-erbB-2 single chain Fv using affinity-driven selection. *J. Mol. Biol.*, **1996b**, *255*, 28–43.

Schier, R., McCall, A., Adams, G. P., Marshall, K. W., Merritt, H., Yim, M., Crawford, R. S., Weiner, L. M., Marks, C., and Marks, J. D. Isolation of picomolar affinity anti-c-erbB-2 single-chain Fv by molecular evolution of the complementarity determining regions in the center of the antibody binding site. *J. Mol. Biol.*, **1996c**, *263*, 551–567.

Schillberg, S., Fischer, R., and Emans, N. Molecular farming of recombinant antibodies in plants. *Cell. Mol. Life Sci.*, **2003**, *60*, 433–445.

Sheedy, C., MacKenzie, C. R., and Hall, J. C. Isolation and affinity maturation of hapten-specific antibodies. *Biotechnol. Adv.*, **2007**, *25*, 333–352.

Sinclair, G. and Choy, F. Y. Synonymous codon usage bias and the expression of human glucocerebrosidase in the methylotrophic yeast *Pichia pastoris*. *Protein Expr. Purif.*, **2002**, *26*, 96–105.

Skerra, A. and Plückthun, A. Assembly of a functional immunoglobulin Fv fragment in *Escherichia coli*. *Science* **1988**, *240*, 1038–1041.

Smith, G. P. Filamentous fusion phage: novel expression vectors that display cloned antigens on the virion surface. *Science* **1985**, *240*, 1038–1041.

Smith, G. P. and Petrenko, V. A. Phage display. *Chem. Rev.*, **1997**, *97*, 391–410.

Songsivilai S., Bye J. M., Marks J. D., and Hughes-Jones N. C. Cloning and sequencing of human lambda immunoglobulin genes by the polymerase chain reaction. *Eur. J. Immunol.*, **1990**, *20*, 2661–2666.

Stave, J. Protein immunoassays methods for detection of biotech crops: applications, limitations and practical considerations. *J. AOAC Int.*, **2002**, *85*, 780–786.

Stemmer, W. P. C. DNA shuffling by random fragmentation and reassembly: *in vitro* recombination for molecular evolution. *Proc. Natl. Acad. Sci. USA*, **1994**, *91*, 10747–10751.

Stemmer, W. P. C., Crameri, A., Ha, K. D., Brennan, T. M., and Heyneker, H. L. Single-step assembly of a gene and entire plasmid from large numbers of oligodeoxyribonucleotides. *Gene* **1995**, *164*, 49–53.

Stoger, E., Vaquero, C., Torres, E., Sack, M., Nicholson, L., Drossard, J., Williams, S., Keen, D., Perrin, Y., Christou, P., and Fischer, R. Cereal crops as viable production and storage systems for pharmaceutical scFv antibodies. *Plant Mol. Biol.*, **2000**, *42*, 583–590.

Stoger, E., Sack, M., Fischer, R., and Christou, P. Plantibodies: applications, advantages and bottlenecks. *Curr. Opin. Biotechnol.*, **2002**, *13*, 161–166.

Stoger, E., Sack, M., Nicholson, L., Fischer, R., and Christou, P. Recent progress in plantibody technology. *Curr. Pharm. Des.*, **2005**, *11*, 2439–2457.

Strachan, G., McElhiney, J., Drever, M. R., McIntosh, F., Lawton, L. A., and Porter, A. J. R. Rapid selection of anti-hapten antibodies isolated from synthetic and semi-synthetic antibody phage display libraries expressed in *Escherichia coli*. *FEMS Microbiol. Lett.*, **2002**, *210*, 257–261.

Sudarshana, M. R., Roy, G., and Falk, B. W. Methods for engineering resistance to plant viruses. *Methods Mol. Biol.*, **2007**, *354*, 183–195.

Tan, W. and Lam, P. H. Expression and purification of a secreted functional mouse/human chimaeric antibody against bacterial endotoxin in baculovirus-infected insect cells. *Biotechnol. Appl. Biochem.*, **1999**, *30*, 59–64.

Tanha, J., Xu, P., Chen, Z., Ni, F., Kaplan, H., Narang, S. A., and MacKenzie, C. R. Optimal design features of camelized human single-domain antibody libraries. *J. Biol. Chem.*, **2001**, *276*, 24774–24780.

Tavlodoraki, P., Benvenuto, E., Trinca, S., de Martinis, D., and Galeffi, P. Transgenic plants expressing a functional single-chain Fv antibody are specifically protected from virus attack. *Nature*, **1993**, *366*, 469–472.

Terpe, K. Overview of bacterial expression systems for heterologous protein production: from molecular and biochemical fundamentals to commercial systems. *Appl. Microbiol. Biotechnol.*, **2006**, *72*, 211–222.

Tomlinson, I. M., Walter, G., Marks, J. D., Llewelyn, M. B., and Winter, G. The repertoire of human germline VH sequences reveals about fifty groups of VH segments with different hypervariable loops. *J. Mol. Biol.*, **1992**, *227*, 776–798.

Tonegawa, S. Somatic generation of antibody diversity. *Nature*, **1983**, *302*, 575–581.

Tripathi, L. Techniques for detecting genetically modified crops and products. *Afr. J. Biotechnol.*, **2005**, *4*, 1472–1479.

Turner, D. J., Ritter, M. A., and George, A. J. Importance of the linker in expression of single-chain Fv antibody fragments: optimisation of peptide sequence using phage display technology. *J. Immunol. Methods*, **1997**, *205*, 43–54.

Valjakka, J., Hemminki, A., Niemi, S., Soderlund, H., Takkinen, K., and Rouvinene, J. Crystal structure of an *in vitro* affinity- and specificity-matured anti-testosterone Fab in complex with testosterone. *J. Biol. Chem.*, **2002**, *277*, 44021–44027.

Van den Bulcke, M., De Schrijver, A., De Bernardi, D., Devos, Y., MbongoMbella, G., Casi, A. L., Moens, W. and Sneyers, M. Detection of genetically modified plant products by protein strip testing: an evaluation of real-life samples. *Eur. Food Res. Technol.*, **2007**, *225*, 49–57.

Vaughan, T. J., Williams, A. J., Pritchard, K., Osbourn, J. K., Pope, A. R., Earnshaw, J. C., McCafferty, J., Hodits, R. A., Wilton, J., and Johnson, K. S. Human antibodies with sub-nanomolar affinities isolated from a large non-immunized phage display library. *Nat. Biotechnol.*, **1996**, *14*, 309–314.

Verch, T., Yusibov, V., and Koprowski, H. Expression and assembly of a full-length monoclonal antibody in plants using a plant virus vector. *J. Immunol. Methods*, **1998**, *220*, 69–75.

Verma, R., Boleti E., and George, A. J. Antibody engineering: comparison of bacterial, yeast, insect and mammalian expression systems. *J. Immunol. Methods*, **1998**, *216*, 165–181.

Wang, S.-H., Du, X.-Y., Huang, Y.-M., Lin, D.-S., Hart, P. L., and Wang, Z.-H. Detection of deoxynivalenol based on a single-chain fragment variable of the antideoxynivalenol antibody. *FEMS Microbiol Lett.*, **2007**, *272*, 214–219.

Wang, S.-H., Du, X.-Y., Lin, L., Huang, Y.-M., and Wang, Z.-H. Zearalenone (ZEA) detection by a single chain fragment variable (scFv) antibody. *World J. Microbiol. Biotechnol.*, **2008**, *24*, 1681–1685.

Weiss, Y., Shulman, A., Ben Shir, I., Keinan, E., and Wolf, S. Herbicide-resistance conferred by expression of a catalytic antibody in *Arabidopsis thaliana*. *Nat. Biotechnol.*, **2006**, *24*, 713–717.

Whitlow, M., Bell, B. A., Feng, S. L., Filpula, D., Hardman, K. D., Hubert, S. L., Rolenence, M. L., Wood, J. F., Schott, M. E., and Milenic, D. E. An improved linker for single-chain Fv with reduced aggregation and enhanced proteolytic stability. *Protein Eng.*, **1993**, *6*, 989–995.

Willats, W. G. Phage display: practicalities and prospects. *Plant Mol. Biol.*, **2002**, *50*, 837–854.

Winter, G. Synthetic human antibodies and a strategy for protein engineering. *FEBS Lett.*, **1998**, *430*, 92–94.

Winter, G. and Milstein, C. Man-made antibodies. *Nature*, **1991**, *349*, 293–299.

Winter, G., Griffiths, A. D., Hawkins, R. E., and Hoogenboom, H. R. Making antibodies by phage display technology. *Annu. Rev. Immunol.*, **1994**, *12*, 433–455.

Woo, J. H., Liu, Y. Y., Mathias, A., Stavrou, S., Wang, Z., Thompson, J., and Neville, D. M., Jr. Gene optimization is necessary to express a bivalent anti-human anti-T cell immunotoxin in *Pichia pastoris*. *Protein Expr. Purif.*, **2002**, *25*, 270–282.

Wu, H., Beuerlein, G., Nie, Y., Smith, H., Lee, B. A., Hensler, M., Huse, W. D., and Watkind, J. D. Stepwise affinity maturation of vitaxin, an $\alpha_v\beta_3$-specific humanized mAb. *Proc. Natl. Acad. Sci. USA*, **1998**, *95*, 6037–6042.

Wu, H., Nie, Y., Huse, W. D., and Watkins, J. D. Humanization of a murine monoclonal antibody by simultaneous optimization of framework and CDR residues. *J. Mol. Biol.*, **1999**, *294*, 151–162.

Yano, A., Maeda, F., and Takekoshi, M. Transgenic tobacco cells producing the human monoclonal antibody to Hepatitis B virus surface antigen. *J. Med. Virol.*, **2004**, *73*, 208–215.

Yau, K. F. Y., Lee, H., and Hall, J. C. H. Emerging trends in the synthesis and improvement of hapten-specific recombinant antibodies. *Biotechnol. Adv.*, **2003**, *21*, 599–637.

Zhang, J., Tanha, J., Hirama, T., Khieu, N. H., To, R., Tong-Sevinc, H., Brisson, J.-R., and MacKenzie, C. R. Pentamerization of single-domain antibodies from phage libraries: a novel strategy for the rapid detection of high-avidity antibody reagents. *J. Mol. Biol.*, **2004**, *335*, 49–56.

Zhao, H., Giver, L., Shao, Z., Affholter, J. A., and Arnold, F. H. Molecular evolution by staggered extension process (StEP) *in vitro* recombination. *Nat. Biotechnol.*, **1998**, *16*, 258–261.

zu Putlitz, J., Kubasek, W. L., Duchêne, M., Marget, M., von Specht, B. U., and Domdey, H. Antibody production in baculovirus-infected insect cells. *Biotechnology*, **1990**, *8*, 651–654.

MICROTITER PLATE ELISA

Michael C. Brown

Immunoassays in Agricultural Biotechnology, edited by Guomin Shan
Copyright © 2011 John Wiley & Sons, Inc.

4.1   ELISA: A HISTORICAL VIEW

Enzyme immunoassay was first described by Engvall in 1971 (Engvall and Perlman, 1971; Van Weemen and Schuurs, 1971). At that time the radioimmunoassay (RIA), using antigens or antibodies labeled with radioactive isotopes, was the predominant immunoassay format in both clinical and research laboratories. Many of these tests were performed in coated plastic tubes. These tubes could fit directly into the instrumentation (gamma counters) used to detect and quantify the isotopes (e.g., I^{125}) predominantly used for these tests. Enzyme immunoassay offered the opportunity to move to a nonradioactive format with simpler, safer handling, and longer shelf life. Early enzyme immunoassays used this same coated tube format and could be transferred to a spectrophotometer for reading or read directly in the tube (Lequin, 2005). It should be noted that 30 years later, many enzyme immunoassays, particularly those done in low numbers, are still read directly in that coated tube format.

Microtiter plates were first described by Takatsy (1951) and later in that decade produced in a commercial form in an 8×12 format by Linbro Corporation and by Cooke Labs in the early 1960s (Manns, 1999). The predominant use of these plates, typically with round or vee bottoms, was in hemagglutination assays with a series of highly specialized pieces of manual equipment used to facilitate dispensing and diluting multiple samples at once. Enzyme immunoassays were first adapted to microtiter plates in 1974 by Voller et al. (1974), researchers well skilled in the art of hemagglutination and aware of the handling advantages these offered for multiple samples. The introduction of the first commercially available reader capable of directly reading absorbance of wavelength filtered light (Clem and Yolken, 1978) a few years later by Dynatech (formerly Cooke Labs and now Dynex) ushered in the era of microtiter plate enzyme-linked immunosorbent assays (ELISAs).

Both the terms ELISA and EIA (enzyme immunoassay) are used to describe these enzyme-based assays. As with any popular format, there are, of course, many spin-offs of these with associated acronyms. In purist terms, ELISA was originally used for assays in a direct bind or competitive format comparable to the old RIA. EIA is a description for sandwich assay, comparable to the old immunoradiometric assay (IRMA). For the sake of simplicity, and because in common usage the terms have become quite blurred, in this chapter both will be referred to as ELISA. Many different enzyme substrates are used in ELISA producing colored, fluorescent, or chemiluminescent end products. These are described in Section 4.2.4, but again, for the sake of simplicity, we will refer to the assay as producing a colorimetric response throughout most of this chapter. Finally, capture reagent will refer to the immunologically specific reagent (antibody or antigen) bound to the surface of the microplate (whether bound directly or through a linker or second antibody) and the detection reagent will refer to the immunologically specific reagent that must be bound for color to be specifically generated.

ELISAs are the assay of choice for quantitative determination of the presence of a target protein in many laboratories throughout the world. With careful development, coefficients of variation on analyte determination can be kept below 10% and in many instances under 5%. The microtiter plate format allows easy handling of a large number of standards, samples, and controls, giving an advantage over other ELISA formats such as coated tubes or coated beads. Most assays can be configured to be performed in well under one workday, allowing acquisition of samples, assay performance, and actionable results in less than 8 h. Assays yielding qualitative results (i.e., the presence or absence of a trait in agricultural applications) can be performed with relatively brief incubations and a large number of samples (in the thousands) can be assayed per day when required. Indeed, the bottleneck in these methods is often the sample preparation rather than the assay itself. Automation of these assays is discussed in a later chapter.

ELISAs can be performed in a field or nonlaboratory environment with dropper bottles and squeeze bottles instead of pipettes and visually read results instead of optical readers. However, these environments are best suited to semiquantitative or qualitative determinations. ELISAs are often useful as a confirmation tool for other test methods. They do require training of the user to be performed. When there are only a very limited number of samples to be analyzed (i.e., less than eight), microplate ELISAs do not offer significant handling advantages over other formats such as coated tube ELISAs.

4.2 ASSAY DESIGN CONSIDERATIONS

Microplate ELISAs lend themselves to a wide variety of formats for a myriad of analytes. Assay design should start with a clear definition of the intended goals desired. Are qualitative or quantitative results required, or perhaps both? What is the required analytical sensitivity in terms of a known standard for quantitative assays or in terms of a detection limit (e.g., detecting 1 seed expressing a protein in a sample of 800 seeds that do not express that protein)? Have similar assays been developed before and what were the most difficult issues associated with them? Will there be potential issues with cross-reactive antigens that may need to be considered up front in antibody selection? What are the types of samples to be tested and how will they be processed? Based on previous experiences, are there likely to be matrix effects, or effects caused by nonanalyte components in the sample (this is usually true)? Finally, there are practical considerations of the user skill and sophistication, the environment in which the user functions (what else they do while running the assay), and the time consideration of how long it takes to get to an actionable result. It is important to define as many of these parameters as possible before one embarks on any experimentation.

In optimizing assay design, one should first carefully consider the building blocks of such assays as described below.

4.2.1 Antibodies

Both polyclonal and monoclonal antibodies can be well suited to be used either as a capture reagent by immobilization on solid substrates or as a detection reagent, either indirectly or conjugated to an enzyme. In most cases, antibodies should be highly purified and free from aggregates that can cause erratic results in most assays. A wide variety of purification techniques, most commonly protein A or G affinity chromatography or ion exchange chromatography (Harlow and Lane, 1988), will give good yields and suitable purity.

If sensitivity is a problem with polyclonal antibodies, using one that has been immunoaffinity purified (on antigen covalently bound to a resin or gel in columns) (Ballon, 2000) can frequently result in dramatic increases in sensitivity. This is due to the removal of nonreactive immunoglobulins and thereby more active antibody can be immobilized onto the limited surface area of well. This can also be true when polyclonal antibodies are used for detection since the presence of nonspecific immunoglobulins may cause a higher background signal. If sensitivity is not an issue and often in competitive assays where the primary limitation on sensitivity may be antibody affinity, immunoaffinity purification may offer less advantage and be unwarranted.

Specificity may also be a consideration in the use of polyclonal antisera. Since polyclonal antibodies can be elicited against any and all components used for immunization, the purity of the immunogen is a major consideration. If the immunogen could be contaminated, it is likely that antibodies will be produced against these contaminants and these may be the cause of false negative or positive results in the final assay. Immunoaffinity purification can cure this problem but it must be kept in mind that if the same material is used to make an immunoaffinity column as was used for immunogen, then these contaminating antibodies will not be removed.

Monoclonal antibodies are free of the specificity concerns as all the immunoglobulins in the preparation are identical. Careful screening and testing can often select the clone that possesses the required specificity and affinity constants. When antibody is used in an indirect format (i.e., detected with an enzyme labeled anti-immunoglobulin from a different species), purification is not necessary and these assays can be done quite effectively with diluted sera, tissue culture supernatant, and so on. Similarly, purification can be avoided if the antibody is bound to a plate by means of an anti-immunoglobulin from a different species or protein A or G that is first immobilized to the plate.

Desired specificity of the ELISA is a strong factor in influencing choice of antibodies. If closely related antigens exist and their reactivity in the ELISA is undesired, monoclonal antibodies may be the antibody of choice. Alternatively, the polyclonal antibody can be cross-absorbed against these related antigens immobilized to a resin to render it specific. The requirement here, of course, is for the availability of suitable quantities of these antigens in purified form, and depending on the extent of the cross-reactivity, a considerable loss of antibody may result (Hermanson et al., 1992). In sandwich ELISA, only the combination of two antibodies used should have the required absolute specificity. For instance, the capture antibody could be reactive with the target antigen and related antigens, but when paired with a detection antibody with specificity that may also have cross-reactivity, but with different antigens than the capture antibody, can yield a specific ELISA. High specificity may not always be desirable. In some instances, a class of related molecules may be the target, in which case high specificity within this class may be detrimental.

In general, high-affinity antibodies are desirable in ELISA methods. High affinity will allow reactions to proceed toward equilibrium and possibly allow less antibody to be used. Less antibody may tend toward lower background that taken with a more complete reaction may result in greater sensitivity. However, most of the time going into a development project one does not have information on affinity, other than perhaps a rough sense from activity of the antibodies in direct bind assays that often are affected by other factors as well.

Fragmentation of antibodies (FAb or FAb_2) (Parham, 1986) rarely offers advantages in agricultural ELISA, unlike in human and veterinary diagnostics where nonspecific binding factors can play a major role. While the Fc portion of antibodies is glycosylated, problems arising from lectins found in many plants are rare. Specificity of most plant lectins is known and addition of the specific free carbohydrate would cure any detrimental effects in an assay.

4.2.2 Antigens

Antigens suitable for use in ELISA may range from small haptens of under 100 Da to large multimeric complexes (>1000 kDa) though strategies must be employed to handle these size differences. Smaller antigens (defined as less than 20 kDa and sometimes even less than 10 kDa) generally require covalent conjugation to a larger carrier molecule to elicit a strong immune response, generally related to preventing them from quickly migrating from the immunization site or eliminated by means other than the immune system. These chemistries must be selected such that they do not bind to or modify the most immunogenic or structurally unique portion(s) of the molecule. In some cases, this may require de novo synthesis of the antigenic derivative for coupling. As mentioned later, on plate coating, small antigens also require covalent linking to the plate or covalent linkage to a carrier that is, in turn, passively absorbed to the plate. It is generally important to use a different carrier and linking chemistry between immunization and preparation for plate binding when the antigen is the immobilized ligand on the plate or a different linking chemistry for binding to the enzyme when the antigen is the labeled ligand used for detection. Larger antigens (>10 kDa) give greater flexibility in that they generally do not require linking chemistries for immunization (which can induce antibodies to the linker) and can be directly bound to the plates (passively or covalently). These larger antigens are also amenable to multiple assay formats.

4.2.3 Enzymes

Suitable enzymes for use in ELISA would be those that are stable in storage and during assay condition, inexpensive, readily conjugated to an antigen or antibody with a variety of controllable conjugation options, and for which a convenient, inexpensive, and stable source of substrate and/or color reagent exists. Enzymes should have a molecular weight that does not pose untoward steric hindrances when conjugated to the antigen or the antibody. Enzymes should also have a sufficiently high turnover number to get rapid conversion of substrate. This converted substrate (or a chromogen participating in a electron donor/acceptor reaction with the enzyme/substrate complex) should be readable in a quantitative fashion in available instrumentation.

TABLE 4.1 Enzymes for ELISA

Enzyme	Molecular weight	Coupling chemistry	Advantages	Disadvantages
Horseradish peroxidase	40 kDa	Periodate oxidation of carbohydrate glutaraldehyde, heterobifunctional linkers	High turnover number, large number of sensitive substrates, lowest MW of enzymes used for ELISA, easiest conjugation	Reaction is highly pH dependant, high endogenous levels in some plant tissues
Alkaline phosphatase	108 kDa	Heterobifunctional cross-linkers	Large number of sensitive substrates	High molecular weight results in large conjugates
ß-Galactosidase	540 kDa	Coupling to high thiol content	Low endogenous background in samples	Few substrates, high molecular weight results in large conjugates
Urease	483 kDa	Heterobifunctional cross-linkers	Broad pH reactivity	Best for visual reading, few substrates, high molecular weight conjugates
Glucose oxidase	153 kDa	Heterobifunctional periodate oxidation of carbohydrate, glutaraldehyde, linkers	Low endogenous background in samples, very stable	Requires coenzyme, few direct substrates, high molecular weight conjugates

There are many possible enzyme candidates for these assays (Table 4.1). Horseradish peroxidase (HRP) and alkaline phosphatase (AP) perhaps come the closest to meeting all these criteria. These two enzymes have been extensively used in human, veterinary, industrial, and agricultural immunoassays for almost 30 years. Because of the popularity of these enzymes, there has also been commercial interest in improvement of substrates, and a wide range of sensitive ones (high yield per reaction event) exist both in chromogenic, chemiluminescent, and fluorescent forms and in precipitating and nonprecipitating forms. Other enzymes, such as beta galactosidase, glucose oxidase, beta glucuronidase, and urease certainly have some applications, but do not, as of this writing, have the same breadth of excellent substrates available.

4.2.3.1 Enzyme Conjugation
When conjugates are of a secondary antibody (i.e., goat anti-rabbit IgG conjugated to HRP) or a secondary detection reagent (i.e., alkaline phosphatase–streptavidin), it is often most effective and quite reliable to purchase these from commercial sources.

When conjugates are required of a specialized polyclonal or monoclonal antibody, it is generally quite easy to prepare them. Preactivated forms of many enzymes are commercially available as are many heterofunctional linkers used to specifically create covalent linkages (Hermanson, 1996).

Peroxidase conjugates, however, are rather easy to prepare by oxidation of the carbohydrates on the molecule with periodate to generate aldehydes and then reaction with the free amino groups on antibodies. These can be made as follows in a variation of the original procedure (Nakane and Kawaoi, 1974):

1. Dissolve enzyme grade horseradish peroxidase at 4 mg/mL or greater in 10 mM acetate buffer at pH 4.4
2. Add 1/10 volume of freshly prepared 100 mM sodium periodate
3. Incubate for 20 min at room temperature in the dark
4. Separate enzyme from free periodate on a small (1–2 mL) Sephadex G25 column equilibrated in 1 mM acetate buffer at pH 4.4. Enzyme will be dark greenish brown
5. Promptly add an appropriate amount (assume 90% recovery from the Sephadex column) to antibody that has been previously dialyzed or diluted into 100 mM carbonate buffer, pH 9.5
6. Incubate in the dark for 2 h at room temperature
7. Chill the reaction on ice
8. Add 10 μL of freshly prepared sodium borohydride (10 mg/mL in 100 mM carbonate buffer, pH 9.5) per milliliter of reaction mixture
9. Incubate in the dark for 2 h on ice
10. Dialyze versus several changes of phosphate buffered saline

Molar ratios of horseradish peroxidase (40 kDa) to antibody (150 kDa) used in assays typically range from 1:1 to 8:1. It is advisable to prepare a range of these ratios and test for optimal activity in ELISA. It is rarely necessary to separate any nonconjugated antibody or enzyme and size fractionation of the conjugates also does not offer any significant advantages over using the whole mixture.

Alkaline phosphatase conjugates are also easily made by introducing a protected sulfhydryl group onto the enzyme, cleaving it with dithiothreitol or mercaptoethanolamine to create a free sulfhydryl. After separation from the reducing agent, the modified enzyme is reacted with a bifunctional reagent with a maleimide (sulfhydryl reactive) at one end and amino reactive (typically a succinimidyl ester) at the other. This reaction proceeds at pH < 6 so that only the maleimide–sulfhydryl reaction proceeds. After separation at pH 6, the activated enzyme is combined with antibody at pH > 7 to complete the reaction (Hermanson, 1996). There are many variations on this general protocol and many of the available kits make some or all these steps simpler for the novice. Some examples of commercially available kits are the EZ-Link kits from Pierce Biotechnology, Inc. (Rockford, IL) or the NH2-reactive kit from Dojindo Molecular Technologies, Inc. (Gaithersburg, MD).

Another strategy that may be employed is the use of antibodies labeled with biotin and combining these with commercial conjugates of streptavidin with enzyme. Labeling with biotin is extremely easy (Kantor, 2008) and lends itself to handling many antibodies at once. This can be useful when many antibodies need evaluation. Once the desired ones have been identified, one can move to direct conjugation and optimization of those. For example, EZ-Link® NHS-Biotin reagents are available from Pierce with varying spacer arm length. These active esters are simply diluted in a miscible organic solvent such as DMF or DMSO and then combined with antibody diluted in PBS for a short reaction period followed by passing on a desalting column or dialysis for removal of free biotin.

4.2.3.2 Stability and Shelf Life

Properly prepared antibody enzyme conjugates at concentrations of antibody in excess of 1 mg/mL are typically quite stable at 4°C for at least 1 year. Stabilizing proteins can be added if desired, but these are often unnecessary with concentrated conjugates. It is however essential to keep these conjugates free of bacterial contamination, and so use of preservatives, compatible with the enzyme, is highly advised. Horseradish peroxidase is sensitive to light and so storage conditions should be in the dark. Conjugates of most enzymes may also be frozen in 50% glycerol for periods of many years.

Dilute conjugates or conjugates in a "ready to use form" will require stabilizers and preservatives to be added. Without these stabilizers, these dilute conjugates may have a shelf life of less than 1 day. These stabilizers may be formulated in-house or, though expensive, conveniently purchased commercially from many vendors. Those stabilizers formulated in-house will benefit from the inclusion of animal serum (typically fetal calf serum) at concentrations ranging from 10% to 50% and a significant lot to lot variability may be observed. Adult animal sera usually have a detrimental effect on the stability and are best avoided. Alternatively, blends of albumin and IgG may be substituted. Trace metals may also be required depending on the enzyme, and chelating agents must generally be avoided. For peroxidase, the addition of cyclic compounds such as luminol, tetramethylbenzidine (TMB), aminonapthylsulfonic acid, or aminopyridine can dramatically improve stability (Schuetz et al., 1997). For low volume use (defined as less than 10 L or 1000 plates per year), the commercial stabilizers, typically used at 20–50% concentration, may be most cost-effective. This 20–50% range will often be compatible with other components (buffers, salts, surfactants, and polymers) required for optimal conjugate activity with the analyte. It would not be atypical for a properly prepared conjugate in an optimized diluent to have shelf life in excess of 2 years at 2–8°C or up to 1 year at room temperature. In choosing stabilizers, one should also consider where the assay is to be used, as export regulations may significantly affect the choice of protein or serum used for stabilization.

As mentioned for conjugate stocks, it is critically important to prevent bacterial growth in dilute conjugates, and most diluents, in the absence of antimicrobials, are quite effective in growing a wide range of microorganisms. Sodium azide (0.1–0.05% effective bacteriostatic concentration) is suitable for alkaline phosphatase but will result in inactivation of horseradish peroxidase. Very popular immunoassay compatible biocides active against bacteria, yeasts, and fungi are isothiazolinones, the ProClin series from Supelco (Eremin et al., 2002). Several notes of caution are in order here. These preservatives slowly react with amine containing buffers above pH 7.0 and may over time lose effectiveness. Similarly, serum components can cause loss of activity for the most common isothiazolone 5-chloro-2-methyl-4-isothiazolin-3-one. In contrast, 2-methyl-4-isothiazolin-3-one is quite stable in serum containing buffers. Antibiotics, such as penicillin, ciprofloxacin, and gentamicin, can help with particularly problematic bacteria; however, their lifetime in solution can be short and heavily influenced by the makeup of the diluent. Fungizone (amphotericin B) may be added for problematic yeasts or molds. The antimicrobial agents all may work very well for a quick kill and to get into containers aseptically. However, when these bottles are opened while using, they are subject to potential recontamination as the active ingredients break down after a few days or weeks. Accordingly, conjugates prepared with these preservatives should be aseptically opened.

Lyophilization is another method for preservation of enzyme conjugates and may be most applicable when material is to be shipped over large distances or to areas where refrigeration is limited, such as to an agricultural field station or other remote locations. Packaging can be arranged in unit doses (single use) or bulk. Optimum formulas usually include protein as a bulking agent, typically in the range of 2–4% w/v prior to lyophilization. Sugars may be included as stabilizers but are best kept in the range of 0.1–1% to avoid a glassy crystalline pellet that may not lyophilize efficiently. Proper formulation and cycle times should allow very high recovery of enzyme and antibody or antigenic activity. With proper lyophilization, an antimicrobial preservative should be unnecessary; however, once reconstituted it may be required, depending on the length of storage. If preservatives are to be incorporated into the lyophilization formula, it is important to ensure that they are not volatile as they will be removed in the process.

4.2.4   Substrates

A wide variety of substrate/color (or chromogen) systems are available commercially and the effort to prepare one's own from scratch is seldom cost-efficient. The ideal substrate/color system should be stable (in the absence of the enzyme), ready to use, nontoxic, compatible with widely available instrumentation, and not extremely light or air sensitive. If the substrate is colored or results in a colored or chromogenic product, it is usually desirable for the product to have a high extinction coefficient at a wavelength in the visible range to use existing instrumentation and light filters. If the substrate is fluorescent, it should be excitable by a wavelength where there is an abundance of light from a given lamp source, exhibit a large stokes shift (difference between the excitation wavelength and the emission wavelength), exhibit a high quantum yield (efficiency of reemitting the absorbed light at the new wavelength), and emit at a wavelength where background fluorescence from the plastic of the well or sample residue is minimal. If a substrate is a chemiluminescent one, it should have a relatively long life in the excited state to maximize the available light for detection. For ELISA, colorimetric substrates are the predominant ones used and are certainly the cheapest. In many cases, they provide excellent sensitivity within the requirements of the assay. When additional sensitivity and assay background are minimal, chemiluminescent ones may be used and increases may be 10–100-fold (Roda et al., 2004). Substrates with fluorescent end products are the least popular, though fluorescence can be used for direct detection (i.e., antibody or antigen labeled with a fluorescent molecule instead of with an enzyme for assay simplification by eliminating one step).

In some substrate systems, the substrate itself is cleaved or acted upon to produce an end product that is read. In others, such as peroxidase systems, the substrate is part of an electron transfer process. For instance, the commonly used chromogen 3,3′,5,5′-tetramethylbenzidine serves as a donor in the breakdown of the substrate for the enzyme, hydrogen peroxide, by peroxidase (Josephy et al., 1982). The resultant oxidized TMB is a dark blue color.

Many of the commercially available single part (no mixing required) substrates have stabilizers added to them to allow this one-part formulation. In some instances, this may result in slower color formation than two-part formulations, mixed immediately before use. Several commercial substrates are further stabilized to allow storage at room temperature. This eliminates the need to thoroughly warm a refrigerated reagent to room temperature (or the intended temperature of the enzyme substrate reaction) prior to use.

Many substrates are somewhat light sensitive. While incubation during the substrate incubation step is usually not an issue, they should be protected from light prior to this and if a two-part formulation (or tablet and liquid) is mixed immediately prior to the substrate incubation. These precautions should eliminate formation of background color prior to the incubation.

Substrate enzyme reactions should be allowed to run at least 10 min (or as recommended by the manufacturer in the case of chemiluminescent substrates). Incubating for shorter periods could result in increased variability (see Section 4.6.3 for more details) In many instances (i.e., with peroxidases), enzymes may decrease in activity during the reaction with substrate. Reaction times longer than 30 min should be used with caution.

4.2.5 Stop Reagents

It may be desired to stop the enzyme substrate reaction prior to reading. This is most commonly done by adding enzyme inhibitors (i.e., sodium fluoride for horseradish peroxidase) or shifting the pH of the reaction. Using an enzyme inhibitor will not require any changes in wavelength for reading the reaction. Using a strategy of shifting pH may shift the wavelength of maximum absorption and may also increase the extinction coefficient, sometimes from twofold to threefold. This will only result in increased sensitivity in assays that have been optimized for extremely low background. In other instances, both the background and the specific signal will increase by the same proportion, and while signal has increased, sensitivity remains the same.

Stopping the reaction will be advantageous when the ELISA assay includes many plates and there will be differences in time of reading during substrate incubation due to the speed of the microplate reader. Nonetheless, these should be read within as short a time frame as practical (with 1 h) as the stop reagents in many cases will not eliminate the light sensitivity of the substrate. In extreme cases (such as a malfunctioning plate reader), the plates could be sealed and stored in the dark.

4.2.6 Solid Phase

Microtiter plates specifically designed for use in ELISA are available from a number of manufacturers. Molding of plates to a uniform surface to get uniform protein binding requires careful control of process parameters (temperature, injection pressure, number of injection gates, number of cavities in the mold, and uniformity of plastic formulation) and plates not specifically controlled in this fashion, while significantly cheaper, can give erratic results, both within a plate and more severely from lot to lot. Reputable manufacturers, with many years of experience behind them, include Nunc (Thermo Fisher Scientific, Rochester, NY), Corning Costar (Lowell, MA), Dynex (Chantilly, VA), and Greiner (Wemmel, Belgium).

Most plates are made of polystyrene or PVC. In many cases, surface treatment with ionizing radiation is done by the manufacturer to enhance binding ("high binding") or derivatized to introduce functional groups for covalent coupling. Many manufacturers will also subject their products to a level of quality control than guaranteeing consistent well to well performance. Most assays in which antibody is the reagent to be bound to the plate will generally benefit from the use of "high-binding" plates. Algorithms for plate selection for other molecules can usually be found on the plate manufacturers' Web sites.

With the availability of instrumentation to read directly in the well, most microplates for this purpose are in a flat-bottom format, and this is the format used almost exclusively at present. Exceptions are minor in nature, such as "C" bottom with rounding of the corners to facilitate washing but essentially flat in the portion read optically.

Most colorimetric assays are performed using clear plates, using a top to bottom or bottom to top read. When the substrate for the enzyme results in luminescence, white plates are typically used to prevent cross talk between wells and to maximize the light measured (reflecting off the white surface) and these are measured from the top. When the substrate for the enzyme results in fluorescence, the choice is typically to minimize scattered or stray light at the excitation wavelength, autofluorescence of the plastic itself, and again to minimize cross talk. For assays run either by instruments or manually, a 96-well configuration is the most popular. 384 well plates, using the same footprint, are also available, but these can be a bit more challenging to work with on a manual basis. Incorporating a dilute but visible neutral dye (phenol red or a food coloring dye) into the reagents allows users to more readily observe their progress as they load the plate at each step of the assay and this can be a real lifesaver with these 384 plates. Also available are 1536-well plates, but they are best suited to only handling with instruments.

Plate configurations are also available from some manufacturers in strips (i.e., 8 or 12 wells molded together). These then fit into a holder, although secureness of fit can sometimes be an issue, particularly when the plates are inverted during or after washing. These wells are quite expensive compared to a solid plate and typically used where coating reagents (antibodies or antigens) are at a premium and the number of samples tested per assay is low. Similarly, wells are available that can be broken apart into individual ones, for a very low number of wells are being run. Again, pricing is higher for these and there can be some challenges putting these into holders such that all wells are in the same plane (very important for washing).

4.3 ASSAY OPTIMIZATION

The first level of assay optimization is to optimize around the basic components of the assay. These basics are common to all microtiter plate ELISAs.

4.3.1 Coating onto the Plate

Far and away the majority of microtiter plate ELISAs utilize passive absorption of antibody onto the surface of the wells. This passive adsorption is due to hydrogen bonding, van der Waals forces, and hydrophobic interactions with the surface (Rowell, 2001). Absorption is typically accomplished by diluting the desired ligand to be coated into a buffer. Over time absorption and desorption to the surface of the plastic will occur. At some point, the ligand will have a sufficient number of physical bonds at contact points to the surface so that it is stable and no longer desorbs. This is the most desired state for reliable, stable ELISAs that can withstand robust washing and incubations. This typically takes at least 2 h to occur (Rowell, 2001) and longer with some proteins. It is convenient to coat plates overnight to reach this steady state. This can be accomplished at room temperature or 2–8°C. It is important that the plates be sealed during this coating step to avoid evaporation, which can cause a host of problems.

At their isoelectric point, most proteins are the least stable in solution and this may be an advantageous pH to coat at. In evaluating optimum coating conditions, it is important to focus on signal, background, and long-term stability (if desired). Oftentimes harsher coating conditions may give greater signal, but at the expense of higher background and less long-term stability. Since the sensitivity of an immunoassay should always be viewed as signal over noise, optimization of coating should have a major focus on the background. If background is clean, then one may take advantage of enhancements to the signal by increasing length of substrate incubation, substrates with higher extinction coefficients, or enhancements to the final colored product (i.e., by an acid stop solution to increase color, usually accompanied by a shift in wavelength). Typical buffers used for coating include phosphate buffered saline, pH 7.2–7.5, and 0.05 M carbonate, pH 9.6. Purity of the buffers used for coating is absolutely critical and it is quite prudent to use dedicated glassware for this to eliminate possible contamination by detergents or trace proteins. A contamination problem here will typically manifest itself as a low-level, nagging, and elusive problems, which are the worst kind to troubleshoot.

Binding capacity for high binding plates recommended for ELISA is typically 400–500 ng/cm^2 for a protein of 150,000 kDa (Gibbs, 2001). A well coated with 100 μL of reagent has approximately 1 cm^2 of surface area in contact with the liquid, depending on the exact configuration. Thus, at equilibrium an antibody solution at 5 μg/mL under optimum conditions of pH, ionic strength, time, and temperature would saturate this well. Often, however, the curve to this saturation point is

TABLE 4.2 Available Microtiter Plate Chemistries for Covalent Linkage

Surface chemistry	Use
Amine	For use with hetero- or homobifunctional cross-linkers to bind to amino, thiol, hydroxyl, or carboxyl groups
Photoactivatable	For linking to amino, thiol, and hydoxyl groups without additional cross-linkers
Maleimide	For covalent binding to sulfhydryl containing molecules, no cross-linker required
Hydrazide	For binding to carbohydrates (after mild oxidation)
Maleic anhydride	For covalent binding to amine containing molecules

nonlinear and near saturating conditions can be achieved with as little as 2–3 µg/mL of antibody.

When direct passive absorption of the analyte is not possible (i.e., when analyte is not a protein, such as a carbohydrate or a hapten), other strategies must be employed. These include coupling the analyte to a protein that will absorb to the plates, such as albumin, IgG, or ovalbumin, and then using passive absorption, or linking directly to the plate covalently. Actual density of antigen that can be achieved may be less with the protein linking strategy; however, this method does space the analyte away from the surface of the plastic and may cut down on steric hindrance that can interfere with antibody binding. Coupling chemistries, linkers, or carrier proteins that have been used to generate the immunogen from the analyte are best avoided here. Frequently, antibodies will be generated to the linking groups or biased toward the analyte linked in that fashion, and this can be a confounding factor when attempting to select antibodies. Once coupled, the protein analyte is typically quite stable and can be used for reproducible plate coating as desired.

Binding capacity for covalent linkage to plates varies considerably depending on the chemistry involved and the manufacturer of the plates. As with a protein carrier, coupling will need to take place through a functional group not involved in the recognition of the molecule by the antibody. Plates available for covalent coupling and their chemistries are listed in Table 4.2.

With proteins, use of a covalent coupling chemistry will often result in a fraction bound passively (sometimes the majority of the analyte are bound in this fashion) and a fraction bound covalently. Stringent washing conditions will be required to remove the passively bound fraction. These conditions however must not be so harsh that they destroy antigen or antibody.

4.3.2 Blocking

For passively adsorbed antibodies or proteins on plates, blocking is usually required. The purpose of the blocking step is twofold: (**1**) bind an agent inert in the immunoassay to any potential binding sites remaining on the plastic and (**2**) remove any antibodies or proteins weakly adhered to the plastic that might otherwise be released during the ELISA.

Blocking agents typically consist of a protein in a buffer and may include surfactants that should be the same or similar to those used in the assay to assist in function 2 described above. Proteins used are typically bovine serum albumin (BSA), casein, skim milk, porcine gelatin, and fish gelatin. In general, one is looking for molecules smaller than the primary coating protein to fill any voids that may have been left between proteins in the coating. When coating with antibody, all blocking reagents mentioned above fulfill this function. When using anti-goat IgG in detection, care must be taken that the blocking agents do not contain IgG that will cross-react. This is true of many grades of BSA, casein, and skim milk, all of which contain bovine IgG that will be cross-reactive. Similarly, when coating with avidin, streptavidin, or related molecules, one must avoid skim milk (as well as some grades of casein), which contains biotin, resulting in severely attenuated activity of the avidin in the assay.

Levels of protein used to block are typically in the range of 0.1–1% (1–10 mg/mL). This is a large excess over the amount of protein used to initially coat the plate (typically 1–10 µg/mL).

Surfactants used for blocking are typically nonionic, such as Tween-20 (polyoxyethylene (20) sorbitan monolaurate) or less commonly Triton X-100 (polyoxyethylene octyl phenyl ether). Typical concentrations are in the range of 0.05–0.2%. These serve to both block hydrophobic groups on the plate and remove loosely bound proteins.

Both proteins and surfactants in this blocking reagent will promote microbial growth, and most proteins used already have some level of microbial contamination in them, so inclusion of a bactericidal or bacteriostatic agent is highly recommended. Sodium azide may be used at levels of 0.05–0.1%. While azide is detrimental to activity of peroxidases, including it at the blocking step has no adverse effects on assay performance, provided it is washed away prior to enzyme exposure. As mentioned with conjugates, isothiazolinones have been used throughout the diagnostic industry for many years and are quite effective at levels of the active ingredient from 15 to 30 ppm (Eremin et al., 2002).

Typically, microtiter plates are coated with 100 µL of coating solution. The walls of the wells above the level of coating however will be exposed to solutions as they are pipetted into, shaken, moved about, or washed. These walls require washing. Blocking volumes should be in the range of 250–300 µL to completely fill the wells.

The effects of blocking are typically complete in a very short amount of time (5 min); however, it is most common to block for 1 h. Frequently leaving for longer times or even storing for brief periods refrigerated in the blocking agent as discussed in Section 4.3.3 is often acceptable. For covalently bound surfaces, it is essential that all covalent reactions be complete and that any remaining linkers be quenched. This is usually accomplished by addition of a small molecule containing an abundance of the same groups used in the coupling. For instance, if the coupling chemistry links to amino groups, addition of low concentrations (10 mM) of ethanolamine would be desirable, effectively complexing with any remaining reactive groups and terminating the reaction phase of coating.

As previously mentioned, many surfaces derivatized for covalent linking are also capable of binding proteins in a passive fashion, so a blocking step of proteins and surfactants might still be in order.

4.3.3   Long-Term Storage

Most coated microplates may be conveniently stored in the presence of the blocking agent at 2–8°C for periods of a week and sometimes longer. It is essential that this blocking agent should have antimicrobials in it. It is also important that evaporation be eliminated by covering the plate. Most available adhesive plate sealers attempt to strike a balance between adhering to and removing from the plastic. Unequal evaporation can occur if the plate is not sealed securely due to ineffective adhesive sealers. This can result in assay noise or even unacceptable background. If storage is to be for a prolonged period (>1 week), drying the plate is highly advised.

Drying imposes some special considerations on the antibody bound to the surface of the plastic. The structure of the antibody (or antigen) is already altered or denatured to some extent by virtue of being bound to the plastic. Drying needs to be designed to not exacerbate this condition. As in lyophilization, preservation of activity during removal of water will require the addition of compounds that can substitute for the bonding that would normally occur between protein and water. Sugars and polymers are quite effective at this at concentrations ranging from 0.1% to 10%. Other stabilizing or bulking reagents, such as protein, may also be desired to maintain activity. Liquid surfactants, which will not dry properly, must be avoided or used at very low concentrations. A stabilizing solution may be added instead or subsequent to the blocking step. After incubation for a brief period, the solution is aspirated or dumped and excess moisture removed from the wells by gentle but firm tapping. The goal of these reagents is to create a very thin but protective hydrating film over the antibody (or antigen) coated plastic. This film should scavenge water quickly and dissolve to allow quick rehydration of the immunoreactive component. As in lyophilization, the bulking agents in the formula will further prevent any trace moisture from interacting with the microplate-bound antibody (or antigen).

The plates are then thoroughly dried overnight in an incubator, hood, or other low humidity environment, followed by which they may be stored in airtight containers with desiccant to maintain relative humidity at less than 10%. Suitable airtight containers would include foil lined bags. Plastic zip lock bags are too permeable and will not offer long-term storage. Properly stabilized plates can be used for up to several years at room temperature, or longer if refrigerated, with no change in performance. As with any assays, however, it is imperative to always run appropriate control materials to validate performance.

4.3.4   Plate Washing

Plate washing is usually required to remove all unbound reagents between steps in an ELISA. Free antibody or antigen that carries over will interfere with the next step in the assay, usually resulting in reduced signal. Similarly, free enzyme will give an erroneous high estimation of the amount of reagent bound. Since error introduced through inefficient or incomplete washing is seldom uniform across the plate, the net result is often a very noisy assay with high background. Increasing washing efficiency is one of the simplest ways to cure problems in assays plagued by this. It is difficult to overwash a plate, but it is very easy to underwash one.

Plate washing can be accomplished in a number of ways. A 500 mL squirt bottle filled with wash solution provides a quick, inexpensive, and an effective wash. It is generally advisable to cut the tip to an opening 3–4 mL in diameter to allow filling the plate quickly and avoid experiencing hand fatigue that comes from prolonged squeezing on a typical squirt bottle. The plate is filled to capacity (or even slightly overflowing), then inverted over a sink, and finally with a flick of the wrist most of the contents are discarded. This process is quite effective and is repeated for the desired number of washes (typically 3 or 4) before tapping the inverted plate on a paper towel to remove most residual moisture. Gentle tapping usually suffices. It is also advisable to dry the backside (bottom) of the plate when washing in this fashion as liquid may end up here as well.

Washing may also be accomplished using a 8- or 12-channel pipette, used both to aspirate and to add wash solution, with pipette tip changes in between. While this will work for a single row of wells, it is a very slow process for an entire plate and is not recommended if other alternatives are available.

Automated plate washers are also available for washing. These work through a series of cycles by filling the wells and then aspirating the contents. It is advisable to be certain that the wells are *completely* filled on each cycle to avoid residue at the top of the well. This may require different adjustments on these readers based on the brand and configuration of the plates being used. Careful maintenance is also required to ensure that none of the probes used for aspirating are plugged with debris. Most of these washers will not work with strip wells or individual wells, unless these are assembled to mimic a full plate. As with manual washing, here also at the end of the wash cycle gentle tapping is advisable. There is rarely a need to vigorously slam the plate to remove this liquid, although this is all too common behavior (perhaps a stress relieving one).

Though die hard squirt bottle advocates may disagree, plate washers offer an advantage when a large number of plates are processed in an assay. Stacking feeders are also available for some of these to further speed up the process for many plates.

Effective washing is critical for uniform responses on the plate and is one of the most frequent causes of random noise (see Chapter 9). Choice of wash buffer can help with the effectiveness. A typical wash buffer will include a buffer and a surfactant and perhaps a preservative if long shelf life is required. Surfactant (typically nonionic ones such as Tween or Triton at concentrations from 0.05% to 0.1%) can be quite helpful in lowering background due to loosely or nonspecifically bound molecules. Tween has also been shown to have a stabilizing effect on peroxidases (Tijssen, 1985). The pH of the wash buffer should generally be matched to the pH the assay is performed at. Incorporation of salts into the wash buffer to physiological strength will also provide best stability of the antibody–antigen complexes. Some assays will employ a "soak" time in the wash to remove loosely bound material. In general, this should be avoided and often there are other improvements in the assay (typically the assay buffers or composition of the wash buffer) to make this unnecessary.

Water also can be used alone as a wash; however, this should be approached with great caution as the low ionic strength environment can be detrimental to some antibody–antigen interactions that are probably quite dependent on the exact nature of the epitope on the antigen. This may be more problematic for monoclonal antibodies than for polyclonal ones, where multiple epitopes are involved. In general, it is best to optimize assays around physiological strength wash buffers.

4.3.5 Temperature

The structure of the typical microplate does not lend itself to uniform cooling and heating. Typically, wells at the periphery equilibrate more rapidly than those at the center, depending, of course, on how the microplate is exposed to changes in temperature. Static incubators typically have hot and cold spots and can produce great variation across the plate. Forced air incubators represent an improvement if they are not overloaded and if care is taken to avoid evaporative effects. Water baths are perhaps the most consistent for equilibrating the plates but bring special challenges in keeping the plate from becoming submerged. A reliable strategy is to bring all reagents and plates to the temperature the assay will be performed at for ~1 h prior to initiating the assay. Avoid adding cold reagents to a plate and never perform an assay with a cold plate. Assays performed at room temperature may show significantly more consistency across the plate than those performed at elevated temperatures, although those done at the higher temperatures (i.e., 37°C) may show higher signal. Room temperature suffers, however, from the drawback in that in many laboratories there is fluctuation of temperature, within and between days, let alone between seasons. The last step in most assays is the enzyme–substrate reaction, and this is the most temperature sensitive of all the steps in an ELISA. With adequate standardization (running truly appropriate control materials), this variability can be factored out.

4.3.6 Covering of Plates

In most 96-well microplate assays, volumes used are 100 µL per well. Over the course of an hour's incubation, evaporation can be considerable. Unfortunately, this effect is rarely uniform across the plate, due to the microenvironment above the plate, and hence results in well to well variability. Evaporation can be effectively minimized by use of removable adhesive plate sealers, plastic plate lids, or stacking of plates with the top plate in the stack sealed or covered with another plate. Plate sealers should not be reused as they may be a source of cross-contamination through splashing onto the sealer.

4.4 FORMATS FOR ELISA

A wide variety of formats are applicable to ELISA. Some of these are illustrated in Figure 4.1.

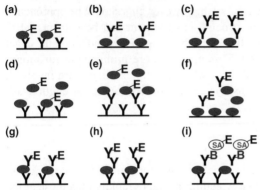

Figure 4.1 ELISA formats. (a–c) Direct bind formats: (a) antibody bound to solid phase reacting with labeled antigen, (b) antigen bound to solid phase reacting with labeled (E) antibody, and (c) antigen bound to solid phase reacting with antibody in turn being detected with secondary labeled antibody reagent. (d–f) Competitive bind formats: (d) unlabeled antigen competes with labeled antigen for binding to antibody bound to solid phase, (e) unlabeled antigen competes with labeled antigen for binding to antibody bound to secondary antibody that in turn is bound to solid phase, (f) antigen competes for binding of labeled antibody that binds to antigen bound to solid phase. (g–i) Sandwich formats: (g) direct labeled antibody binds to antigen to complete sandwich, (h) labeled secondary antibody binds to antibody bound to antigen to complete sandwich, and (i) labeled streptavidin (SA) binds to biotinylated (B) antibody bound to antigen to complete sandwich.

4.4.1 Direct Bind Format

Direct bind assays are most useful for selection and evaluation of antibodies or for selection and optimization of hapten conjugates when antibodies have been already identified in preliminary screening.

For antibody evaluation, plates are coated with analyte in one of the manners described in Section 4.3.1. Amount of analyte required to saturate the plate will vary with the molecular weight and composition of the analyte. Following blocking, the primary antibody putatively directed against the antigen is serially diluted and added to the wells. The wells are covered and incubated, followed by washing and addition of a conjugated antibody directed against the immunoglobulin of the species of the primary antibody. For instance, if the primary antibody were a rabbit antibody, this step might use a conjugated goat anti-rabbit IgG. Anti-immunoglobulin antibodies may be directed against the whole molecule or portions thereof (i.e., anti-Fc region), and for this application, it makes little difference. As noted in Section 4.3.2, one should be sure that all components are fully compatible and noncross-reactive with the blocking agents or nonspecific background may occur. After a second incubation, washing is performed followed by substrate addition and color development.

Not all analytes are capable of binding to plates efficiently. Molecules smaller than 10,000 molecular weight may benefit greatly from using either covalent coupling to an activated plate, or coupling to a carrier protein that may then be passively absorbed to the plate. Most likely these analytes were also conjugated to a carrier for immunization to elicit a strong local response and prevent rapid clearing.

Both the carrier and the chemical structure used for linking will have also elicited an immune response, sometimes much stronger than the antigen itself. It is highly desirable to use a different carrier for screening in the direct bind format as well as a different linking chemistry in preparing these molecules for binding to microplates. The same holds true when one is preparing enzyme conjugates of these molecules when antibody rather than antigen is the immobilized portion of the assay.

The direct bind format is quite useful for evaluating a cycle of immunizations to judge both the strength and the maturity of the immune response. As the response matures, one should find that the titrations reach a plateau. This method is also quite useful for evaluation of hybridoma supernatants for preliminary identification of antibodies directed against the target antigen.

In rare instances, an analyte may have such strong preferential orientation when bound to the surface that potential epitopes are hidden. Examples of this include epitopes on the alpha chain of human chorionic gonadotropin (Norman et al., 1987). In these instances, secondary screens employing sandwich immunoassays, described later, may be required. It is difficult to know if this is the case, so one should always consider secondary screening for the most comprehensive evaluation and certainly in cases where one is selecting hybridomas.

For direct bind evaluation of conjugated analytes, antibody is instead coated to the wells of the microplate. Conjugates of the analyte are diluted over the wells, incubated, washed, and color developed. Most frequently, this is done in combination with varying antibody concentration on the plate as a preliminary to establish conditions for a competitive immunoassay.

4.4.2   Competitive Formats

Competitive formats may be done in either of two orientations. While applicable to analytes of any molecular weight, they are most commonly performed with analytes having molecular weights of 10 kDa or less, where a sandwich format may prove challenging. The goal of constructing such an assay is to determine the conditions under which a change in free analyte, either standards or unknowns, will produce a predictable change in color formation. In the absence of analyte, color formation will be maximal.

In the first orientation, antibody is coated onto the plate as previously described. Antigen with an enzyme label is subsequently added that after appropriate incubations, washing, and substrate addition results in detectable signal. When unlabeled antigen is added either prior to or coincident with the labeled antigen, this material competes for binding to the antibody, resulting in a decrease of signal.

In the second format orientation, antigen is coated onto the plate and labeled antibody is added at the second step. When unlabeled antigen is added either prior to or coincident with the labeled antibody, this material competes for binding to the antibody resulting in less antibody available to bind to the antigen on the plate.

Either of these two formats is quite applicable to competitive ELISA. Choice between them may be based on availability of antigen and/or antibody, labeling, or binding characteristics of antigen or antibody. If both antigen and antibody are equally available, it may be desirable to set up both assays in a preliminary fashion and make choice based on performance.

To establish the conditions for such an assay, checkerboard titrations are initially done. For the antibody-coated plate, primary antibody is serially diluted in coating buffer and then added to the microplate in one direction, across the columns of the plate in this example. After incubation on the plate, these are treated with blocking agent and washed. Dilutions of conjugated analyte are then added to coated wells in the 90° direction to the dilutions of the antibody. After incubation and washing, color is developed. Where antibody is in excess, conjugated analyte binding will be strong. As antibody approaches limiting conditions, the binding will decrease. Similarly, at high levels of conjugated analyte, binding will be strong (Figure 4.2a). In

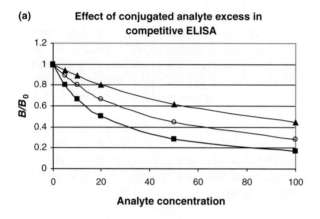

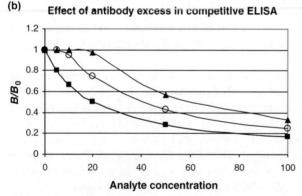

Figure 4.2   (a) Effect of excess conjugated analyte in a competitive ELISA. B/B_0 is the binding that occurs in the presence of a given concentration of sample analyte divided by the binding that occurs in the absence of sample analyte. ■ represents an optimized level of conjugated analyte for maximum sensitivity, ○ represents an analyte concentration twice the optimum level, and ▲ represents an analyte concentration four times the optimum level. (b) Effect of excess antibody in a competitive ELISA. B/B_0 is the binding that occurs in the presence of a given concentration of sample analyte divided by the binding that occurs in the absence of sample analyte. ■ represents an optimized level of antibody for maximum sensitivity, ○ represents an antibody concentration twice the optimum level, and ▲ represents an antibody concentration four times the optimum level.

general, the most sensitive assay will result when minimal concentrations of both antibody and conjugated analyte required to give a reasonable signal at zero analyte concentration are used.

In the second format, the microtiter plate is coated with antigen at a range of dilutions in coating buffer and then added to the microtiter plate in one direction, across the columns of the plate in this example. After incubation on the plate, these are treated with blocking agent and washed. Dilutions of conjugated antibody are then added to coated wells in the 90° direction to the dilutions of the antigen. After incubation and washing, color is developed. As antigen coating levels decrease, one should observe a decrease in antibody binding (as measured by color formation) at some portion of the dilution curve. Similarly, as the conjugated antibody decreases, one should observe a decrease in color as the amount of antibody becomes limiting (Figure 4.2b).

Plate coating at very low levels of antibody or antigen (less than 1.0 μg/mL) may suffer serious problems with reproducibility between lots. This may manifest itself in coating on different days, coating with different lots of reagents, or coating with different lots of plates. For antibody, this can readily be addressed by instead coating the plate under saturating conditions with anti-immunoglobulin, typically directed against the Fc portion of the molecule to provide an orientation. Following a blocking step and a wash to remove any unbound antibody, the antigen-specific antibody is added to the plate, incubated, and is bound in an oriented fashion. Plates can then be washed again and used for the assay or stabilized and stored as previously described. Using this strategy, while more time-consuming, will typically result in good lot to lot reproducibility.

For coating with antigen, it may be advantageous to add low concentrations of a filler or bulking protein to the antigen. For an antigen such as a hapten coupled to a carrier, the filler would consist of the carrier alone. For antigen not requiring a carrier, a molecule of similar composition, charge, and molecular weight would be most suitable.

Competitive assays may be performed using either a simultaneous or a sequential incubation. In a simultaneous incubation, both the labeled and unlabeled antigens (standards or unknowns) are present together with the antibody at the same time. In a sequential incubation, first the unlabeled antigen is added, incubated, and washed and then labeled antigen is added. Alternatively, the labeled antigen can be added to the mixture in the absence of a wash at that stage. Following further incubation, washing is performed and substrate added. For assays with short incubation times (usually because the time to an actionable result needs to be short), the antibody–antigen reaction does not reach equilibrium. In these instances, the simultaneous assay format is the format of choice. This is also true when the affinity of the antibody for the antigen is low. For longer assays, an advantage may be conferred when the unlabeled antigen is first added. Allowing those complexes to preferentially form could significantly shift the equilibrium such that a later addition of labeled reagent (with or without a wash) may shift the standard curve to create a more sensitive assay.

Time to result is always an important immunoassay specification. The best way to select the configuration (simultaneous or sequential) for a competitive

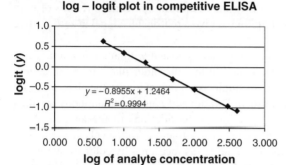

Figure 4.3 An optimized competitive ELISA, yielding a straight line in log–logit format. X-axis is the log of the analyte concentration. Y-axis is logit(y) or $\log_{10}(((B/B_0)/(1 - (B/B_0))))$, where B_0 represents the binding with no analyte present and B represents the binding with analyte present.

immunoassay is to explore the effect of time constraints and then optimize around those limits.

Like most assays, a key part of optimization of competitive assays is optimizing the shape of the standard curve. This is best done by attempting to match the actual performance to the theoretical performance as described by Rodbard and Lewald 1970, where both labeled and unlabeled ligands compete equally for binding (i.e., there is no preferential binding based on differences in chemical structure). This can be accomplished by applying the following formula for a log/logit analysis: Maximum binding (B_0) is defined as the signal (absorbance in the case of a chromogenic substrate) in the presence of no analyte. B is defined as the binding or signal in the presence of x quantity of analyte. Logit(y) is defined as $\log_{10}$ of $(((B/B_0)/(1 - (B/B_0))))$ and is plotted on the y axis. The x axis represents the log of the concentration of analyte. Linear regression analysis of logit(y) versus log(x) should give a straight line (Figure 4.3). The point at which the line intersects the x axis (logit(y) $= 0$) will be the point at which 50% binding occurs. Optimizing the competitive assay around a straight line and the desired midpoint of the curve will create the most reproducible and robust competitive assays. Slope of the line will be indicative of the relative affinity of the antibody for the analyte and the conjugated analyte. Using this fit as the data analysis of choice will greatly simplify optimization.

4.4.3 Sandwich Format

When antigen exceeds 10 kDa, sandwich assays should always be considered as a potential format. A sandwich assay is defined as an antibody immobilized to a surface, complexed to an antigen, and complexed in turn to a labeled antibody. The sandwich format requires two antibodies be able to bind to an antigen, either a repeating epitope or an antigenic structure on the molecule, or two discrete antigenic determinants. In the former, a single antibody could serve as both halves of the sandwich. The requirement for an antigen suitable for a sandwich assay would be one in which these two epitopes are sufficiently spatially separated that two different antibody molecules can bind at the same time with minimum steric hindrance. As antigens increase in size, the availability of multiple epitopes without steric hindrance increases.

There are multiple variants on this assay. Instead of detection with enzyme directly linked to the detection antibody, a third antibody may be used wherein this antibody is directed against the species of antibody (or isotype of the antibody) that forms the top half of the sandwich. In this scenario, the species or isotype of the antibody at the top half of the sandwich must vary from the antibody used for capture. For instance, a goat antibody directed against antigen "X" might be used to coat the plate. After blocking, washing, and incubating with standards or unknowns, the plate is washed and incubated with rabbit anti-"X" antibody. After washing, the plate is incubated with enzyme-conjugated goat anti-rabbit IgG, washed, and color developed. It is very important that the species of the enzyme-conjugated antibody match that of the plate-coated antibody or that it be extensively absorbed against that species on the plate to prevent background from species cross-reactivity. Some amplification over the direct sandwich assay may be seen using this format, but it is usually less than twofold, hardly worth the extra step required. More often this format is used when it is impractical to label the antibody at the top half of the sandwich, or when that antibody is in an impure form (such as sera) or in short supply. For instance, this format could be used for screening culture supernatants during hybridoma development. A polyclonal antibody directed against the antigen is coated on the plate, blocked, washed, and a source of antigen added. Hybridoma supernatants are added, incubated, and washed and enzyme-conjugated anti-murine IgG (same species as polyclonal) is added, incubated, washed, and developed with substrate. This has a further advantage in that very crude extracts with antigen in them may be used such as plant extracts so that the exact form of the antigen as expressed in the plant is the target antigen.

When the desired coating antibody is also impure or relatively scarce, the plate can be coated with anti-immunoglobulin of a third species, highly absorbed for reactivity to the targeted species. The plate is then blocked, washed, and the capture antibody to antigen X added. The assay would proceed as described previously. Again, the detection antibody can be conjugated for direct binding or unlabeled and subsequently detected by an anti-IgG from a third species. This conjugate must be highly adsorbed to guarantee specificity. As an example, coat with donkey anti-goat antibody (highly absorbed against rabbit) to capture goat anti-antigen X antibody then detect antigen X using rabbit anti-antigen X antibody, followed by enzyme-labeled donkey anti-rabbit IgG (highly absorbed against goat). Careful antibody selection and titration of all reagents to minimize background are the key to getting this format to work well. It may also be advantageous to incorporate IgG of the nontargeted species (goat in the above example) into the diluent so that the enzyme conjugate is diluted to tie up any remaining activity. It should be kept in mind, in the case of using anti-IgG antibodies to immobilize capture antibodies, that the anti IgG should be carefully selected for high affinity as, in the final assay, performance will be determined by the lesser of the affinity constants of the anti-IgG and the capture Ab.

Avidin (or related molecules)/biotin systems may also be incorporated into these sandwiches. The anti-antigen antibody that forms the upper portion of the sandwich could be biotin labeled and detection accomplished using enzyme-conjugated streptavidin. Again, there may be some amplification over direct detection, but this is minimized by the steric constraints of enzyme-conjugated streptavidin. One could also incorporate the avidin/biotin system into the capture phase of the sandwich assays as previously described, in place of anti-immunoglobulins. Biotinylation is more amenable to very small-scale conjugations when antibody quantity is limiting than direct enzyme conjugation. Generally, proteins with similar activity to avidin, such as streptavidin or neutravidin (Pierce), are used instead of avidin because of the high nonspecific binding that can sometimes be observed due to the high pI of avidin. While the affinity of these for biotin may be less than that of avidin, this is of little practical consequence as all are quite high.

Optimization of sandwich assays is significantly easier when the format is simpler. The more complex formats may serve as effective screening methodologies but are more difficult to develop into robust quantitative methods. Initial experiments should be checkerboard-type titrations where one reagent is diluted across the rows while a second reagent is diluted across columns. It is generally advisable, assuming adequate quantities of antibody are available, to test all potential pairs. Test each antibody both in the capture mode and in the detection mode. If antibody quantities are restricted, the detector antibody typically requires less antibody (10–100-fold less) than the capture antibody.

Solid-phase optimization was described previously. Effective sandwich assays will have ample capacity on the solid phase to effectively capture antigen in the range of interest. Titration of enzyme-conjugated antibodies over the solid phase with antigen should focus on two issues: background and the shape and slope of the standard curves.

Typically, in any immunoassay, sensitivity is defined as signal over noise. The limit of detection can be defined as the background plus 2 or 3 standard deviations for 95% and 99% confidence that the signal is different from background. If an assay has minimal or no background, the detection limits can be increased by longer substrate incubations, chemical conversion of formed substrates, or moving to more sensitive substrates. Conversely, with noticeable background, sensitivity will be identical regardless of substrate sensitivity. Minimizing background can be achieved through conjugate titrations, ratios of enzyme to antibody used to make the conjugate, diluents, wash composition, and wash efficiency. Keeping assays simple, as with direct antibody labeling, will also assist in minimizing the background variation.

Specific binding is a phenomenon where there is a point (reagents, time, temperature, etc.) at which no additional specific binding will occur. In contrast, nonspecific binding is a phenomenon where increasing reagents, time, or other parameters will simply continue to increase nonspecific binding.

A wide variety of sophisticated software programs exist that allow fitting curves to ELISA results. Many of these are incorporated into programs that come with microplate readers. Given algorithms of increasing complexity, even poorly optimized ELISAs can generate results that the user may attempt to use in a quantitative fashion. This deceptively easy data reduction unfortunately contributes to the use of assays outside their reliably quantitative range. Rather, in an optimized sandwich

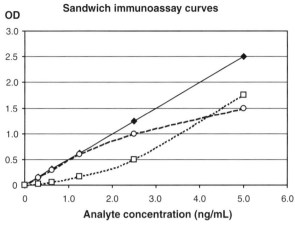

Figure 4.4   Possible immunoassay curves. ♦ represents an optimized sandwich immunoassay with all reagents, except analyte, present in excess to generate a linear curve with good range, ○ represents an assay where one or two reagents are not in sufficient excess, leading to a reduced linear portion of the curve, and □ represents an assay with multiple components (including perhaps the affinity of the reagents) at limiting concentration.

assay, the linear portion of the curve gives the most reliable quantitation. The only factor affecting the response should be the dose of antigen. One should optimize conditions around producing a linear response in the portion of the standard curve most useful for quantitating the samples to be tested. Standard curves and the causes of the deviations from linear are illustrated in Figure 4.4.

As sandwich ELISAs become more linear, the net result will be a reduction in variability throughout the curve. This will result in a lower coefficient of variation (CV) on determination of analyte levels. When plotted on a linear scale with analyte on the x axis and response (OD, fluorescence, light output, etc.) on the y axis, a slope of 1 is often an ideal compromise between accurate determination of analyte concentration and dynamic range of the assay. Slopes above 2 will limit useful range of the assay. Slopes below 0.5, or working in a portion of the curve where some component in the assay is saturating, may produce unacceptable CVs.

4.5   FURTHER PERFORMANCE OPTIMIZATION

The goal of performance optimization is to create a robust assay that can withstand small, sometimes inevitable variations in procedure and still produce reliable results. These conditions can be found by careful experimentation, often in a factorial fashion to identify the assay parameters creating the largest effects and then taking steps to minimize these. Evaluation of performance or method validation is covered in Chapter 6, but following the guidelines outlined previously in this chapter combined with the ones listed below should assure robust assays.

As a note, it is relatively easy to get ELISAs to perform well in laboratory buffers with purified proteins or antigens. It is quite another matter to get them to perform well with real-world samples. During optimization, one should move to include diverse real-world samples as early in the process as possible.

4.5.1   Assay Kinetics

Optimization of assay kinetics is sometimes a balance between the ideal immunoassay conditions, where reagents move toward equilibrium, and the reality of the timing in a workday. Nonetheless, a careful exploration of incubation times (even toward times that would not prove practical in real use) will help define where the ELISA sits relative to ideal conditions. This should be done for incubation times of all steps, with the exception perhaps of the enzyme substrate step where one can rely on previous experience or suggestions of the substrate vendor. One should avoid choosing incubation times where the response of the assay is changing rapidly. This knowledge may later prove useful in troubleshooting if required.

4.5.2   Simultaneous Versus Sequential Incubations

Sandwich assays may be performed with each step done sequentially or in some instances with sample and detection antibody added simultaneously. The advantage of simultaneous incubation is often one of shortening overall length of the assay and certainly eliminating one set of wash steps. Often, when the affinity of the antibodies, particularly the detection antibody, is low, simultaneous incubations will increase assay sensitivity since at least one of the reactions (analyte with detection antibody) can take place more rapidly in the liquid phase as opposed to at the surface of the microplate where diffusion rates will limit opportunities for interaction. This format requires additional optimization beyond just applying conditions worked out for a sequential format. Reoptimization of reagent concentrations and kinetics will be required, but elimination of the wash and second incubation will make these experiments go quickly. When time to result is critical, simultaneous condition should always be evaluated.

There are several precautions with this format. First, sample and sample diluent must be fully compatible with the enzyme used as the label. Second, this format is prone to an effect known as "high-dose hook effect" (Fernando and Wilson, 1992). This phenomenon occurs when analyte exceeds the binding capacity of the detection antibody. As a consequence, a significant portion of the analyte bound to the plate does not have detector antibody bound. The net result is an increase in response to a point where this phenomenon starts to occur followed by a decrease in response as analyte continues to increase. More detail about hook effect will be discussed in Chapter 9. There are two solutions to this: The first is to optimize assay conditions around the absolute greatest concentration of analyte that would ever be expected. This can usually be accomplished by increasing the concentration of detector antibody. A side effect of this, assuming optimal conjugates and plate coating so that background does not occur, will be more favorable kinetics and increased sensitivity. The second solution, although more laborious and usually unacceptable (if the goal of using simultaneous incubation was to reduce labor), is to run all samples at several dilutions so that if this effect occurs, it may be detected. Issues with simultaneous versus sequential considerations have been covered in Section 4.4.2.

4.5.3 Matrix Effects

In any immunoassay, the matrix (or overall composition of the sample) may exert a profound influence on the immunoassay. This will often manifest itself as reduced availability of the analyte to the antibody(ies). This is a problem for both quantitative and qualitative immunoassays, although somewhat lesser for the latter. Sample dilution invariably reduces the influence of sample matrix and should be the first approach tried, but this may unacceptably reduce assay sensitivity and certainly add an additional step by requiring the dilution of the sample into assay buffer. The primary goal of the assay buffer if required should be to minimize these matrix effects. Oftentimes, the properties required of a sample buffer may be formulated into the sample extraction buffer. Matrix effects may be readily assessed by spiking a known quantity of purified or highly concentrated analyte into matrix, allowing equilibration to occur (1 h would not be unreasonable), and then measuring quantitatively in the immunoassay. Generally, values in the range of 80–120% ((expected value/observed value) $\times$ 100) are acceptable in quantitative ELISA. Outside this range, it will be necessary to either alter the sample extraction buffers and/or adjust the assay buffer. Generally, an extraction buffer should effectively solubilize the analyte, but minimize other compounds extracted that might interfere with the assay. Water, for instance, will solubilize some analytes but often leaves interferents behind. If the analyte is not water soluble, a minimalist approach should be taken to optimize the composition of the extraction buffer. Both the formulation of the extraction and the assay buffer may involve the use of buffers to manipulate pH, salts to manipulate the ionic environment, and detergents and/or polymers to free analyte and/or antibodies from potential interfering compounds.. Kits containing a large number of surfactants and those with a large number of polymers are available commercially (RDI, 2005). Effective concentrations of surfactants are generally in the range of 0.1–1.0% and concentrations of polymers, if used, between 0.1% and 5%. Within plant-derived materials, polyvinylpyrrolidone is particularly effective at complexing with polyphenols often found in these samples and reducing matrix problems (Bruyns et al., 1998).

A second way to assess matrix effects would be to evaluate the linearity, relative to the standard curve, of sample dilution. Samples should be serially diluted, quantitated in the ELISA, and then the values plotted. Plotting expected value versus observed values should yield a straight line with slope of 0.85–1.15. If the curve is not a straight line, then the linear portion of the curve will correspond to the dilutions where the assay has no or minimal matrix effects. Fine-tuning the assay in this fashion is also possible, but this is significantly more time-consuming than the "spike and recovery" method previously described (although perhaps more accurate).

There is another type of matrix effect with relatively insidious effects that is often overlooked. With undiluted samples of ground plant seed, which generally contain a high percentage of undissolved solids, a thick layer may form on the bottom of the wells during incubations. This obviously will inhibit diffusion of analyte. One solution is to keep the microplate vigorously agitated during the sample incubation. Washing should be done with sufficient stream strength to remove any sludge that may settle between incubation and wash. Another more time-consuming but safer alternative is to gently centrifuge (1–2 min at $500 \times g$ will suffice) or filter samples prior to assay. More detail about matrix effect will be discussed in Chapters 6 and 9.

4.6 TROUBLESHOOTING

During development some common phenomena described below may be encountered. It is essential to understand their sources and apply corrective action so that they do not continue to haunt the assay as it moves to more routine use. In many cases, these corrective actions will be more carefully written protocols. Phenomena that are consistently present are relatively easy to troubleshoot. Unfortunately, all too many of these effects may come and go, making experimental troubleshooting difficult and often frustrating, so the fundamentals behind preventing them become all the more important. Other than the techniques described here, more troubleshooting tips will be discussed in Chapter 9.

4.6.1 High and Erratic Background

If background is high and erratic, it is often from inefficient washing. It is essential to be sure that wells are completely filled during wash steps and that sufficient number of washes have been performed. Other causes may be allowing the plate to dry after washing or may be due to aggregates present in the conjugate. Overall high background may result from inappropriate dilution of conjugate, bacterial contamination or other debris in reagents, unanticipated cross-reactivity of reagents (improper previous evaluations), substrate contamination, or light exposure.

4.6.2 Edge Effects

Edge effects have three origins. One is inconsistency of the microplate molding process that results in surfaces with variable binding capacity. Plates not sold for ELISA should not be used for quantitative assays. A second source is discrepancy of reagent temperatures with incubation temperatures. Wells of the microplate will not equilibrate to temperature at the same rate, with edges equilibrating quickly. All reagents should be brought to the incubation temperature prior to use. Using an incubator can exacerbate this effect as incubator walls are often significantly warmer than the interior. If incubators are used, plates should be placed in the middle and the incubator not overloaded to the point that airflow is impeded. The third source is evaporative effects. The microenvironment above the plate edges is more susceptible to evaporation than wells in the middle. Careful covering or sealing throughout preparation of plates and assay will eliminate this. The latter two sources generally result in edge wells with more color than interior wells, which is the most common complaint when edge effects are observed.

4.6.3 Front to Back Effects

These effects are observed when there is considerable variation in degree of equilibrium from the first well compared to the last well in the assay. This effect is further exacerbated when assay incubation times are brief. This can be eliminated by use of multichannel pipettes and associated specialized containers (8- or 12-well troughs, deep well plates for preparing dilutions, etc.). Also, paying careful attendance to reagent addition and lengthening the incubation times so that variation in reagent addition represents a small fraction of the total time will minimize or eliminate this effect. All sample preparation and dilution should be done prior to loading the assay plate. It is also essential that all reagents be brought to temperature before use.

4.6.4 Poor Intra-Assay Precision

Poor CVs can come from poor washing efficiency, variations in plate coating, pipetting error, edge effects, and front to back effects, or incubations times not at equilibrium. Optimizing plate coating, reevaluating kinetics, and working at slight excess of reagents in sandwich assays will all improve CVs and the precision with which analyte concentration may be determined. CVs of sandwich assays should be in the range of 2–10% in the working portion of the curve. CVs of competitive assays will be higher (since using reagent in excess is not possible), but should be in the range of 10–15% in the working portion of the curve.

A few wells with erratic color generally come from poor washing (insufficient filling again) or splashing of reagents during plate setup.

4.6.5 Poor Interassay Precision

One of the major sources of poor CVs between assays in plant matrix ELISA is variation in sample extraction. This may come from both efficiency of extraction and stability of the analyte once extracted. Liquid controls (which could be stored frozen in aliquots) should be run with each assay, as well as control samples to be extracted to distinguish the source of the variation. Standards should be prepared in sets (again that may be frozen if stability is an issue) rather than fresh for each assay to eliminate pipetting error. While interassay CVs are typically higher than intra-assay CVs, they should still be less than 20% within the working range of the assay.

4.6.6 Day to Day Variation in Absorbance (or Light Emission)

This is usually attributable to temperature variations in incubations from day to day. This is particularly true for the enzyme substrate step that will be strictly temperature dependent, whereas the immunoreactions may, depending on assay optimization, still reach the same degree of equilibrium even with temperature variation. Assays run at "room temperature" can be notorious for this effect. It is also important to read assays at precisely the same time of substrate development (or use a stop reagent) for consistent results. In most instances, the standard curves and controls, which should be run on every plate, will factor out the normal day to day variation. It is generally recommended to set a range in response where one expects some of the standards to fall and use this as a criterion for deciding whether to accept or reject the assay.

4.7 FUTURE APPLICATIONS OF MICROTITER PLATE ELISA

For over 35 years, microtiter plate ELISA has remained a standard method in laboratories throughout the world for accurately analyzing many samples at once. While the 8×12 format remains the same, the number of possible wells in the same footprint has increased to 384 and 1536. Automation and robotic technology have facilitated handling these more complex plates and processing larger numbers of samples. More information on automation of ELISAs is covered in Chapter 8. Instrumentation is available to measure a variety of substrates (colorimetric, fluorescent, and luminescent) on a single reader. In many areas, the recent trends have been toward multiplexing. With robotic spotters, it is now possible to construct arrays of 25 (5×5) antibodies immobilized onto each well in a plate (Roda et al., 2002). A single sample can be tested with a mixture of 25 conjugates (and precipitating substrates to keep the response localized to the spot) and read with CCD imagers to maximize the data per sample. Whether this offers advantages in agricultural immunoassay remains to be seen. Regardless, the principles outlined in this chapter can be applied to most microplate ELISAs, no matter how small, how fast, or how many per well.

REFERENCES

Ballon, P. (Ed.), *Affinity Chromatography Methods and Protocols*, Humana Press Inc., New York, 2000.

Bruyns, A. M., De Neve, M., De Jaeger, G., De Wilde, C., Rouzé, P., and Depicker, A. Quantification of heterologous protein levels in transgenic plants by ELISA. In Cunningham, C. and Porter, A. J. R. (Eds.), *Recombinant Proteins from Plants: Production and Isolation of Clinically Useful Compounds*, Humana Press Inc., New York, 1998, pp. 251–269.

Clem, T. R. and Yolken, R. H. Practical colorimeter for direct measurement of microplates in enzyme immunoassay systems. *J. Clin. Microbiol.*, **1978**, *7*, 55–58.

Engvall, E. and Perlman, P. Enzyme-linked immunosorbent assay (ELISA). Quantitative assay of immunoglobulin G. *Immunochemistry*, **1971**, *8*(9): 871–874.

Eremin, A. N., Budnikova, L. P., Sviridov, O. V., and Metelitsa, D. I. Stabilization of diluted aqueous solutions of horseradish peroxidase. *Appl. Biochem. Microbiol.*, **2002**, *38*, 151–158.

Fernando, S. A. and Wilson, G. S. Studies on the "hook" effect in the one step sandwich immunoassay. *J. Immunol. Methods*, **1992**, *151*, 47–66.

Gibbs, J. *Immobilization principles—selecting the surface*, Elisa technical Bulletin 3, Corning Incorporated, Kennebunk, ME, **2001**.

Harlow, E. and Lane, D. *Antibodies: A Laboratory Manual*, Cold Spring Harbor Laboratory Press, Woodbury, NY, 1988.

Hermanson, G. T. *Bioconjugate Techniques*, Academic Press, San Diego, CA, 1996.

Hermanson, G. T., Mallia, A. K.,and Smith, P. K. (Eds.). *Immobilized Ligand Techniques*, Academic Press, San Diego, CA, 1992.

Josephy, P. D., Eling, T., and Mason, R. P. The horseradish peroxidase-catalyzed oxidation of 3,5,3',5'-tetramethylbenzidine. Free radical and charge-transfer complex intermediates. *J. Biol. Chem.*, **1982**, *257*, 3669–3675.

Kantor, A. *Biotinylation of antibodies*, available at http://www.drmr.com/abcon/Biotin.html (accessed December 2, 2008).

Lequin, R. E. Enzyme immunoassay(EIA)/enzyme-linked immunosorbent assay (ELISA). *Clin. Chem.*, **2005**, *51*, 2415–2418.

Manns, R. *Microplate history*, **1999**, available at http://www.microplate.org/history/det_hist.htm (accessed December 2, 2008).

Nakane, P. K. and Kawaoi, A. Peroxidase labeled antibody: a new method of conjugation. *J. Histochem. Cytochem.*, **1974**, *22*, 1084–1091.

Norman, R. J., Haneef, R., Buck, R. H., and Joubert, S. M. Measurement of the free alpha subunit of human glycoprotein hormones by monoclonal-based immunoradiometric assay, and further exploration of antigenic sites on the choriogonadotropin molecule. *Clin. Chem.*, **1987**, *33*, 1147–1151.

Parham, P. Preparation and purification of active fragments of mouse monoclonal antibodies. Weir, D. M. (Ed.), *Cellular Immunology*, 4th edition, Blackwell Scientific Publications, California, 1986.

RDI, 2005, available at http://www.researchd.com/rdioem/oemindex.htm (accessed 2008).

Roda, A., Mirasoli, M., Venturoli, S., Cricca, M., Bonvicini, F., Baraldini, M., Pasini, P., Zerbini, M., and Musiani, M. Microtiter format for simultaneous multianalyte detection and development of a PCR-chemiluminescent enzyme immunoassay for typing human papillomavirus DNAs. *Clin. Chem.*, **2002**, *48*, 1654–1660.

Roda, A., Pasini, P., Mirasoli, M., Michelini, E., and Guardigli, M. Biotechnological applications of bioluminescence and chemiluminescence. *Trends Biotechnol.*, **2004**, *22*, 295–303.

Rodbard, D. and Lewald, J. E. Computer analysis of radioligand assay and radioimmunoassay data. *Acta Endocrinol. Suppl. (Copenh.)*, **1970**, *147*, 79–103.

Rowell, V. (Ed.). *Nunc Guide to Solid Phase*, Nunc A/S, Roskilde, Denmark, 2001.

Schuetz, A. J., Winklmair, M., Weller, M. G., and Niessner, R. Stabilization of horseradish peroxidase (HRP) for use in immunochemical sensors. *Proc. SPIE*, **1997**, *3105*, 332.

Takatsy, G. *Kiserl. Orvostud.*, **1951**, *4*, 60.

Tijssen P. *Practice and Theory of Enzyme Immunoassays*. Elsevier, Amsterdam, 1985.

Van Weemen, B. K. and Schuurs, A. H. Immunoassay using antigen-enzyme conjugates. *FEBS Lett.*, **1971**, *15*, 232–236.

Voller, A., Bidwell, D., Huldt, G., and Engvall, E. A microplate method of enzyme-linked immunosorbent assay and its application to malaria. *Bull. World Health Organ.*, **1974**, *51*, 209–211.

CHAPTER 5

LATERAL FLOW DEVICES

Murali Bandla
Rick Thompson
Guomin Shan

5.1   INTRODUCTION

Lateral flow immunoassay is a rapid membrane-based assay where the analyte and antibody reagents are chromatographed on a porous membrane. These devices are also known as "dipstick immunoassays," "strip tests," "immunochromatographic devices," and "lateral flow devices." For simplicity, in this chapter these will be referred as LFDs.

The LFDs are used for the specific qualitative or semiquantitative detection of many analytes including small inorganic molecules, peptides, proteins, and even the products of nucleic acid amplification tests. The basic concept of LFDs was first described in the 1960s, and the first commercial application was Unipath's Clearview

Immunoassays in Agricultural Biotechnology, edited by Guomin Shan
Copyright © 2011 John Wiley & Sons, Inc.

home pregnancy test commercialized in 1988. Since then, it has been a popular platform for rapid tests for clinical, veterinary, agricultural, food safety, and environmental applications. One or multiple analytes can be tested simultaneously on the same strip (O'Farrell, 2009). These devices are portable, relatively inexpensive to manufacture, and do not require any laboratory equipment or training for the operator. Lateral flow immunoassays are very stable and robust, have a long shelf life, and do not usually require refrigeration during storage. They generally produce a result within 10–20 min. However, lateral flow strips are usually less sensitive compared with enzyme-linked immunosorbent assay (ELISA). LFD tests are extremely versatile and are available for an enormous range of analytes from blood proteins to mycotoxins and from viral pathogens to bacterial toxins. In the past decade, LFD has become a dominant analytical tool for the detection of transgenic proteins in genetically engineered crops, grain, and foods. The assay formats of lateral flow immunoassays are comparable to the plate format ELISA. The concept of LFD was originated from the enzyme immunoassay by applying the same assay components on a chromatography paper strip (Litman et al., 1980). The first reported LFD used an enzyme as a label in the system (Pappas et al., 1983) and it was soon replaced by more sensitive labels. The current LFD technology is more sophisticated with the contribution of advanced membrane materials and broader selection of labeling particles from colloidal gold (Leuvering et al., 1980) and latex particles (Gussenhoven et al., 1997) to quantum dots (Goldman et al., 2004) and upconverting phosphor particles (Corstjens et al., 2001).

Lateral flow immunoassays are usually used only for qualitative testing purposes; however, one can obtain quantitative or semiquantitative result by measuring the amount of conjugate bound to the test zone. Several LFD manufacturers are offering readers to quantify the signal and are available commercially. In this chapter, we will discuss the principle of LFD, its advantages and limitations, manufacturing and validation, quality control, and troubleshootings.

5.2 PRINCIPLE OF LATERAL FLOW DEVICES

Lateral flow immunoassays are essentially immunoassays adapted to operate along a single axis to suit the test strip format. A number of variations of the technology have been developed into commercial products, but they all operate according to the same basic principle. The function of antibody to detect the analyte in LFDs is the same as that of ELISA except the detection antibody is labeled with a particle that gives a visible signal.

A typical LFD consists of a sample application pad, a conjugate or reagent pad, a reaction membrane, and a wick or waste reservoir pad (Figure 5.1). The sample application pad is an absorbent pad located at one end of the strip onto which the test sample is applied and it facilitates filtration of unwanted particulates from the sample. Usually, the sample pad is made of cellulose or cross-linked silica. The conjugate or reagent pad is adjacent to and in contact with the sample pad and contains labeled analyte or antibodies specific to the target analyte depending on the application. When the liquid sample passes through the reagent pad, the reagent is released, mixes

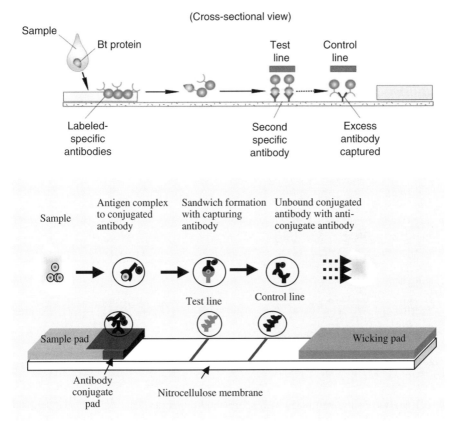

Figure 5.1 Diagram of an immunochromatographic device. (*Source:* Envirologix, Inc. Portland, ME.)

with the sample, and the mixture enters the reaction membrane. The reaction membrane is usually a hydrophobic nitrocellulose or cellulose acetate membrane. There are two zones located within the result window, the test zone and the control zone. In a typical antibody–antigen–antibody sandwich assay, an analyte-specific capture antibody is immobilized on the membrane at the test line, and a detection antibody-specific antibody is deposited in the control line. The test line indicates the presence or absence of analyte in the sample and the control line serves as an internal control to ensure proper functionality of the device (Figure 5.1). More test lines may be applied for multianalyte testing (Snowden and Hommel, 1991; Oku et al., 2001; Kolosova et al., 2007; Leung et al., 2005). The waste reservoir is an absorbent pad designed to draw the reaction mixture across the membrane by capillary action and then collect it.

In general, there are two main types of lateral flow immunoassay: double antibody sandwich format and competitive format (Posthuma-Trumpie et al., 2009). In the double antibody sandwich format, the sample migrates from the sample pad through the conjugate pad where labeled analyte-specific antibody is present and the

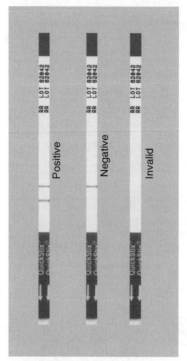

Figure 5.2 Visualized test results by immunochromatographic devices. (*Source:* Envirologix, Inc. Portland, ME.)

target analyte in the sample will bind to the conjugate. This antigen and labeled antibody complex then continues to migrate across the membrane until it reaches the test zone where the antigen and antibody complex will bind to the immobilized antigen-specific antibodies and form an antibody–antigen–antibody sandwich and produce a visible line (test line) on the membrane. The unbound labeled detection antibodies will pass through the test zone and continue to move upward. It will then be trapped by a detection antibody-specific antibody in the control zone and form a visible line on the strip (control line). If no analyte is present in the test solution, only the control line appears in the result window and the result is negative. This band indicates that the liquid flowed properly up the strip. If two bands appear in the result window (test line and control line), the result is positive (Figure 5.2). The device may also be enclosed in a plastic cassette with a window for reading the signal. The double antibody sandwich format is most suitable for larger analytes with multiple epitopes. This is the primary diagnostic format for modern agricultural biotechnology products and it will be the main focus in this chapter. A similar format is also used for nucleic acid detection (Posthuma-Trumpie et al., 2009). In this case, the analyte is an amplified double-stranded DNA using primers with two different tags. An antibody specific to the tag is immobilized in the test zone. In such a typical test, the target DNA is amplified using PCR with two tagged primers. The resulting double-stranded amplicon contains one strand labeled with biotin and the other labeled with a tag such as digoxigenin. When amplified target DNA sequence moves through the conjugate pad containing avidin-labeled detector reagent, the biotin will bind to the avidin conjugate and form tag-DNA-biotin/avidin-reporter complex. In the test zone, the tag will bind to the anti-tag antibody, resulting in the colored signal (Van Amerongen and Koets, 2005).

The competitive immunoassay format is primarily used for testing small molecules and antigenic molecules with only one antigenic epitope. In this format, there are two possible configurations. In one configuration, the conjugate pad contains labeled antigen and the test line is composed of immobilized antigen-specific antibody. The antigen in the sample analyte and the labeled antigen compete for binding sites on the antibody at the test zone. In another configuration, the conjugate pad contains labeled antibody, and a protein conjugate of the purified antigen is immobilized at the test line. The antigen in the sample binds with antibody and the complex migrates along the membrane. In the competitive format, the detection signal is negatively correlated to analyte concentration. Two visible lines in both test and control zones is a negative result, and a single control line on the membrane without a visible line in test zone is a positive result.

The choices of membrane and detector reagents are two critical factors for the LFD assay system. Nitrocellulose is the most popular material for the LFD membrane (Leung et al., 2003; Van Dam et al., 2007; Jin et al., 2005). Other polymeric materials used for LFDs include nylon (Buechler et al., 1992), polyethersulfone (Edwards and Baermner, 2006; Kalogianni et al., 2007), polyethylene (Fernandez-Sanchez et al., 2005), polyvinylidene fluoride, and fused silica. Capillary flow characteristics and capacity for protein binding are the two key parameters for the membrane selection. Millipore's technical publication on lateral flow strips (Millipore, 2008) thoroughly describes the properties and specification of various membrane materials and provides an excellent guide for membrane selection during product development. The capillary flow properties affect reagent deposition, assay sensitivity, and test line consistency. For lateral flow devices, the membrane must irreversibly bind capture reagents (proteins) at the test and control zones. A membrane's protein binding capacity is governed by the polymer surface area that is determined by pore size, porosity, and thickness. In general, proteins bind to membrane polymers primarily through two mechanisms: electrostatic and hydrophobic interactions. Nitrocellulose membrane binds proteins through electrostatic interaction between the strong dipole of the nitrate ester and the strong dipole of the protein–peptide bonds (Millipore, 2008). The capillary flow rate refers to the speed of a sample front moving along a membrane strip when liquid is introduced at the sample pad and is expressed in seconds per centimeter. For any given membrane, its flow rate is directly determined by the cumulative physical properties such as pore size, pore size distribution, and porosity—the three important parameters that need careful consideration while choosing a membrane for assay development. The pore size of a membrane is a measure of the diameter of the largest pore in the filtration direction, and pore distribution is a measure of the range of pore sizes on a membrane. Membrane porosity is the volume of air in a three-dimensional membrane structure and usually expressed as a percentage of the membrane's total volume (Millipore, 2008). Commercially available nitrocellulose membranes have various pore sizes ranging from 0.05 to 12 µm and pores are typically not randomly distributed due to the manufacturing process. In contrast, pore sizes can be controlled precisely in poly-ethylene membrane (Posthuma-Trumpie et al., 2009).

The detector reagent is another key component of the LFD system. There are a few choices of particles for labeling or conjugation. The most popular choice is colloidal gold. Colloidal gold has extensively been used in immunoelectron micros-copy. It is made by reducing gold tetrachloric acid. The resultant gold particles are negatively charged and stay in colloidal state by repelling each other. Proteins readily bind to the gold by interacting mostly through tryptophan, lysine, and cysteine residues (Leuvering et al., 1980; Nielsen et al., 2008). Another popular labeling material is colored polystyrene beads, also known as "latex beads," (Gussenhoven et al., 1997; Greenwald et al., 2003). Latex beads are available in different sizes and colors and offer a wide choice of surface chemistries ranging from simple passive absorption of antibodies to covalent linking. Other materials being evaluated for LFD use include carbon, selenium, chemiluminescent, and fluorescent nanoparticles (Zuiderwjijk et al., 2003; Edwards and Baermner, 2006). Paramagnetic particles may be useful for developing quantitative LFDs.

Particle size is one of the most important factors that influences the flow rate and sensitivity of the LFD. Studies showed that larger particles may result in a better assay sensitivity; however, the colloidal stability decreases with increase in size (Laitinen and Vuento, 1996). Also, the number of antibody molecules per particle decreases with increase in gold particle size. Usually, an optimum gold nanoparticle size for LFD is 80 nm or smaller (Aveyard et al., 2007).

Other than membrane and detector type, there are other components such as sample pad, conjugate pad, and buffer reagents that directly affect the quality of LFD. More detailed specifications and consideration of each parameter in LFD develop-ment will be discussed later in this chapter.

5.3 ADVANTAGES AND LIMITATIONS

5.3.1 Advantages of LFDs

There are several advantages of LFDs used in immunodiagnostic procedures for the Ag Biotech applications.

Portability: Most of the diagnostic applications in agricultural biotechnology are field based and, therefore, do not have access to laboratory facilities. These devices are portable and do not require any equipment for use in the field.

Stability: Due to the dry chemistry used in manufacturing the reagents, these devices are more stable at room temperature. This provides an added advantage to the portability and the robustness, important considerations for field or on-site applications.

Easy to Use: The LFDs are very easy to use with minimal or no sample preparation. The buffers required for optimization of antigen–antibody binding can be introduced through the sample pad and thus require only samples extracted in simple buffers or even tap or potable water. These devices are simple to use and do not require trained personnel.

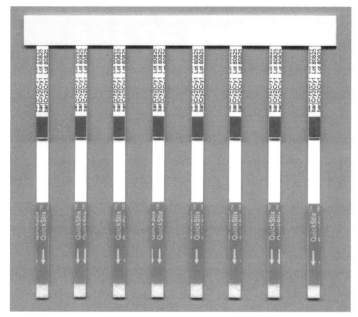

Figure 5.3 Comb format lateral flow devices. (*Source:* Envirologix, Inc. Portland, ME.)

Speed: LFDs are rapid tests and should produce a qualitative answer in less than 10 min. The speed is ideal for situations where rapid process and turnover of samples are critical.

Flexibility: LFDs can more easily be configured for multianalyte testing compared to ELISA. The membrane and the conjugate pad can accommodate more than one antibody, and as such three to four analytes can be measured from one sample (Figure 5.3). By combining individual LFDs into combs (Figure 5.4), multianalyte combs can be assembled quickly according to the diagnostic needs. The flexibility is even much greater with multianalyte LFD combs. Also, the LFDs can be formed into combs that can suit 96, 48, or 24 well formats that greatly enhance the efficiency of testing. In addition, the developed LFDs can be stored long term without losing the data and can be archived or digitally scanned.

5.3.2 Disadvantages or Limitations

Until recently, LFDs have not been useful as quantitative tests. Currently, several test strip readers are marketed for agricultural immunoassays. The readers can provide accurate and reproducible quantitative results. A few manufacturers have developed the ability to produce LFDs with sufficient consistency to provide results with acceptable reproducibility. Some of the newer readers support quantitative LFD tests that can produce results with reproducibility approaching that of ELISA methods. GIPSA has issued program notices that define performance criteria for mycotoxins such as deoxynivalenol (DON) and aflatoxin.

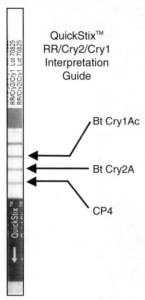

Figure 5.4 Multiplex analyst format LFD. (*Source:* Envirologix, Inc. Portland, ME.)

Sensitivity: Typical sensitivity of about 1 ng/mL can be achieved using high-affinity antibodies. This compares to about 10 pg/mL sensitivity typically obtained in ELISA. However, a more practical limit of 2–5 ng/mL is considered a reasonable compromise for the speed and ease of use compared to ELISA.

Robustness: Due to variability in manufacturing of the raw materials and the nature of dry chemistry in the reagents, LFDs are not as robust as ELISA. They can be affected by the humidity, high temperature, and variable sample matrix. Due to the use of high volume of antibodies for achieving faster results, false positives can often occur. Sufficient validation practices should be incorporated into the development process to address these potential issues.

Cost: Due to the high cost of materials and manufacturing, LFDs cost more compared to other immunoassays. However, none of these disadvantages limit the use of the LFDs and are easily outweighed by the advantages.

5.4 DEVELOPMENT AND VALIDATION

5.4.1 Steps Involved from Concept to Development of LFD

Developers working within the framework of quality systems use a defined design control process for product development. The product design, efficacy of immunoreagents, and effectiveness of the development plan will determine the time required to complete product development. The basic LFD design includes characteristics important in applications of the test. A good understanding of regulatory and diagnostic requirements will provide a basis from which performance criteria for the LFD are established. Critical performance parameters include speed, sensitivity, specificity, stability, and reproducibility. Some important factors, including the nature of sample matrices and the configuration of cassettes or other enclosures, influence the design and should be considered carefully before beginning application. These factors will define the constraints for physical dimensions of the test strip backing material, sample pad, conjugate pad, and top (absorbent) pad. The self-contained, compact nature of LFDs allows them to be used as test strips *per se* or to be incorporated into a wide variety of housings and other devices. LFDs may be designed to dip directly into the sample matrix or to accept sample delivered by pipette, dropper, or other devices. It is important to recognize the likelihood that simple test strips may react differently to samples than test strips incorporated into cassettes or other devices. If the test strip is to be housed in a cassette or other device, it is important for final development work to be conducted with test strips mounted in such devices.

Various reaction schemes may be used, depending on characteristics of the analyte. The most common approaches to developing rapid tests involve either direct double antibody "sandwich" or competitive inhibition reaction schemes (Figure 5.1). These are well known, extensively documented in the literature, and discussed elsewhere in this chapter. In general practice, the antibody sandwich scheme is used for large molecules such as proteins. The competitive inhibition scheme is used for smaller molecules that provide few or limited antigenic sites resulting in increased potential for steric interference from more than one antibody as in the antibody sandwich scheme (Figure 5.2).

Antibodies are the most important and critical reagents for the development of LFDs. The importance of the antibodies selected for the immunoassay sometimes requires the analysis of multiple antibodies. Commercially available "catalog" antibodies may provide satisfactory results in some cases. In other cases, it is necessary to produce and screen a large number of antibodies to identify those most useful for an assay. This process can be expensive and time-consuming, but will provide the best chance for developing a highly effective LFD.

5.4.1.1 Detector Reagents

The most commonly used detector reagents for LFDs are colloidal gold and colored microspheres. Passive adsorption techniques are commonly used to attach antibodies to these reagents. Colloidal gold may be purchased from numerous sources or may also be prepared by the developer. Microspheres composed of various polymers with colored dyes, fluorescent dyes, or magnetic characteristics are available from commercial suppliers. Each of these materials offers advantages for which they may be selected, depending on the requirements of the application. Regardless of the detector reagent selected, it is important that the particles are uniform in shape, size, and physical and electro-chemical characteristics. The detector reagent must be capable of flowing uniformly through the porous membrane. Particles of different size, shape, and polarity proceed at different rates through the membrane. Antibodies or other reagents such as protein A and protein G (for subsequent primary antibody attachment) are commonly linked by conjugation to the detector reagent. Protein conjugates of small molecules may also be attached to the detector reagent.

5.4.1.2 Colloidal Gold Conjugates Colloidal gold is readily available, highly visible, and capable of binding most immunoglobulins and other proteins by passive adsorption to produce a stable labeled colloidal suspension (Figure 5.3). Colloidal gold particles may be purchased or produced in various sizes by varying the ratio of reactants (Frens, 1973). The particles used most commonly for LFDs range from 20 to 40 nm. It is generally believed that the optimal pH for adsorption of a protein is at or near the protein isoelectric point (PI) (Norde, 1986). Polyclonal antibodies exhibit isoelectric points within a relatively wide range of pH values. An average PI for a polyclonal antibody is often about pH 8, whereas PI for monoclonal antibodies varies with the isotype, and the pI for monoclonal antibodies may be much different from polyclonal antibodies and may occur in a narrower range of pH. For purposes of attaching proteins to colloidal gold, it is a general practice to adjust the pH of the reaction buffer to slightly more basic than the isoelectric point of the protein or antibody to be conjugated as this may maximize protein density on the particle surface (Geoghegan, 1988). The Fc and Fab fragments of IgG adsorb differently at different pH values. This allows developers to select conditions for optimal adsorption of the Fc region to colloidal gold particles, while minimizing adsorption of the Fab region by choosing a slightly alkaline pH (Kawaguchi et al., 1989).

5.4.1.3 Latex Microspheres Latex microspheres for diagnostic applications are commonly produced from polystyrene or copolymers of styrene. Most common methods for attaching antibodies and other proteins to particles are passive adsorption and covalent binding. The process of passive adsorption of biomolecules is a simple but effective method of attachment. The binding of protein onto hydrophobic particles occurs by strong interactions of nonpolar or aromatic amino acid residues with the surface polymer chains on the particle. While this method is convenient, it may also result in varying degrees of denaturation of the protein (Hermanson, 2008).

A typical procedure for passive adsorption of proteins to latex microparticles involves diluting the protein in a reaction buffer at or near the pI of the protein. An appropriate quantity of purified protein is dissolved in reaction buffer and then added to a suspension of particles in a ratio calculated by formulas provided by the manufacturers of detector reagents. After incubating the protein/particle suspension for a prescribed period, the particle conjugate is washed by centrifugation to remove unconjugated protein, usually with a storage buffer containing a blocking reagent such as bovine serum albumin (BSA), a nonionic surfactant, and preservative such as sodium azide (Bangs, 2008).

Proteins may also be covalently coupled to microspheres to provide a stronger attachment that may confer greater stability to the detector reagent–protein probe and its resultant signal. Covalent coupling to particles may be accomplished using a carbodiimide to couple carboxyl functional microspheres to available amines or by glutaraldehyde to couple amino functional microspheres to available amines. Proteins are attached via carboxylic acid or primary amine groups. Detector reagents may also be coupled by single-step coupling protocols using activated particles available from commercial suppliers. Various suppliers offer a broad range of microspheres opti-mized with various surface chemistries for immobilizing peptides, small antigens, antibodies, and other proteins and nucleic acids.

5.4.1.4   Test Line Capture Reagents   When using a double antibody sandwich reaction scheme, one of the antibodies is applied to the test line region on the membrane. For competitive inhibition schemes, a protein conjugate of a small molecule, homologous to the analyte, is applied at the test line. In this case, it is often necessary to screen conjugates made from different carrier proteins and with different hapten loading ratios to find the optimal reagent for the test system.

In a review of protein binding to lateral flow membranes, Jones discusses the importance of maintaining the appropriate balance of hydrophobic and electrostatic interactions between proteins and membranes and also emphasizes the importance of adequate drying of membranes after application of protein to ensure long-term stability of the protein–membrane bond. The nature of binding of proteins to nitrocellulose is not well understood. It is generally accepted that hydrophobic interactions, hydrogen bonding, and electrostatic interactions have a role in the binding mechanism (Jones, 1999a, 1999b).

This information suggests that factors such as ambient humidity and compo-sition and properties of the capture reagent application buffer may affect binding characteristics. It is generally advantageous to use low concentrations of buffers, at approximately the isoelectric point of the capture protein. It is often useful to include a solvent such as alcohol to enhance binding to the membrane. The alcohol destabilizes the protein and increases hydrophobicity. The alcohol may confer a further advantage by enhancement of the orientation of immunoglobulins on the membrane surface since the Fc and FAb regions exhibit different stabilities to the alcohol. The structure of the Fc region is more likely to be destabilized by alcohol than the FAb regions, and exposure of more hydrophobic groups favors binding to the membrane (Tijssen, 1993).

5.4.1.5   Control Line Capture Reagents   LFDs usually include a control line that indicates the test has been performed properly after the specified period for test completion. Formation of signal at the control line depends on excess anti-analyte antibody–detector reagent migrating past the test line and continuing to the control line, where it binds to a second (different) capture reagent. Species-specific anti-immunoglobulin against the detector reagent is used most commonly as the control line capture reagent. Other reagents such as protein A and G may also be used to bind the anti-analyte antibody–detector reagent complex. High concentrations of analyte may deplete the detector reagent in some LFD systems and adversely affect the formation of the control line. Since the control line should appear uniform, regardless of analyte concentration, it is sometimes desirable to use a separate, independent reagent system to produce the control line.

5.4.1.6   Membrane   Immunoreagents typically used for the test and control lines are dispensed on a membrane and the most commonly used immunoreagent is nitrocellulose. The test line reagent must effectively "capture" or bind to the detector reagent as it migrates up the membrane after sample is added to the device. Initial experiments should be conducted to determine conditions for optimal binding. Most membrane-based immunoassays use either nylon or nitrocellulose. Both materials bind proteins noncovalently. Nylon binds proteins via electrostatic charge interactions, while the interaction of nitrocellulose with proteins appears to be primarily hydrophobic. Both polymers bind most immunoreagents such as antibodies to make them good candidates for use in lateral flow immunoassays. These membranes generally do not require covalent attachment of the immobilized reactant (Millipore, 2008).

Membranes are made from various polymers such as nitrocellulose, nylon, polysulfones, and other materials. These membranes are manufactured with various pore sizes to allow different flow rates. This is often expressed as rate of capillary rise (s/cm) when tested in the vertical orientation. This measure is inversely related to the flow rate, so that a membrane with rate of capillary rise of 180 is slower than membrane with rate of 120. This is an important consideration when selecting membranes as the flow rate affects both time to completion and assay sensitivity. This is illustrated in *Rapid lateral flow test strips: considerations for product development* (Millipore, 2008) in which the following information is provided for Millipore's Hi-Flow Plus membranes (Table 5.1).

When considering capillary flow rate, it is important to recognize that the effective concentration of analyte increases with the square of the change in flow rate. Lateral flow tests are designed to create a unidirectional flow of reagents across the test strip. In the case of direct double antibody sandwich schemes, the development of visual or other signal on the test line depends on the formation of a complex resulting from binding of analyte–detector reagent to capture reagent immobilized on the test line. Since the formation of the analyte–detector reagent complex occurs before reactants enter the membrane, the formation of this analyte–signal–antibody sandwich complex depends on the length of time the reactants are close enough to interact and also depends on the concentration of antigen and antibody. This relationship has been described by the following formula:

$$R = k[\text{Ag}][\text{Ab}]$$

where R is the reactant (complex of antibody/antigen) and k is a rate constant determined by affinity of the antibody for the antigen.

TABLE 5.1 Relationship Between Membrane Flow Rate and Sensitivity

Membrane	Capillary flow time (s/4 cm)	Flow rate	Sensitivity
HF240	240	Slowest	Most sensitive
HF180	180		
HF135	135		
HF120	120		
HF090	90		
HF075	75	Fastest	Least sensitive

From *Rapid lateral flow test strips: considerations for product development* (Millipore, 2008).

This relationship may be used to demonstrate the effect of flow rate on assay sensitivity in the following example. Consider a case in which membrane flow rate is doubled and therefore allows only half the time for reactants to interact. In this case,

$$R = k(0.5 \times [Ag])(0.5 \times [Ab]) = 0.25k[Ag][Ag]$$

It is apparent that doubling the flow rate will reduce formation of complex by a factor of 4, or the square of the increase in the flow rate. Given the range of flow rates of commercially available membranes, it follows that assay sensitivity may be affected by approximately one order of magnitude, depending on choice of membrane flow rate (Millipore, 2008).

5.4.1.7 *Blocking the Membrane*   It is often advantageous to block the membrane after test and control line reagents have been applied and dried to introduce blocking reagents to reduce nonspecific background signal. The need for membrane blocking should be determined empirically for each test. From a practical perspective, if sufficient differentiation of positive and negative samples can be achieved without blocking the membrane, then the additional time and expense of doing so may not be justified. It is common, however, that additional blocking of the membrane is required to achieve sufficient differentiation and appropriate flow characteristics. The hydrophobicity of the membrane can produce nonspecific binding of detector reagent and may also interfere with uniform and efficient flow of reagents through the membrane. Some types of samples are extracted and/or diluted in reagents that may also improve differentiation characteristics of the test. If membrane blocking is necessary, a broad variety of reagents may be used, including buffers, proteins, and synthetic compounds. A reagent selected from these categories may simultaneously improve and impair different performance characteristics. For example, a protein such as gelatin, BSA, casein, fish gelatin, or various synthetic polymers, such as PVP, may reduce nonspecific binding of detector reagent to capture reagent, while also reducing the rate of flow of sample along the membrane. For this reason, it is often useful to consider combinations of reagents selected from the general categories mentioned previously. It may be useful to incorporate additional reagents, such as surfactants, into the sample pad or conjugate pad. Addition of a surfactant such as Tween 20 or Triton X-100 to the sample pad may offset a reduction of flow rate caused by blocking the membrane with a protein or a synthetic polymer. The blocking reagents with or without surfactants could be introduced through sample extraction of diluent buffer, sample pad, or conjugation pad, thus eliminating the need for blocking step. The choice of blocking agents often requires a compromise between nonspecific background, flow rate, and test line intensity.

5.4.1.8 *Conjugation of Detector Reagent*   Particles of colloidal gold or various styrene microspheres are most often used as detector reagents for LFDs and are both referred to as "particles" for the following discussion. Proteins, and in particular immunoglobulin, may be either passively or covalently absorbed to particles.

Preparing stable protein complexes depends on several interactions: (a) the electronic attraction between the negatively charged gold particles and the positively charged sites on the protein molecule, (b) adsorption between hydrophobic pockets on the protein and the gold metal surface, and covalent binding of gold to free sulfhydryl groups by dative binding (Hermanson, 2008).

It is a common practice in such conjugations to coat the surface of the spheres with a monolayer of protein. To maximize surface coverage, the pH of the conjugation buffer should be at or slightly more basic than the isoelectric point of the protein. This theoretically exposes hydrophobic groups for efficient binding. It is useful to conduct experiments to determine the optimal pH for conjugation.

Using the optimal pH, the ratio of protein to particles may then be optimized by titration. The working ratio of protein to particles is achieved at approximately 5–10% greater than the ratio found to stabilize particles on a small-scale experimental level. Using the optimal conditions, a small pilot batch of conjugate may be prepared and used for further experimentation in prototype test strips.

The detector reagent conjugate may be dispensed to a conjugate pad using an appropriate reagent dispenser or by dipping conjugate pad in a suspension of the detector reagent conjugate. After drying, the conjugate pad may be tested by assembling test strips configured as described in the Manufacturing and Quality Control section of this chapter.

5.4.1.9 Sample Pad Materials and Treatment

Selection of sample pad material and treatment are critical aspects of LFD development. When developing tests for agricultural diagnostics, woven glass fiber material is often selected. It is necessary to treat and condition the sample pad material with proteins, surfactants, and buffer reagents to control several important functions including (a) modifying sample viscosity and related flow properties, (b)improving the solubilization and release of detector reagent, (c) reducing nonspecific binding of detector reagent conjugate, and (d) modifying the nature of the sample to enhance the development of signal at the test line (Millipore, 2008). As described Section 5.4.1.7, it is sometimes unnecessary to block the membrane directly due to the flexibility to incorporate one or more reagents into the sample pad or sample diluent. Test samples may exhibit great variation in characteristics such as pH and concentrations of substances that may affect test results. It is often possible to design the reagent used to treat sample pads to accommodate such sample variations.

5.4.1.10 Conjugate Pad

The conjugate pad performs a critical function as it bridges the sample pad to the membrane and contains detector reagent. It is desirable for the detector reagent to become solubilized and release readily from the conjugate pad as sample flows onto the membrane. Releasing the detector reagent without delay is advantageous as its concentration influences the formation of signal at the test line and, therefore, sensitivity of the test. Conjugate pads are composed of materials such as spun fibers, glass, cellulose, polyester, and polypropylene. These materials may be pretreated with buffers, surfactants, and blocking reagents to reduce nonspecific binding and enhance release of conjugate and flow characteristics.

5.4.2 Factors Affecting Sensitivity and Specificity

Among the most important factors affecting LFD sensitivity and specificity is the nature of the immunoreagents. Assuming that immunoreagents with necessary affinity and specificity are available, the other test strip parameters can usually be optimized to achieve the required performance specifications. The sample matrix may contribute detrimental or enhancing effects on test performance; therefore, it is important to optimize test parameters to accommodate all anticipated sample matrices. Early in the development process, this should be investigated by testing negative matrix and also negative matrix spiked with analyte at concentrations within the anticipated range of detection of the assay. As discussed previously, it is possible to incorporate reagents into the sample pad to enhance the signal and differentiation of the assay. Diluents may also be used to mitigate the effects of different matrices on test response.

5.4.3 Validation of LFDs

Consistent with guidelines for the validation of analytical methods (e.g., Codex, 2004; Horwtiz, 1995; ICH, 2005), U.S. Pharmacopeia states "Validation of an analytical method is the process that establishes, by laboratory studies, that the performance characteristics of the method meet the requirements for the intended analytical applications" (U.S. Pharmacopeia). Westgard 2008 states that the following steps are required to conduct a good method validation study:

1. Define a quality requirement for the test in the form of the amount of error that is allowable
2. Select appropriate experiments to reveal the expected types of analytical errors
3. Collect the necessary experimental data
4. Perform statistical calculations on the data to estimate the size of analytical errors
5. Compare the observed errors with the defined allowable error
6. Judge the acceptability of the observed method performance

5.4.4 Validation Protocol

Companies operating under quality systems follow a design control process that includes definition of performance criteria and a validation protocol for LFDs. An adequate validation study should be established and documented to include the intended use of the test, principles of operation, and pertinent parameters to be studied. The validation procedure should include criteria for acceptance of performance and also a description of the test method. The validation protocol should be written in advance, reviewed for scientific soundness, and approved by qualified individuals. The protocol should describe the procedure in detail, including acceptance criteria and statistical methods to be incorporated. The data produced during the validation protocol are analyzed with results, conclusions, and deviations and recorded in the validation summary report. If the results indicate that the predefined acceptance criteria are met, then a method can be considered valid and the statement of its validity is included in the validation report (Westgard, 2008).

5.4.5 Threshold Testing (Screening Tests)

LFDs are widely used for on-site or field testing due to convenience, speed, accuracy, and low cost. A group of industry experts comprised of members of the Analytical Environmental Immunochemical Consortium (AEIC) proposed standards for validation of immunoassays for genetically modified crops and derived food ingredients (Lipton et al., 2000). USDA's Grain Inspection, Packers and Stockyards Administration (GIPSA) defines criteria for Performance Verification of Qualitative Mycotoxin Tests and Biotech Rapid Test Kits in Directive 9181.2 3-29-04 (current update). This program was established by GIPSA "for verifying the performance of qualitative rapid tests to detect the presence of mycotoxins and/or biotechnology events present in grains and oilseeds" (USDA GIPSA, 2004). The Rapid Test Evaluation Program involves the following four-step process:

1. The manufacturer submits a data package supporting performance claims. The data include the following:

   a. Results of testing 120 negative control samples and 120 fortified (currently naturally contaminated) samples using three different lots. All test results must coincide with predetermined positive or negative results for the mycotoxin or biotechnology event of interest.

   b. Temperature sensitivity testing is not required for LFDs.

   c. Cross-reaction with other biotechnology events (biotech test kits only).

   d. Data from five independent tests of all other biotechnology events in commercial production in the United States demonstrating that the test does not detect proteins produced by other biotechnology events. All tests must be negative.

2. GIPSA reviews the data and the performance verification submitted by the manufacturer.

3. If data are complete and support claims of the rapid test manufacturer, GIPSA conducts an in-house performance verification of the rapid test.

4. If the manufacturer's claims are verified by GIPSA in-house performance testing, a certificate of performance is issued to the manufacturer.

5.4.6 Quantitative Tests

Over the past 5 years, quantitative tests for LFDs have become more prevalent. LFD readers capable of detecting light reflectance, fluorescence, and magnetic fields are currently available. Several companies are currently marketing LFDs developed for use with these instruments. GIPSA has published directives for performance verification of quantitative LFDs for several analytes. For example, Program Notice

TABLE 5.2 Acceptable Limits

Conc. (ppb)	Max. RSD$_1$ (%)	S.D.	Range^a (ppb)
5.0	30	1.5	2.0–8.0
10	30	3.0	4.0–16
50	25	5.0	10–30
100	25	25	50–150

^{a} For naturally contaminated samples, this range will be recalculated using the actual meaning, and the maximum RSD values at each concentration is listed. The mean will be determined by using HPLC values obtained using GIPSA's Aflatoxin Reference Method. At least 95% of the individual values at each concentration must be within the applicable range.

FGIS-PN-01-15, Performance Evaluation Criteria for Aflatoxin Test Kits, 2004, outlines the following requirements:

1. The time required for analysis, including extraction and dilution, must be less than 30 min.

2. The limit of detection must be less than or equal to 3.0 ppb.

3. The test must meet the acceptable limits provided in Table 5.2 using four corn samples naturally contaminated with aflatoxins at concentrations of 5, 10, 20, and 100 ppb.

GIPSA suggests approximately 17 additional commodities for verification testing.

Developers include sufficient testing during the development process to ensure that the LFD will perform to such specifications.

5.4.7 Estimation of Shelf Life

The shelf life of a product may be defined as the time span in which critical performance characteristics are maintained under specific conditions of storage and handling. Manufacturers' quality assurance criteria typically require that a product must recover at least 90% of the initial value throughout its shelf life. Applying this performance criterion to the results of stability testing thereby determines the period of the shelf life. Although shelf life is commonly estimated by accelerated stability testing, real-time stability testing is necessary to validate stability claims for clinical chemistry reagents and reference materials (Anderson and Scott, 1991).

A precedent for agricultural diagnostic tests is established by FDA guidelines for stability studies and expiration dating. When IVD stability is a design concern, appropriate procedures such as stability studies are conducted and an expiration period, supported by the studies, is established to define the period in which stability is assured. The expiration period is included as part of the product specifications for the IVD and its components as required by Section 820.181(a). Currently, FDA accepts only real-time data for supporting an expiration period. If real-time data are insufficient to support the full expiration period claimed, FDA may, on a case by case basis, accept accelerated data with the understanding that the data will be supported by real-time data, or the shelf life is adjusted to reflect the real time expiry.

It has become common practice for manufacturers to initiate real-time and accelerated stability testing simultaneously and to release new products with shelf life claims based on results of accelerated stability testing. Although nonclinical diagnostic tests are not controlled by the FDA, it is prudent practice to continue real-time stability testing beyond the length of time assigned as shelf life of the product, until significant degradation is observed.

5.4.7.1 Real-Time Stability Testing

The length of time for real-time stability testing should be sufficient to allow significant product performance degradation under recommended and storage conditions. This is necessary in order to distinguish between merely interesting variations and significant performance degradation. A good control for such studies involves including a single lot of the reference material with established stability characteristics in each assay. By normalizing assay results to those obtained with this reference, systematic and interassay imprecision may be minimized.

5.4.7.2 Accelerated Stability Testing

Accelerated stability testing can provide some early indication of shelf life without waiting for results of real-time stability testing to be completed. This is accomplished by subjecting product to higher temperatures and determining the amount of heat input required to cause product failure. Data from elevated temperature testing may then be used to predict product shelf life. The best results are obtained from studies designed to avoid temperatures that will denature labile components such as antibodies. It is considered good practice to use internal controls comprised of product not subjected to elevated temperatures. Activity of product subjected to higher temperatures may be expressed as a ratio to activity of unstressed product (Kirkwood, 1977).

5.4.7.3 The Arrhenius Equation and Activation Energy

Accelerated stability testing is based upon the Arrhenius relationship (Conners et al., 1973). The Arrhenius equation establishes the functional relationship between time and stability of product stored under constant conditions. The speed of the reaction depends on the order of reaction at a rate constant. The logarithm of the reaction rate is a linear function of the reciprocal of absolute temperature and can be expressed by the following equation:

$$\log k_2/k_1 = (-E_a/2.0303R)(1/T_2 - 1/T_1)$$

where k_1 and k_2 are rate constants at temperatures T_1 and T_2, respectively. E_a is the activation energy and R is the gas constant.

This equation describes the relationship between storage temperature and degradation rate. The use of the Arrhenius equation permits a projection of stability from the degradation rates observed at elevated temperatures (Lachman and DeLuca, 1976). Activation energy, the independent variable in the equation, is equal to the energy barrier that must be exceeded for the degradation reaction to occur. When the activation energy is known (or assumed), the degradation rate at low temperatures may be projected from that observed at "stress" temperatures.

TABLE 5.3 Typical Accelerated Stability Testing Schedule

Temperature	Testing interval (once per indicated interval) (day)
37°C	5
45°C	4
50°C	3
60°C	2
80°C	Daily

Manufacturers in pharmaceutical and diagnostic reagent industries use various shortcuts, such as the Q rule (Conners et al., 1973) and bracket tables (Porterfield and Capone, 1984), to estimate product shelf life. These techniques offer a reasonable means by which manufacturers can make shelf life decisions by analyzing relatively few stressed samples. However, they are based on assumptions about the activation energy of product components and are valid only insofar as these assumptions are accurate. Whatever method is chosen, the validity of product stability projections depends on analytical precision, the use of appropriate controls, the assumptions embodied in the mathematical model, and the assumed or measured activation energy of product components. Minor variations in complex chemical systems can significantly affect lot-to-lot stability due to variations in activation energy of product degradation. Due to these uncertainties, shelf life projected from accelerated studies must be validated by appropriate real-time stability testing.

The typical procedure for performing accelerated stability testing is as follows. Samples are stored at three or four elevated temperatures. The highest temperature used is limited by the denaturation characteristics of antibodies used in the immunoassay. Storage temperature should be held constant at $\pm 1°C$. In accelerated stability testing, it is useful to follow the stability up to 50% decomposition because this allows the rate constant to be determined with relative accuracy. This would typically be on the order of 4–8 weeks. A common protocol allows testing at 20, 37, and 50°C. It is important that test materials used for accelerated stability testing are packaged in a manner similar to that employed with commercial product (Table 5.3).

Deshpande 1996 suggests that the time intervals between 37 and 50°C may be considered to produce reasonable room temperature shelf life equivalents (Table 5.4).

These guidelines are considered generally acceptable and reduce some of the necessities of performing the complex experimental and mathematical analyses

TABLE 5.4 Approximate Shelf Life Equivalence

Months	1	2	3	6
37°C			1 year	2 years
50°C	1 year	2 years		

described previously. Real-time stability studies at the storage conditions recommended in the product insert sheet are also necessary and will serve to validate the accelerated stability data.

5.5 APPLICATION AND TROUBLESHOOTING

5.5.1 Application of LFD

In the past 15 years, LFD have been used as important analytical tools for agricultural biotechnology for purposes including product development, event sorting, seed production, product stewardship, and grain channel. LFDs are routinely used for testing both seed and leaf samples. To ensure the repeatability, applicability, and adaptability of LFD method in target tissues, a thorough validation is necessary before use. Guidelines and process of method validation of LFD can be found in this chapter and in Chapter 6. During seed production and quality control, LFD is often used to determine seed purity to confirm the high level of the target trait and evaluate low level presence of unintended traits. A more detailed discussion of the LFD application in this field is presented in Chapter 10. Finally, in food supply chain and quality management systems, LFD is mainly utilized for non-GM (nongenetically modified) identity preservation such as testing for unapproved events and for the purpose of non-GM labeling. Similarly, LFD is applied to test for the presence of target traits such as high-value GM commodities and validate label claims (Chapter 11). LFDs are extensively used in the food and feed supply chain including testing at the grain elevator, testing of raw material for food processing, and testing of food fraction during processing (Grothaus et al., 2006).

5.5.2 Troubleshooting the Devices

Interpretation of data with LFDs is primarily qualitative in nature. The LFDs signal can also be read with the aid of a scanner to get semiquantitative data. Qualitative interpretation of LFDs gives results as either a positive or a negative reading. The LFDs contain a control line, the absence of which makes the test invalid. If an LFD kit is validated under strict SOPs and quality assurance, there are several sources that can cause either false-positive (FP) or false-negative results (FN). Grothaus et al. 2006 has listed sources of errors for LFD applications. Discussion here is intended to help the analyst to troubleshoot during application. Main factors causing FP and FN results include precipitation of conjugate on the strip, LFD stability, matrix effects from samples, and operator error.

5.5.2.1 Precipitation of the Conjugate The FPs and FNs usually appear when an unexpected precipitation of gold conjugate occurs during the development of the strip. The gold conjugate can be precipitated by extreme salt content, presence of high concentration of hydrophobic molecules such as bacterial extracellular polysaccharides in the sample. These reactions can be caused by the sample, state of the sample at testing time (e.g., rotten tissue), or extraction medium used to extract the analyte from the sample. If the precipitation is slow and gradual, the test line appears during the course of development (FP) even though the sample is negative. On the contrary, if the precipitation is instantaneous, the gold complexes do not enter into the membrane. However, the latter case also gives rise to a faint or no control line.

5.5.2.2   *Sample Matrix Effect*

The samples and the sample matrix sometimes contribute to errors by nonspecific protein binding, precipitating the gold conjugates, inhibiting capillary flow, and extreme pH condition of the extract that inhibits binding and promotes desorption of membrane-bound antibody.

Nonspecific protein binding may result from nonspecific interaction between the sample, conjugate, and capture line that can be caused by protein attachment such as charge attraction, hydrophobic adsorption, or nonspecific immunogenic binding. The precipitation of gold conjugate may be due to the nonspecific binding as explained previously. This may also happen when LFDs are used with a matrix that has not been optimized or validated. For example, an LFD validated only for cottonseed should not be used for soybean. Inhibition of flow contributes to false negatives with a faint or no control line and may occur when the extraction ratio of sample to buffer is lower than the theoretical flow tolerance of the strip. Other factors that reduce the flow are fineness of grind in case of seed and grain, oil content in the material, and, rarely, complex sugars from leaf and other plant materials that form into mucus-type material.

Sometimes, the seeds to be tested are treated with pesticides, bleach, or acid to control seedborne diseases and the residual presence of these compounds may lower the pH of the extracted matrix. Any pH below 5.0 may not favor the optimal binding of the antigen to the antibody and thus may give rise to false negatives. Certain leaf tissues have high chlorophyll content that may be released by maceration. The chlorophyll may stick to the test line resulting in a light visible "green line" that may be misinterpreted as a positive test result. Vigilant observation by the operator could avoid such interpretive error.

5.5.2.3   *Operator Errors*

Operator factors that contribute to errors are usually associated with improper handling of the test devices. Lateral flow test strips and cassettes typically have certain limitations on sample volume that may be used and for test strips, the depth to which these may be inserted into the sample. In these devices, the gold conjugate is dried into a pad at the end of the sample pad and requires the lateral flow of the sample liquid to release the gold conjugate–antigen complex onto the membrane. If the conjugate pad directly comes in contact with the sample extract, the gold conjugate may be released into the sample, so that the test and control lines do not develop completely or at all.

Also, due to the dry chemistry involved in manufacturing these devices, LFDs must always be stored under dry conditions. Leaving test strips exposed to high humidity may compromise performance resulting in poor or no capillary flow.

REFERENCES

Anderson, G. and Scott, M. Determination of product shelf life and activation energy for five drugs of abuse. *Clin. Chem.*, **1991**, *37*, 398–402.

Aveyard, J., Mehrabi, M., Cossins, A., Braven, H., and Wilson, R. One step visual detection of PCR products with gold nanoparticles and a nucleic acid lateral flow (NALF) device. *Chem. Commun.*, **2007**, 4251–4253.

Bangs Laboratories, Inc. *Adsorption to microspheres*. Tech Note 204, **2008**.

Buechler, K. F., Moi, S., Noar, B., McGrath, D., Villela, J., Clancy, M., Chenhav, A., Colleymore, A., Valkirs, G., and Lee, T. Simultaneous detection of seven drugs of abuse by the Triage panel for drugs of abuse. *Clin. Chem.*, **1992**, *38*, 1678–1684.

Codex Procedural Manual 13th Ed. and Consideration of the Methods for the Detection and Identification of Foods Derived from Biotechnology; General Approach and Criteria for Methods Codex, Alimentarius, Rome, Italy, 2004, http://www.codexalimentarius.net.

Conners, K. A., Amidon, G. L., and Kennon, L. *Chemical Stability of Pharmaceuticals: A Handbook for Pharmacists*, Wiley, New York, 1973, pp., 8–119.

Corstjens, P., Zuiderwijk, M., Mrink, A., Li, S., Feindt, H., Niedbala, R. S., and Tanke, H. Use of up-converting phosphor reporters in lateral-flow assays to detect specific nucleic acid sequences: a rapid, sensitive DNA test to identify human papillomavirus type 16 infection. *Clin. Chem.*, **2001**, *47*, 1885–1893.

Deshpande, S. S. *Immunoassays: From Concept to Product Development*, Springer, 1996.

Edwards, K. A. and Baermner, A. J. Optimization of DNA-tagged dye-encapsulating liposomes for lateral-flow assays based on sandwich hybridization. *Anal. Bioanal. Chem.*, **2006**, *386*, 1335–1343.

Fernandez-Sanchez, C., McNeil, C. J., Rawson, K., Nilsson, O., Leung, H. Y., and Gnanapragasam, V. One-step immunostrip test for the simultaneous detection of free and total prostate specific antigen in serum. *J. Immunol. Methods*, **2005**, *307*, 1–12.

Food and Drug Administration. Center for Devices and Radiological Health Office Compliance. *Guideline for the manufacture of in vitro diagnostic products* (January 10, **1994**).

Frens, G. Controlled nucleation for the regulation of the particle size in monodisperse gold solution. *Nat. Phys. Sci.*, **1973**, *241*, 20–22.

Geoghegan, W., The effect of three variables on adsorption of rabbit IgG to colloidal gold. *J. Histochem. Cytochem.*, **1988**, *36*, 401.

Goldman, E. R., Clapp, A. R., Anderson, G. P., Uyeda, H. T., Mauro, J. M., Medintz, I. L., and Mattoussi, H. Multiplexed toxin analysis using four colors of quantum dot fluororeagents. *Anal. Chem.*, **2004**, *76*, 684–688.

Greenwald, R., Esfandiari, J., Lesellier, S., Houghton, R., Pollock, J., Aagaard, C., Anderson, P., Hewinson, R. G., Chambers, M., and Lyashchenko, K. Improved serodetection of *Mycobacterium bovis* infection in badgers (*Meles meles*) using multiantigen test formats. *Diagn. Microbiol. Infect. Dis.*, **2003**, *46*, 197–203.

Grothaus, G. D., Bandla, M., Currier, T., Giroux, R., Jenkins, G. R., Lipp, M., Shan, G., Stave, J. W., and Pantella, V. Immunoassay as an analytical tool in agricultural biotechnology. *J. AOAC Int.*, **2006**, *89*, 913–928.

Gussenhoven, G. C., van der Hoom, M. A., Goris, M. G., Terpstra, W. J., Hartskeerl, R. A., Mol, B. W., van Ingen, C. W., and Smits, H. L. LEPTO dipstick, a dipstick assay for detection of leptospira-specific immunoglobulin M antibodies in human sera. *J. Clin. Microbiol.*, **1997**, *35*, 92–97.

Hermanson, G. T. *Bioconjugate Techniques*, 2nd edition, Elsevier, 2008.

Horwitz, W. Protocol for the design, conduct, and interpretation of method-performance studies. *Pure Appl. Chem. revised*, **1995**, *67*, 331–343.

ICH Harmonized Tripartite Guideline: Validation of Analytical Procedures: Text and Methodology (2005) http://www.ich.org.

Jin, S., Chang, Z. Y., Ming, X., Min, C. L., Wei, H., Sheng, L. Y., and Hong, G. X. Fast dipstick dye immunoassay for detection of immunoglobulin G (IgG) and IgM antibodies of human toxoplasmosis. *Clin. Diagn. Lab. Immunol.*, **2005**, *12*, 198–201.

Jones, K. D. Troubleshooting protein binding in nitrocellulose membranes. Part 1: principles. *IVD Technol.*, **1999a**, 32–41.

Jones, K. D. Troubleshooting protein binding in nitrocellulose membranes. Part 2: common problems. *IVD Technol.*, **1999b**, 26.

Kalogianni, D. P., Goura, S., Aletras, A. J., Christopoulos, T. K., Chanos, M. G., Christofidou, M., Skoutelis, A., Ioannou, P. C., and Panagiotopoulos, E. Dry reagent dipstick test combined with 23S rRNA PCR for molecular diagnosis of bacterial infection in arthroplasty. *Anal. Biochem.*, **2007**, *361*, 169–175.

Kawaguchi, H. K., Sakamoto, Y., Ohtsuka, T., Ohtake, H., and Sekiguchi, H. I. Fundamental study on latex reagents for agglutination tests. *Biomaterials*, **1989**, *10*, 225–229.

Kirkwood, T. B. L. Predicting the stability of biological standards and products. *Biometrics*, **1977**, *33*, 736–742.

Kolosova, A., De Saeger, S., Sibanda, L., Verheijen, R., and van Peteghem, C. LEPTO dipstick, a dipstick assay for detection of leptospira-specific immunoglobulin M antibodies in human sera. *Anal. Bioanal. Chem.*, **2007**, *389*, 2103–2107.

Lachman, L. and DeLuca, P. Kinetic principle and stability testing. *The theory and Practice of Industrial Pharmacy*, 2nd ed., Lea and Febiger, Philadelphia, 1976, pp. 32–69.

Laitinen, M. P. A. and Vuento, M. Affinity immunosensor for milk progester- one: identification of critical parameters. *Biosens. Bioelectron.*, **1996**, *11*, 1207–1214.

Leung, W., Chan, C. P. Y., Bosgoed, F., Lehmann, K., Renneberg, I., Lehmann, M., and Renneberg, R. One-step quantitative cortisol dipstick with proportional reading. *J. Immunol. Methods*, **2003**, *281*, 109–118.

Leung, W., Chan, C. P. Y., Leung, M., Lehmann, K., Renneberg, I., Lehmann, M., Hempel, A., Glatz, J. F. C., and Renneberg, R. Novel digital-style rapid test simultaneously detecting heart attack and predicting cardiovascular disease risk. *Anal. Lett.*, **2005**, *38*, 423–439.

Leuvering, J. H. W., Thal, P. J. H. M., van der Waart, M., and Schuurs, A. H. W. M. Sol particle immunoassay (SPIA). *J. Immunoassay Immunochem.*, **1980**, *1*, 77–91.

Lipton, C. R., Dautlick, J. X., Grothaus, G. D., Hunst, P. L., Magin, K. M., Mihaliak, C. A., Rubio, F. M., and Stave, J. W. Guidelines for the validation and use of immunoassays for determination of introduced proteins in biotechnology enhanced crops and derived food ingredients. *Food Agric. Immunol.*, **2000**, *12*, 156.

Litman, D. J., Hanlon, T. M., and Ullman, E. F. Enzyme channeling immunoassay: a new homogeneous enzyme immunoassay technique. *Anal. Biochem.*, **1980**, *106*, 223–229.

Millipore Corporation. *Rapid lateral flow test strips: considerations for product development*, **2008**, http://www.millipore.com/techpublications/tech1/tb500en00.

Nielsen, K., Yu, W. L., Kelly, L., Bermudez, R., Renteria, T., Dajer, A., Gutierrez, E., Williams, J., Algire, J., and de Eschaide, S. T. Development of a lateral flow assay for rapid detection of bovine antibody to anaplasma marginale. *J. Immunoass. Immunochem.*, **2008**, *29*, 10–18.

Norde, W. Adsorption of proteins from solution at the solid–liquid interface. *Adv. Colloid Interface Sci.*, **1986**, *4*, 267–340.

O'Farrell, B. Evolution in lateral flow-based immunoassay systems. In Wong, R. and Tse, H. (Eds.), *Lateral Flow Immunoassay*, Humana Press, New York, NY, 2009, pp. 1–34.

Oku, Y., Kamiya, K., Kamiya, H., Shibanda, L., Ii, T., and Uesaka, Y. Development of oligonucleotide lateral-flow immunoassay for multi-parameter detection. *J. Immunol. Methods*, **2001**, *258*, 73–84.

Pappas, M. G., Hajkowski, R., and Hockmeyer, W. T. Dot-enzyme-linked immunosorbent assay (Dot-ELISA): a micro technoque for the rapid diagnosis of visceral leishmaniasis. *J. Immunol. Methods*, **1983**, *64*, 205–214.

Porterfield, R. I. and Capone, J. J. Application of kinetic models and Arrhenius methods to product stability evaluation. *Med. Devices Diagn. Ind.*, **1984**, 45–50.

Posthuma-Trumpie, G. A., Korf, J., and van Amerongen, A. Lateral flow (immuno)assay: its strengths, weaknesses, opportunities and threats. A literature survey. *Anal. Bioanal. Chem.*, **2009**, *393*, 569–582.

Snowden, K. and Hommel, M. Antigen detector immunoassay using dipsticks and colloidal dyes. *J. Immunol. Methods*, **1991**, *140*, 57–66.

Tijssen P. *Practice and Theory of Immunoassays*, 8th edition, Elsevier, Amsterdam, The Netherlands, 1993.

United States Department of Agriculture/Grain Inspection, Packers and Stockyards Administration, *Performance verification of qualitative mycotoxin and biotech rapid test kits, Directive 9181.2*, **2004**, http://archive.gipsa.usda.gov/reference-library/directives/9181-2.pdf.

U.S. Pharmacopeia. Chapter 1225, Validation of Compendial Methods. http://www.pharmacopeia.cn/v29240/usp29nf24s0_c1225.html. (Accessed on Oct 14, 2010)

Van Amerongen, A. and Koets, M.In van Amerongen A., Barug, D., Lauwaars, M. (Eds.), *Rapid Methods for Biological and Chemical Contaminants in Food and Feed*, Wageningen Academic Publishers, Wageningen, 2005, pp. 105–126.

Van Dam, G. J.,Wichers, J. H., Ferreira, T. M. F., Ghati, D., van Amerongen, A., and Deelder, A. M. Diagnosis of Schistosomiasis by Reagent Strip Test for Detection of Circulating Cathodic Antigen. *J. Clin. Microbiol.*, **2004**, *42*, 5458–5461.

Virusys Corporation, *Stability Testing*, http://www.virusys.com/stability-testing/. (Accessed on Oct 14, 2010)

Wallis, C., Melnick, J. L., and Gerba, C. P. Concentration of viruses from water by membrane chromatography. *Annu. Rev. Microbiol.*, **1979**, *33*, 413–437.

Westgard, J.O. *Basic method validation*, 3rd ed. Madison, WI: Westgard QC, Inc, 2008.

Wang, X., Li, K., Shi, D., Xiong, N., Jin, X., Yi, J., and Bi, D. Development of an Immunochromatographic lateral-flow test strip for rapid detection of sulfonamides in eggs and chicken muscles. *J. Agric Food Chem.*, **2007**, *55*, 2072–2078.

Zuiderwijk, M., Tanke, H. J., Sam Niedbala, R., and Corstjens, P. L. An amplification-free hybridization-based DNA assay to detect *Streptococcus pneumoniae* utilizing the up-converting phosphor technology. *Clin. Biochem.*, **2003**, *36*, 401–403.

CHAPTER 6

IMMUNOASSAY METHOD VALIDATION

Jean Schmidt
Clara Alarcon

6.1   INTRODUCTION

Other chapters in this book provide the background theory and considerations involved in the development and use of immunoassay methods for products of biotechnology (Lipton et al., 2000; Grothaus et al., 2006). The primary focus of this chapter will be on the validation of developed methods prior to implementing as

Immunoassays in Agricultural Biotechnology, edited by Guomin Shan
Copyright © 2011 John Wiley & Sons, Inc.

an analytical tool in the laboratory. Validation studies are conducted to document that an analytical method, in this case an immunoassay, performs as specified for the purpose intended. There are several guidelines for the validation of analytical methods (Codex, 2004; FDA, 2001; Horwitz, 1988, 1995; ICH, 2005) that may be applied to immunoassays used in biotechnology. Validation experiments are designed to demonstrate that the method has the sensitivity, specificity, accuracy, and precision necessary to reproducibly detect and/or quantify the target analyte in a specific matrix or matrices. The scope of the validation study and acceptance criteria applied can vary depending on the intended use of the analytical method. For example, an assay that is used to determine plus (positive) or minus (negative) calls for samples by researchers would be expected to have different validation requirements from an assay that is used to quantify the amount of insect-resistant protein present in grain samples as a part of the risk assessment for product registration purposes. Generally, to be considered a fully validated method, the validation will include a collaborative interlaboratory trial, with testing by multiple independent laboratories (Codex, 2004). However, single laboratory validation studies are commonly conducted to support the implementation of a specific immunoassay for analysis within a laboratory. Descriptions of several validation studies that vary in scope are provided below.

> *Combined Development and Validation Studies*: Similar experiments are used in the assay development process and in assay validation testing (see Chapters 4 and 5). When the validation study is conducted by the laboratory that developed the method prior to releasing the assay, the scope of the analysis tends to be broader and often includes experiments that helped to determine/define the final assay parameters. For example, the study may include development data used to determine the range of the curve or the limit of detection (LOD), as well as verification of the accuracy of the final range and/or LOD.
>
> *Single Laboratory Validation of Defined Method*: In many cases, a laboratory will obtain an assay developed by a third party for use in sample analysis. The method is fully defined and generally has previously been validated by the development laboratory prior to release. In this case, the validation study conducted by the laboratory receiving the assay might be limited to verification of the performance parameters described in the finalized method.
>
> *Independent Laboratory Validation (ILV)*: The U.S. EPA requires an ILV for methods that will be used for product registration of crops expressing insect-resistant traits. The validation analysis is conducted by an independent third party according to Good Laboratory Practice (GLP) Standards (40 CFR 160). Additional details of the EPA requirements for an ILV study will be presented later in this chapter.
>
> *Performance Verification Testing of Rapid Tests*: Qualitative immunoassay kits (lateral flow devices (LFDs) or microtiter ELISA kits) can be submitted for performance verification testing by USDA/GIPSA. If the GIPSA in-house verification procedure substantiates the manufacturer's claims, a certificate of performance is issued. Additional details on the GIPSA verification testing will be presented later.

Interlaboratory Validations: A collaborative interlaboratory validation for an analytical method includes testing done by multiple independent labs (at least eight). Often a smaller number of labs will be used to verify performance prior to initiating the more extensive collaborative study to ensure that the assay will perform well and to allow for implementation of any adjustments to the method that might be required to attain the desired reproducibility.

This chapter will primarily focus on single laboratory validations of immunoassay methods designed for quantification of a transgenic protein in plant tissues. Validation of assays for the qualitative determination of a positive or negative sample or for a plus/minus call for a specified level of the expressed protein in a sample will also be discussed. In some cases, the same assay will be used to quantify the protein in a positive sample and to qualitatively distinguish positive samples from negative samples. The validation experiments should be designed to cover each of the intended uses.

6.2   QUANTITATIVE ELISA VALIDATION

The same experiments that are done as a part of the method validation are also conducted as the assay is being developed to determine the curve range, verify acceptable performance, or identify areas that need further optimization. As already mentioned, the validation experiments and development can be done simultaneously; however, it is important that the actual validation experiments are performed on the finalized method under the conditions to be used for sample analysis. If, for example, modifications to the buffer composition or the extraction method are made to improve extraction efficiency or minimize matrix effects (see Chapter 9), the validation experiments will all need to be conducted with the modification incorporated.

To conduct the validation study, a written method, the necessary assay reagents, a standard protein, and applicable tissues are required. The written analytical method should include descriptions of the equipment, labware, and reagents required for the analysis. The steps of the procedures should be detailed enough to allow appropriately trained personnel to carry out the analysis from sample preparation to data reporting without additional instruction. Critical parameters and/or limitations of the method should be emphasized. A characterized reference standard with documented identity, concentration, purity, and storage stability information should be available for standard curve generation and fortification recovery testing (see Chapter 9 for discussion on characterization and demonstration of equivalence for reference standards). A characterized set of positive and negative samples representative of the matrices to be analyzed by the method is also required.

The extraction protocol should be detailed and include information on tissue types, applicable sample processing (e.g., lyophilized or fresh frozen), sample size, buffer composition, extraction volume, extraction temperature, tissue maceration method and timing, extract storage conditions, and any minimum recommended dilutions that may be applied to specific matrix types. Minimum recommended dilutions may be imposed by the assay developer, if applicable, to avoid matrix effects that affect quantification.

The format of the assay and the curve fit applied to the standard curve can vary (see Chapters 4 and 9). This discussion will primarily focus on sandwich ELISA formats, which are commonly used in quantification of transgenic proteins in plants. The standard curve should have a defined quantitative range that generally includes five to seven nonzero standard concentration levels. For sigmoidal curve fitting, anchor points outside the quantitative range may also be included in the standard curve but are not required to meet acceptability criteria. The lower limit of quantification (LLOQ) and the upper limit of quantification (ULOQ) are the lowest and highest points, respectively, on the standard curve that can be measured with acceptable precision and accuracy.

6.2.1   Sensitivity

As described above, the standard curve must be defined prior to the validation study. For a quantitative assay, determining the LLOQ during development can be done in a variety of ways and will depend on the needs of the assay and the quality of the reagents. Generally, range finding experiments using a broad range of concentrations and fewer standard levels can be done using a checkerboard design in conjunction with pair testing of antibodies (see Chapter 4). Once the pair is chosen, extended curves that incorporate more points near the targeted LLOQ and ULOQ can be used to further optimize the curve range. By evaluating the interpolated values for each of the points in the extended curve, one can determine where the accuracy falls below an acceptable level. Additional testing using fortification and recovery testing with the matrices of choice can be used to further optimize the standard curve and to validate the standard curve range once it is defined (see Section 6.2.3.2).

An AAPS/FDA Bioanalytical Workshop recommended the following for immunoassay standard curve acceptance criteria: a matrix-matched standard curve should be able to predict at least 75% of the standard points in the quantitative range within 20% of the theoretical value except at LLOQ and ULOQ, which should fall within 25% of the theoretical value (Viswanathan et al., 2007). Depending on the curve fit and the quality of the antibodies used, it is not unusual to have standard curves (without added matrix) that predict all of the standards within 10% or better of the theoretical value. Tighter acceptance criteria may be applied to the standard curve in development and prevalidation testing to allow the greater variability that is expected when an assay is used in a high-throughput analytical setting. For example, using a quadratic fit for a curve that has a 20× range (e.g., 0.25–5 ng/mL) with low background, it is often possible to apply the criteria that the curve predict at least six out of seven nonzero points in the quantitative range within 10% of the theoretical value (see Table 6.1 and Figure 6.1 for an example of a standard curve with good curve fit and agreement between theoretical and observed concentrations).

Sensitivity for a qualitative assay is defined as the LOD and is further described later. The LOD can be defined as the lowest concentration at which a positive sample can be reliably distinguished from a negative sample. The LOD is often incorporated

TABLE 6.1 Standard Curve Example with Good Prediction of Standard Concentration

Standard	Concentration (ng/mL)	Back calculated concentration (ng/mL)	Mean back calculated concentration (ng/mL)	OD values	Percent error	Mean percent error
Std1	0	0.02	0.01	0.01	Inf	Inf
		0.01		0.01	Inf	
		0.01		0.01	Inf	
Std2	5.00	4.96	4.99	1.92	0.9	0.4
		5.01		1.94	0.1	
		5.01		1.94	0.2	
Std3	4.00	4.04	4.00	1.60	0.9	0.5
		3.98		1.58	0.6	
		4.00		1.59	0.1	
Std4	3.00	3.03	3.02	1.24	1.1	0.8
		3.03		1.24	1.0	
		3.01		1.23	0.2	
Std5	2.00	2.05	1.99	0.86	2.4	2.1
		1.98		0.83	0.8	
		1.94		0.81	3.2	
Std6	1.00	0.97	0.99	0.42	3.3	1.7
		0.99		0.42	1.4	
		1.00		0.43	0.5	
Std7	0.50	0.49	0.49	0.22	1.5	2.5
		0.49		0.21	2.3	
		0.48		0.21	3.7	
Std8	0.25	0.26	0.25	0.11	3.2	1.7
		0.25		0.11	1.3	
		0.25		0.11	0.5	

Note: More significant figures than shown were used in calculations of percent error; percent error = (back calculated concentration − concentration)/concentration × 100.

into quantitative assays, especially when positive samples are likely to fall below the LLOQ. In many cases, the LOD is incorporated as the lowest point in the standard curve, although it is not expected to meet the accuracy criteria set for the LLOQ. The LOD may also be included as a separate sample in the assay. Sometimes the LOD is defined based on the mean value of the absorbance reading for a negative control sample plus two or three standard deviations. Similar experiments with extended curves below the LLOQ can be used to estimate the LOD empirically. The validation of the LOD consists of analyzing a set of positive samples that fall near the LOD (or negative sample fortified at the LOD concentration) and a set of known negative samples and determining the number of false negatives and false positives, respectively, for each. Additional information on validation of the LOD can be found in Section 6.5.

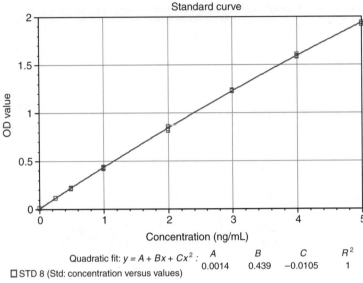

Quadratic fit: $y = A + Bx + Cx^2$:

	A	B	C	R^2
	0.0014	0.439	−0.0105	1

□ STD 8 (Std: concentration versus values)

Figure 6.1 Graph of standard curve example shown in Table 6.1.

6.2.2 Specificity/Selectivity

The potential for cross-reactivity of the antibodies used in the assay with other similar proteins (endogenous or transgenic) should be an important consideration very early in the assay development process. Different antibody production and screening strategies may be employed in cases where cross-reactivity with other proteins is anticipated (see Chapters 4, 5, and 9). In some cases, complete specificity to the introduced transgenic protein can be difficult or even impossible (e.g., overexpression of an endogenous protein) to obtain. It is important to characterize the effect of other proteins and/or matrix components, if any, on the ability of the assay to quantify the target protein.

6.2.2.1 Cross-Reactivity/Interference of Purified Proteins The validation experiments used to demonstrate specificity of assay for the target protein should include, to the extent possible, other relevant transgenic proteins that may be present in samples. This would include other transgenic proteins expressed in the same plant (e.g., coexpressed or stacked products), as well as transgenic proteins that may be expected to be present in the same crop or across crops that may incorporate the target trait protein. Also, when applicable, it is important to understand the relative cross-reactivity of different forms of the target protein that may be found in the samples (i.e., full length versus truncated). If both forms are likely to be present in plant samples and both need to be quantified by the method, it is important to demonstrate that the relationship between OD response and concentration of the two (or more) forms is the same and the final result is the sum of all forms. For example, in validating the specificity of an ELISA designed to quantify the truncated form of the Cry1Ac protein in transgenic cotton, the following purified proteins were tested in the assay:

Cry1F, Cry1Ac (full-length and truncated forms), Cry1Ab, Cry34Ab1, Cry35Ab1, BAR, and PAT (Shan et al., 2007). In addition, if the protein is known to be closely related to an endogenous protein or proteins, an understanding of the level of cross-reactivity with the endogenous counterpart(s) is relevant.

Generally, in testing purified proteins for cross-reactivity, analysis of a dilution series starting with relatively concentrated amounts of the proteins provides a high degree of confidence that the proteins do not cross-react. In the Cry1Ac assay validation example cited above, concentrations of the test proteins ranged from 10 mg/L down to zero in the assay buffer (Shan et al., 2007). The range and number of concentrations may vary depending on the assay format and the expected cross-reactivity. For sandwich assays that use a coincubation of samples/standards with both the coating and the detecting antibody, caution needs to be applied if testing high concentrations relative to potential hook effects (i.e., high amounts of binding protein resulting in a lower absorbance response). For a sequential sandwich assay format (e.g., there is a wash step between the standards/samples and the addition of the conjugated detection antibody), a single high concentration of the test protein may be sufficient to demonstrate a lack of cross-reactivity (e.g., 1000 μg/L). Lower test concentrations may also be justified based on the highest level of the protein that may be expected to be found in samples. For example, in an ELISA designed to measure Cry1Ab protein in the blood plasma of cows fed MON810 containing maize (Paul et al., 2008), the concentrations used to test cross-reactivity of other Cry proteins in the assay (in this case, Cry3Bb1, Cry9C, Cry1C, Cry2A, and Cry3A) were the same as the concentrations used to construct the Cry1Ab standard curve.

The criteria used to define whether a protein is cross-reactive can vary. For example, a protein may be considered cross-reactive if the OD response is more than two times the background OD, or greater than the mean background OD + three standard deviations. When the LOD has been determined, a protein giving a value above the LOD may be considered cross-reactive. Cross-reactivity would obviously be considered significant in cases where the OD response of the protein is greater than that of LLOQ of the assay (i.e., produces a quantifiable result). In cases where the extent of the cross-reactivity is significant and consistent (e.g., homologous protein and/or endogenous counterpart), the level of cross-reactivity may be described as a percentage of the interpolated values relative to the known concentration of the test protein.

In addition to determining whether the purified proteins cross-react in the assay, determination of potential interference by the purified proteins with the ability of the assay to accurately quantify the target protein may be relevant, especially in cases where there are homologies between the test and the target proteins. For example, in a sandwich style ELISA, cross-reactive binding of the test protein to the coating antibody without concomitant binding to the detecting antibody may result in a decrease in the ability of the assay to quantify the target protein due to competition for binding sites. In a coincubation style assay, the same situation may occur if the detecting antibody cross-reacts with the test protein but the coating antibody does not. It is also possible that binding between the test protein and the protein of interest may interfere with quantification either by blocking the specific protein–antibody binding, or through modification of the conformation of the target protein such that the antibody binding is affected.

To determine whether a purified test protein interferes with quantification of the target protein, concentrations of the target protein near the LLOQ (e.g., 20% greater than the LLOQ) and near the ULOQ (e.g., 20% less than the ULOQ) of the assay can be combined with specified amounts of the test protein. The criteria used to define whether a protein interferes with the quantification of the target protein may be modified depending on the expected precision of the assay. In general, if the value of the target protein in the presence of the test protein differs by more than 25% from the value of the target protein alone analyzed on the same plate, the test protein may be considered to interfere with quantification in the assay. However, it is also important to understand if there is a biological significance to the level of interference. For example, if a test protein that has low expression levels in transgenic tissues shows a 30% decrease in the interpolated concentration of the target protein when present at 10,000 µg/L, but at the 1000 µg/L and lower concentrations does not demonstrate the potential for interference, it would be unlikely that interference would be observed in samples containing that protein.

6.2.2.2 Cross-Reactivity/Interference of Matrix

The matrix (e.g., tissue extract) can also affect the ability of the method to quantify the target protein. In some cases, there may be a homologous endogenous counterpart present in negative matrices that may cause a cross-reactive OD response. In other cases, nonspecific binding of matrix components to the antibodies may occur. Alternatively, matrix components may decrease the ability of the method to quantify the protein of interest by interfering with the specific antibody–target protein interactions as described above for the potential for interference by purified proteins.

Commonly, to test for matrix effects in the assay, different levels of the negative tissue extract (this could include testing of nontransgenic and of nontarget transgenic tissue) are used to prepare a standard curve. The response of the matrix spiked curve can be compared to the response of a curve prepared in assay buffer (no matrix) to determine whether there is an effect. Generally, several levels of matrix are tested in this manner during development and the results may be used to determine whether a minimum dilution should be applied to avoid the effect. If the method recommends that a minimum dilution be applied to avoid matrix effects, the validation testing can be limited to verifying that the curve prepared with negative tissue extract at the minimum recommended dilution is equivalent to the curve prepared in the assay buffer.

For testing an undiluted extract for matrix effects, each negative matrix tissue type is extracted according to the method extraction procedure and, as necessary to provide sufficient volume of matrix, extracts from replicate samples are pooled. The pooled extracts are then used in place of the assay buffer for preparation of the standard curve. Each matrix prepared curve is compared to a curve prepared in assay buffer (no matrix) analyzed on the same plate.

A simple approach to testing for matrix effects at 2× dilutions and higher is to prepare a standard curve that is twice as concentrated at each standard curve level as the standard curve used in the method. As described previously, the negative tissue matrix in question is extracted according to the extraction procedure described in the method and extracts of replicate samples are pooled as necessary to get sufficient sample. To test a 2× dilution of matrix, equal volumes of the undiluted matrix extract and each level of the standard curve (at 2× the normal concentration) are mixed, resulting in a standard curve at the 1× standard concentrations fortified with a 2× dilution of matrix. The same 2× concentrated standard curve may be diluted with equal volumes of assay buffer to prepare the reference standard curve. This approach not only simplifies curve preparation but also minimizes the potential for prep to prep standard curve differences that would not be related to an actual matrix effect. This same approach can be used to test other matrix concentrations; for example, adding equal volumes of a 3× dilution of matrix will result in matrix at a final concentration equal to a matrix dilution of 6×.

The difference between the theoretical values for the curve and the observed values for the points on the matrix spiked curve can be calculated using the standard curve prepared in assay buffer without matrix as the reference. Criteria for determining whether the matrix effects are significant can vary and may depend, in part, on the expected precision of the standard curve. A difference of 15% or more for each point on the standard curve has been applied in a recent method validation study (Shan et al., 2007). However, in some cases the matrix effect may be more pronounced in the upper or the lower portion of the curve and criteria of more than 20% difference for two consecutive points may be applied, especially if samples will be interpolated from that level of the standard curve.

The matrix testing described above can also be used to demonstrate that the negative extract in the absence of standard did not cross-react in the assay. Verification of the zero standard level for the matrix spiked curve does not show significant cross-reactivity. The same approach can be applied to assessing the cross-reactivity of the matrix in absence of the standard protein as was applied for the purified proteins (e.g., if the OD response is more than two times the background OD, or a protein giving a value above the LOD, or when the OD response is greater than that of LLOQ).

If a significant matrix effect is observed at a given dilution, it is sometimes possible to apply a minimum dilution of the particular matrix extract that does not exhibit matrix effects. For example, if leaf extract at a 5× final matrix dilution does affect the curve beyond the acceptance criteria but at a 10× dilution of matrix the effect is not significant, a minimum dilution of 10× may be applied for leaf sample extracts. Sometimes, it is not possible or practical to dilute significant matrix effects (e.g., matrix effects remain significant at higher dilutions and/or samples are expected to have target protein concentrations that cannot be quantified at higher dilutions). Although a less desirable option, a matrix-matched curve may be used to correct the matrix effects. In the above example, instead of applying a minimum dilution of 10× to leaf extracts, the standard curve would be prepared in a 2× dilution of negative leaf extract. This presents the limitation that different sample types would each need their own matrix-matched curves for analysis at each dilution tested. In addition, each matrix-matched curve would need to be tested for accuracy independently (see below for further discussion).

6.2.3 Accuracy

6.2.3.1 Extraction Efficiency

The extraction efficiency is determined by dividing the amount of target protein in the extract when the extraction is conducted as described by the analytical method by the total amount of target protein in the sample. Efforts should be made to attain an efficient extraction of the target protein from each matrix (ideally within 70–100% of the total target protein present); however, lower efficiencies may be considered acceptable if the relative amount extracted by the method is consistent (e.g., within 20% CV). The extraction efficiency can be difficult to assess without the use of a complementary analytical method. Usually extraction efficiency is determined through serial extractions of the same sample using the extraction buffer and the extraction procedure described in the method and then the amount of protein detected in the first extract relative to the sum of the target protein detected in all the extracts is determined. The extraction efficiency determined by serial extractions in the same buffer can be further supported by acceptable fortification recovery results that demonstrate the protein is soluble and recoverable from the tissue matrix in the assay buffer using the extraction procedure described in the analytical method.

When available, Western blot analysis can be used to verify extraction efficiency by the inclusion of a final "complete" extraction step using a buffer that contains detergents and reducing agents (e.g., laemmli sample buffer) (Laemmli, 1970) to solubilize any target protein that may not have been extracted by the method extraction buffer. The extracts that were obtained using the assay buffer could be analyzed by the ELISA method and the final extract could be analyzed by Western blot to ensure that no additional significant amounts of target protein were extracted by the complete extraction buffer. If the amount of target protein in the final extract is undetectable by the Western blot procedure, the extraction efficiency can be calculated from the ELISA values as the amount of protein in the first extract divided by the sum of the protein concentrations determined for each of the serial extracts. If the Western procedure is quantitative, each of the extracts could be quantified by Western blot and the extraction efficiency could be calculated by dividing the concentration of protein in the first extract by the total concentration of protein in all extracts. Alternatively, the concentration of a sample extracted in the assay buffer could be compared to a replicate sample extracted in the final extraction buffer (e.g., laemmli) by Western blot. In this case, the relative extraction efficiency is determined by dividing the concentration of the protein in the assay buffer by the concentration of the target protein extracted in the buffer containing detergents and reducing agents.

6.2.3.2 Fortification and Recovery Testing

Fortification recovery (also known as recovery or spike recovery) testing is conducted by adding the purified standard reference protein at known concentrations to negative tissues and extracting and analyzing according to the procedure in the method. Generally, the three fortification concentrations used for this testing are calculated to give final values that interpolate at, or near, the ULOQ, the midpoint of the standard curve, and the LLOQ of the standard curve after accounting for extraction volumes and minimum recommended dilutions (if applicable). The results are expressed as percent recovery, which is calculated by dividing the observed result for the fortified sample by the known amount or the theoretical value added to the sample. During the validation testing, at least three independent measurements of each of the three levels should be obtained (ICH, 2005). Ideally, the mean recovery value at each level should fall between 70% and 120% with a CV of less than 20% (Lipton et al., 2000). If during assay development and/or prevalidation testing the method gives results outside the acceptable range for one or more matrices, attempts should be made to improve the recoveries by modifying buffer components, extraction procedures, and/or antibody reagents (discussed further in Chapters 4 and 9). It should be noted that lower recovery values (below 70%) may be considered acceptable if they are consistent (e.g., CV <20%). Recovery testing should include each of the matrices that are to be validated in the assay. For matrix-matched curves, the matrix level of the fortified sample should correspond to the matrix level of the standard curve (i.e., if the matrix level of the curve is a $2\times$ dilution, the fortification level should be adjusted to allow a $2\times$ dilution of the fortified sample extract).

The fortification concentrations are calculated using the following information: concentrations of the ULOQ, midpoint, and LLOQ of the standard curve, extraction volume, minimum recommended dilution or matrix-matched dilution (if applicable), and the acceptable accuracy at the ULOQ and the LLOQ of the assay. If allowances are made for quantification above and/or below the LLOQ (e.g., can quantify samples if they are within $\pm20\%$ of ULOQ/LLOQ), then the values of the LLOQ and the ULOQ may be used for the fortification testing. If no allowances are made for extrapolation of results above the ULOQ and below the LLOQ, and the acceptable accuracy criteria is within 70% and 120% of the known target protein concentration, then the amount of protein used to fortify the negative matrix sample can be adjusted such that acceptable results can be accurately interpolated within the quantitative range of the assay. For example, the fortification concentration expected to fall near the LLOQ can be adjusted to allow the interpolation of acceptable values at 70% of the fortified amount (e.g., LLOQ/70%). Likewise, the fortified concentration expected to fall near the ULOQ of the assay should be adjusted such that acceptable values for recovery of 120% could be interpolated within the quantitative range of the curve (e.g., ULOQ/120%).

Example Calculation: If the ULOQ is 12 ng/mL, the midpoint is 6 ng/mL, the LLOQ is 0.6 ng/mL, the extraction volume is 0.5 mL, and a minimum recommended dilution of 10 is applied for the tissue of interest (or if a matrix level of $10\times$ dilution is used in a matrix-matched curve), then the amount of fortified protein in the extraction volume can be calculated as follows:

$$\text{Adjusted LOQ} = \text{ULOQ}/120\% \text{ or } \text{LLOQ}/70\%$$

$$\text{Near ULOQ} = (12\,\text{ng/mL}/120\%) \times 0.5\,\text{mL} \times 10\,\text{dilution factor}$$
$$= 50\,\text{ng of target protein}$$

$$\text{Near midpoint} = 6\,\text{ng/mL} \times 0.5\,\text{mL} \times 10\,\text{dilution factor}$$
$$= 30\,\text{ng of target protein}$$

$$\text{Near LLOQ} = (0.6\,\text{ng/mL}/70\%) \times 0.5\,\text{mL} \times 10\,\text{dilution factor}$$
$$= 4\,\text{ng of target protein}$$

The known concentrations of target protein in the extract (i.e., ng target protein divided by the extraction volume) at each fortification level would be 100, 60, and 8 ng/mL, respectively. The percent recovery is calculated as the concentration of the target protein determined for the extract divided by the known amount of protein in the extract multiplied by 100. If the adjusted results obtained for the "known" fortification concentration of 100 ng/mL were 110, 98, and 87 ng/mL, then the percent recoveries (adjusted result of fortified sample/known target fortification concentration $\times$ 100) for each would be 110%, 98%, and 87%, respectively, and the mean percent recovery would be 98% with a 12% CV, and therefore meeting the acceptance requirements set for accuracy.

If the validation is done as a part of the assay development, additional fortification levels are generally incorporated to help define the LLOQ and the ULOQ, as well as to help determine whether a minimum dilution should be recommended. For example, recovery concentrations incorporating 1:2, 1:5, and 1:10 dilution for each level could be analyzed and if the $2\times$ dilution yielded mean recovery results outside the acceptable criteria but the $5\times$ and $10\times$ dilution produced acceptable results, a minimum dilution of at least $5\times$ may be recommended. The minimum recommended dilution may be increased to $10\times$ or perhaps higher depending on the results of matrix testing and/or dilution agreement discussed later.

6.2.3.3 Verification by Independent Method Comparison of results obtained from an independent quantitative method can provide a verification of the method accuracy when the two methods agree (see also Chapter 9). Quantitative Western blot, activity assay, or mass spectroscopy methods may all be used; however, often these methods are not available or not fully quantitative. Another limitation might be the accuracy and/or precision of the independent method. When comparing results between two different methods, it is important to understand and account for the limitations and potential biases of both the methods.

6.2.4 Precision

6.2.4.1 Dilution Agreement Dilution agreement experiments are designed to demonstrate that the assay is capable of giving equivalent results regardless of where the sample OD interpolates in the quantitative range of the standard curve. The CV of the adjusted results from several dilutions of a single sample extract should be less than or equal to 20%. To conduct these experiments, ideally, samples that are positive for the target protein are diluted such that at least three of the dilutions result in values that span the quantitative range of the curve (i.e., near the ULOQ, the midpoint, and the LLOQ), and, if applicable, to have at least one dilution exceed the ULOQ of the assay. If a minimum dilution is recommended, any dilution with results falling in the quantitative range of the assay that is less than the minimum dilution can be excluded from the CV measurements. If the development and validation studies are combined, then the dilution agreement results may be considered in the determination of a minimum recommended dilution. See Table 6.2 for an example of good dilution agreement results. See Table 6.3 for an example where a minimum dilution may need to be applied to avoid matrix effects for a particular tissue type. Note that when a

TABLE 6.2 Dilution Agreement Example with Good Precision

Dilution factor	Mean result (ng/mL)	Adjusted result (ng/mL)	Quantitative range 8.0 to 0.4 ng/mL ($\pm 10\%$)
5	*Range*	*Range*	*Above ULOQ*
10	8.29	82.9	Within range
20	4.03	80.6	Within range
40	2.03	81.4	Within range
80	1.07	85.5	Within range
160	0.52	83.1	Within range
320	*0.27*	*86.9*	*Below LLOQ*
640	*0.15*	*97.1*	*Below LLOQ*
Mean adjusted result		83	
Standard deviation		1.9	
CV		2.3	
Minimum mean result		0.52	
Maximum mean result		8.3	

Texts in italic are excluded.

minimum dilution appears to be indicated in a dilution agreement experiment, matrix testing and recovery experiments usually support the data. For the example in Table 6.3, in matrix testing, the effect of that level of matrix would be expected to increase standard ODs relative to the reference curve, and the recovery of the tissue sample at the 1:10 dilution would likely be higher than 100%.

TABLE 6.3 Dilution Agreement Example without and with a Minimum Dilution Applied

Dilution factor	Mean result (ng/mL)	Adjusted result (ng/mL)	Applying minimum dilution of 1:20	Quantitative range 8.0 to 0.4 ng/mL ($\pm 10\%$)
5	*Range*	*Range*	*Range*	*Above ULOQ*
10	8.80	88.0	*Excluded*	Within range
20	2.87	57.4	57.4	Within range
40	1.48	59.2	59.2	Within range
80	0.73	58.4	58.4	Within range
160	0.36	57.6	57.6	Within range
320	*0.18*	*57.6*	*57.6*	*Below LLOQ*
640	*0.11*	*70.4*	*70.4*	*Below LLOQ*
Mean adjusted result		64	58	
Standard deviation		13.4	0.8	
CV		20.9	1.4	
Minimum mean result		0.36	0.36	
Maximum mean result		8.8	2.87	

Texts in italic are excluded.

6.2.4.2 *Within Plate Precision* Within plate precision should be a routine part of plate quality control (QC) when stabilized coated plates are used in a quantitative sandwich assay and should be conducted for each new plate lot. If plates are coated on an as needed basis, the within plate precision may be monitored during development and validation and then verified periodically to ensure acceptable assay performance. To test for within plate precision, either a homogeneous extract or a standard diluted to fall in the midrange of the standard curve can be analyzed across all wells of a plate. Generally, if there were no issues during the plate coating process and the assay is robust, then within plate variability is low and CVs of the OD values are well below 10% (within plate precision results less than 5% are common). If large lots of stabilized plates to be used over time are produced, then within plate precision testing should be conducted as part of the QC testing on plates pulled from near the beginning, middle, and end of the plate coating production run.

6.2.4.3 *Across Plate Precision* To validate that the precision of the assay is acceptable for implementation of the method for sample analysis within a lab, intralaboratory analysis can be performed using a characterized set of samples analyzed over several assay days by more than one analyst (when possible). When using samples with known stability, a single set of extracts could be generated at the beginning of the validation testing and then separated into single use aliquots and analyzed across days and across analysts. In cases where the sample extracts are not expected to be stable over the required time of testing, or do not respond well to a freeze–thaw cycle, pooled tissue sample replicates may be freshly extracted and diluted as applicable and then analyzed on each plate run for a given analysis day. QC samples with values approaching the ULOQ, the midpoint, and the LLOQ should also be included on each plate. Ideally, the acceptance criteria applied to the precision of the results across days and analysts using previously described extracts should be within 20% CV (see Table 6.4).

6.2.4.4 *Standard Curve Precision* The precision of the back calculated values of each point on the standard curve can also be determined within days and across days. Generally, for standard curves without added matrix, a CV less than 10% is common with the possible exception of the ULOQ and LLOQ levels (especially for curves with sigmoidal fits).

6.2.4.5 *Robustness* Robustness may be defined as the ability of the assay to remain unaffected by small differences in the method parameters and is generally included during the development phase of the assay. To assess the robustness of an assay, generally small, defined changes that might be reflective of variations that could be expected to occur during normal use are purposely incorporated and the effect on quantification is measured (generally CV of interpolated results and fortification recovery testing). Things to consider in testing assay robustness may include variations in the following: buffer composition/pH, incubation times/temperatures, samples weights/extraction volumes, antibody concentrations, and lot-to-lot testing. Robustness testing is most commonly done during method development and is generally included in validation studies conducted by the method developer. If the

TABLE 6.4 Example of Results of Precision Analysis Across Days and Analysts

(a) Example of interpolated results of positive samples					
	Tissue 1	Tissue 2	Tissue 3	Tissue 4	Tissue 5	Tissue 6
n^a	16	16	16	16	16	16
Mean ng/mL	920	120	250	82	160	140
SD	130	13	16	8.3	18	12
CV	14	11	6.4	10	11	8.6

(b) Example of % recovery values for two tissue types fortified at a high level (near ULOQ) and low level (near LLOQ)

	Fortified Buffer high	Fortified Buffer low	Tissue 1% recovery high	Tissue 1% recovery low	Tissue 2% recovery high	Tissue 2% recovery low
n^a	16	16	16	16	16	16
Mean percent recovery	99	110	96	89	100	94
SD	4.9	10	7.6	5.7	2.8	7.1
CV	4.9	9.1	7.9	6.4	2.8	7.6

a = Data collected from 16 independent plates/curves over three separate days by at least two analysts per day.

results of the robustness testing demonstrate that certain parameters need to be tightly controlled for acceptable results, those critical parameters/limitations should be emphasized in the written method. Robustness is also verified in a more indirect fashion during single laboratory validation testing by measuring the precision of interpolated results over several days using different analysts and different buffer/reagent lots as applicable (see above for precision).

6.2.4.6 Stability The stability of the immunoreactivity of standard protein used in the assay is generally investigated during assay development and is verified by demonstrating acceptable performance during the validation process. Standard stability testing is also a part of the characterization and certification process for the reference standard (see Chapter 9) and is monitored periodically as long as the standard is in use.

Early stability of the target protein in the tissue and the tissue extracts is also determined during development. Unstable proteins may require the addition of stabilizing buffer components (see Chapter 9 for some examples) and/or cofactors, as well as special extraction conditions and/or storage conditions that would need to be incorporated into the method prior to validation. Tissue sample/tissue extract stability may be verified as a part of the validation process or followed in separate stability studies. Tissue stability experiments are conducted by measuring the protein in a positive sample at a "time zero" (i.e., minimal time between sample collection and analysis), and then reanalyzing the sample periodically after storage under conditions typically used prior to routine sample analysis. Often to obtain enough tissue for the repeated analysis over time, collected samples are pooled, and it is important that enough replicates are analyzed at each time point to ensure that the results are representative of the entire sample. Stability testing should cover the longest period of time that samples might be expected to be stored prior to analysis. Extract stability testing may also be conducted using a "time zero" (i.e., minimal time between extraction and analysis), followed by analysis at set time points under typical storage conditions. For example, extract stability testing might include refrigerated short-term storage (1–24 h) and/or frozen storage (4–48 h or longer). The method should include any limitations to the stability of the target protein in the extract (e.g., analyze extract within 20 min after extraction or extracts may be stored frozen for up to 24 h following extraction).

Extracts that are stable when stored for prolonged periods of time (i.e., stored in frozen single use aliquots) may be used to monitor assay performance. Quality control samples are included on each plate during routine analysis and stable extracts are often used for this purpose. Data acceptance criteria for the plate would include a range of acceptable values for control samples and may cover the high, middle, and low ranges of the standard curve. Interpolated control sample values should fall within established ranges, for example, within 25% of the determined concentration (Codex, 2004), or within three standard deviations of the established mean value.

Reagent stability testing is conducted by the method developer to assign expiration dates to the assay test kits that are provided to other laboratories. Often accelerated stability testing is conducted by storing reagents at elevated temperatures (i.e., 37°C), followed up with real-time verification of kit performance under recommended storage conditions. If stability testing has been conducted by the kit manufacturer, then it is generally not reconfirmed during a single laboratory validation as long as the reagents are not used beyond the established expiration dates.

6.3 INDEPENDENT LABORATORY VALIDATIONS

The EPA requires that an ILV be performed for methods that will be used for product registration of crops expressing insect-resistant traits. The requirements for an ILV are detailed in Pesticide Registration (PR) Notice 96-1: Tolerance Enforcement Methods—Independent Laboratory Validation by Petitioner (EPA, 1996a) and paraphrased below. An ILV study must accompany submissions for a first tolerance petition for pesticide residues in (e.g., new insect-resistant transgenic trait) or on a raw agricultural commodity or processed food or feed. An ILV is also required if a new method is proposed for enforcement of any new tolerance for residues of a pesticide with previously established tolerances, or if significant modifications have been made to a previously approved method to accommodate the new commodity. However, an ILV is usually not required for confirmatory methods or if the analytical method is judged superior to the currently accepted method by the agency.

The laboratory chosen to do the ILV trail should be capable of conducting the analysis according to FIFRA Good Laboratory Practices as specified in 40 CFR 160 and must be unfamiliar with the method and independent of the lab that developed the method. The ILV lab must not use the same equipment, instruments, and supplies that were used for the original method development, nor should involve personnel who conducted the trial report to a study director that was involved in the development, validation, or subsequent use of the method.

The method should be detailed enough so that the trial can be conducted (according to GLPs) as written without significant modifications. Prior to conducting the first set of sample analysis, the ILV lab may contact the developer or others who have used the method; however, all such communication must be logged and reported to the EPA. Personnel familiar with the method are prohibited from visiting the ILV lab to offer assistance in the procedure.

If the method is to be used for more than one commodity, each commodity will be tested as a separate ILV. The sample set used for ILV trial includes two control samples, two control samples fortified at the proposed tolerance level, and two control samples fortified at the LLOQ. In cases where the proposed tolerance is at the LLOQ, a second concentration fortified at $2\times$ the LLOQ will be included. One additional fortification level may also be included at the discretion of the petitioner. The results for the fortified samples must fall between 70% and 120% of the known fortified value and interference should be negligible relative to the proposed tolerance level. At least one of up to three sample sets must produce results that meet the required criteria for the ILV to be considered acceptable. If the first or the second set of samples does not provide acceptable results, the laboratory conducting the ILV may contact the developer or others familiar with the method, but all communications need to be logged and reported. Any modifications to the method resulting from the communications should be incorporated into the method. If none of the three sample sets provides acceptable results, the method must be revised and another ILV is conducted on the revised method using a different laboratory. The requirements with respect to the reporting of the data are detailed in Environmental Protection Agency PR Notice 96-1 (http://www.epa.gov/opppmsd1/PR_Notices/pr96-1.html) and in Residue Chemistry Test Guidelines. OPPTS 860.1340, Residue Analytical Method (EPA, 1996a, 1996b).

6.4   WESTERN BLOT METHOD VALIDATION

Western blot analysis involves the electrophoretic separation of proteins followed by transfer to a membrane support and detection by antibodies specific for the target protein that are also linked directly or indirectly to a reporter system (Towbin et al., 1979). Western blots may also be used for both qualitative and quantitative analyses and despite limitations have some benefits over a sandwich ELISA. One such benefit is that denaturing conditions can be used to more completely extract less soluble proteins. In addition, the target protein is at least partially separated from matrix/buffer components that may present problems in an ELISA. Western blot analysis also provides additional information on the size of the target protein. However, Western blot methods usually require more sample handling, are lower throughput, and generally have a higher inherent level of imprecision associated with the results compared to ELISA methods. Also, the quantification may require more direct involvement and subjectivity from the analyst, depending on the image analysis software used and the level and consistency of nonspecific background staining. Additional discussion on Western blot analysis can be found in Chapter 10.

The same types of experiments that are done for validating a quantitative ELISA method can also be conducted to validate a quantitative Western Blot method. The standard curve generally contains fewer points due to the limited number of wells on each gel, but still consists of a ULOQ and a LLOQ. Specificity can be investigated using both purified proteins and negative tissue samples (nontransgenic and nontarget transgenic) to determine whether there are cross-reacting bands. Serial extractions can be used for determining extraction efficiency. Fortification recovery testing can be done using fortification concentrations covering the high, midpoint, and low portions of the standard curve. Precision within a blot can be determined by running a single sample in each well of the gel and determining the CV of the densitometry values. Across blot precision can be determined by running samples according to the method and determining the CV of the interpolated results across analysts, across blots, and across days. Due to the limitations of the method, it is especially important that clear criteria based on the validation results should be developed for acceptance or rejection of data during routine analysis.

6.5 QUALITATIVE IMMUNOASSAY VALIDATIONS

Qualitative immunoassay analysis is commonly conducted using either the microtiter ELISA format or LFDs, although Western blots and dot blots are also used. The qualitative ELISA is particularly useful when large numbers of samples (hundreds to thousands) are being evaluated for a positive/negative call in a high-throughput situation. LFDs are commonly used when testing samples on-site (e.g., in the field or at a grain elevator), or when a smaller number of samples are tested in the lab. A Western blot or dot blot could be conducted if an ELISA or an LFD is not available and/or if incompatible buffer conditions are required to extract the protein from the sample.

In a qualitative assay, the emphasis is placed on reliably distinguishing a positive sample from a sample that does not contain the target analyte. The required sensitivity of the method and the potential for cross-reactivity with other matrix components are important considerations in the method development process (see Chapter 5). The experiments done during qualitative method development are also conducted for the validation and generally include a measurement of the specificity, sensitivity (verification of the LOD/threshold level), accuracy (false-positive/false-negative rates), and robustness (especially in development-based validations). The sensitivity of the method is defined by the LOD or threshold level. The LOD is the value above which samples are determined to be positive and below which samples are considered to be negative, or more correctly, nondetectable by the method. For a microtiter ELISA method, the LOD is based on the OD response of positive samples relative to the OD response of negative samples and can be estimated by a variety of methods (ICH, 2005). For

Western blots and dot blots, it is the concentration at which the specific protein band or dot can be distinguished from background. For an LFD, the LOD is the concentration at which positive band is reliably observed. This discussion will focus primarily on the validation of lateral flow devices.

The most significant advantage of the LFD is its simplicity. It provides a visual detection of the presence/absence of the target protein within a few minutes. Most LFDs use a double antibody sandwich format. Antibodies specific to the target protein are coupled to a color reagent. The membrane contains two capture zones, one specific to the target protein and the other specific for the antibodies coupled to the color reagent. The appearance of one line on the strip indicates a negative result (control line); the appearance of two lines on the strip indicates a positive result. For additional information on LFD theory and development, see Chapter 5.

The results of the validation process establish the performance characteristics and limitations of the analytical method, and are used to develop recommendations regarding criteria and quality control measures. The validation is performed in the same way as the method is intended to be used.

The analytical method refers to all experimental procedures needed to detect the analyte in a particular matrix; this includes a protocol describing the extraction method, interference by other analytes, or inapplicability to certain situations. The method should also provide testing conditions, indication of timing, the use of applicable controls, stability of the device and sample extracts, and any other factors that could be of importance to the operator (see Section 6.2.5.5).

The method should be applicable to the matrix of concern, for example, leaf, grain, processed food, and so on. Proteins may be easily degraded, and therefore the measurable concentration will likely decrease during grain processing; this needs to be considered in the performance criteria assessment of the method. This discussion will focus on the identification of transgenic protein in leaf sample, single seed sample, or bulk grain sample of a given crop. The parameters discussed below refer to a single laboratory validation and not a collaborative study. Guidelines for validating characteristics and minimum criteria in a collaborative study require more than 15 laboratories analyzing at least two analyte levels (ICH, 2005; AOAC, 2005).

6.5.1 Specificity

Validation testing should include method specificity, which is the ability to unequivocally detect the presence of the analyte in the sample extract by obtaining positive results from samples containing the analyte, coupled with obtaining negative results from samples that do not contain the analyte, including nontransgenic plants and nontarget transgenic plants. Specific recommendations for specificity testing are provided in Section 6.6.

6.5.2 Sensitivity

Method sensitivity involves the determination and verification of the LOD. For pooled grain samples, the LOD can be described as the concentration at which the method can reliably detect one positive seed in a pool of known negative seeds, without giving rise to a significant number of false positives from known negative samples. The measures of precision and accuracy are determined by the frequencies of false-negative and/or false-positive results at the threshold level as defined by the minimum analyte concentration at which a positive sample gives a positive result at least 95% of the time (Codex, 2004).

The sensitivity of a LFD can be determined by analyzing known low amounts of the target protein in assay buffer or negative plant tissue extract that are expected to fall near the estimated LOD. Strips are placed in different concentrations of the standard preparation and the minimum protein concentration level at which the target protein can reliably be detected is the sensitivity of the strip.

For use in testing bulk or pooled grain samples, a more practical application for determining the LOD is the analysis of mixtures of ground non-GM grain with ground target grain by weight at different levels (e.g., 10%, 5%, 1%, 0.5%, 0.1%, 0.05%, 0.02%, and 0% GM). Protein expression levels in combination with the ability to extract the target of interest can vary significantly; thus, the sensitivity or limit of detection has to be consistent with the lowest expression level. The number of individual grain in a weight unit depends on the variety tested. It is recommended that 100 individual grains of the variety being tested are counted and weighed to the nearest 0.01 g to determine the average weight of an individual grain. This average weight is used to determine the weight of samples at different levels of GM grain. The lowest GM% that reliably gives a clear positive signal is the practical limit of detection.

The false-negative rate of an assay is determined by analyzing a series of known positive samples of leaf, single seed, and/or seed pools with a constant known concentration of positive material (consistent with the determined LOD or threshold level of the method) and evaluating the result. The percent false negatives is the number of known positive samples that give a negative result divided by the total number of positive samples tested multiplied by 100. Confidence intervals and statistical uncertainty should be considered and applied to the risk of false-positive and/or false-negative results as well. The desired level of confidence determines the size and number of pools that need to be tested (Codex, 2004). Additional discussion on the importance of sampling for qualitative analysis of pooled or bulk grain samples can be found in Chapter 11 (Grothaus et al., 2006).

6.5.3   Storage Stability Testing

Storage stability of the strips is commonly determined by storage of the strips for 3 months at room temperature or at 37°C for 11 days (condition design to mimic storage for ~134 days at room temperature), followed by evaluation using known negative samples, positive samples, and negative samples fortified at the limit of detection (e.g., 15 independent samples of each). All negative samples need to give a negative result and all GM and fortified samples need to give a positive result, ensuring that the reactivity of the strip was not diminished when compared to a new lot of strips.

6.5.4   Testing for High-Dose Hook Effect

A high-dose hook effect refers to the false-negative results seen with immunochromatographic strip tests when very high levels of target are present in the tested sample. It is possible that unbound GM protein reaches the test line before the labeled antibody-bound GM protein, resulting in a false-negative result. To assess the threshold of high-dose hook effect, increasing amounts of GM standard protein are evaluated, as well as samples at 0.1%, 1%, 10%, and 100% (w/w). Strong positive results for the 100% sample ensure that no high-dose hook effect is present. High-dose hook effects are also discussed in Chapters 4, 5, and 9.

6.5.5   Performance on Leaf and Single Seed Control Samples

One hundred twenty independent analyses comprised of 20 individual GM and 20 negative control samples are analyzed using three lots of LFDs. All test results should give strong positive results for the target protein in the GM samples and negative results for the negative samples within the time frame of the test. Such result corresponds to less than 1% error rate ((# false positives + # false negative)/total number tested). Performance testing of bulk grain samples is described next as a part of the data required prior to GIPSA verification testing.

Screening grain at very low levels can be accomplished by using a large sample size. LFDs can be used to test multiple subsamples, the size of which does not exceed the sensitivity of the strip test. Although the test does not determine the exact percent of GM protein in the grain, it determines the probability that a sample contains greater or less than a specified threshold concentration (Laffont et al., 2005). The results of a determination are expressed in terms of percent of GM material in the sample.

6.6   GIPSA PERFORMANCE VERIFICATION OF RAPID TEST KITS

The practical applications outlined below are Grain Inspection, Packers and Stockyards Administration (GIPSA) recommendations for verifying biotechnology rapid test performance (Directive 9181.2) for pooled grain samples. Tests should be performed using three different test lots of LFDs (or ELISA kits). The following data are provided to GIPSA by the manufacturer prior to consideration for performance verification and certification and are recommended as part of validation testing for LFDs or qualitative microtiter ELISA tests.

6.6.1   Performance Testing

One hundred twenty analyses comprised of 40 fortified samples prepared by weight at the established limit of detection are evaluated using the three test kit lots. All fortified samples must provide a clear positive result for the GM protein of interest in the selected sample size. In addition, 120 independent analyses of control samples that do not contain the target analyte each must give a negative result.

6.6.2 Cross-Reactivity from Other Biotechnology Crops

All commercially available biotechnology products expressing proteins other than the test protein need to be evaluated for cross-reactivity in the assay. Data from five independent samples of all other biotechnology products, each at 100% of the biotech trait, are analyzed using the three testing lots. All 15 results need to be negative.

6.6.3 Temperature Sensitivity Testing

Although not required for LFD kits testing, temperature sensitivity is required for microtiter ELISA kits. To test temperature sensitivity of the qualitative microtiter ELISA method, 15 independent negative and 15 independent positive GM control samples and 15 fortified grain samples at the limit of detection are analyzed at both 18 and 30°C. All kit components, extracts, and equipment are equilibrated for 1 h at the test temperature prior to performing analyses using three separate ELISA kit test lots. All control samples need to give a negative result at both temperatures tested. All GM and fortified samples need to give clear positive results at both temperatures.

6.6.4 GIPSA In-House Verification Testing

Once the submitted data are reviewed by GIPSA, the manufacturer will be notified whether or not the test met the performance specifications. If the data do meet the specifications, GIPSA will perform an in-house verification with 30 independent analyses of control samples and of samples fortified at the specified LOD using three different kit lots. All control samples must be negative and all fortified samples must be positive by the test. If the verification testing is successful, a certificate of performance will be issued. The certification will expire in 3 years and can be renewed for an additional 3 years by providing additional performance data to GIPSA prior to the expiration date (USDA/GIPSA, 2004).

6.7 VALIDATION RESULTS

As stated at the beginning of this chapter, validation studies are designed to demonstrate that the method has the sensitivity, specificity, accuracy, and precision necessary to reproducibly detect and/or quantify the target analyte in a specific matrix or matrices. There are guidelines for the criteria that can be applied to evaluate the method performance (Lipton et al., 2000; Codex, 2004; FDA, 2001; ICH, 2005; Grothaus et al., 2006). Ideally, for quantitative assays, the extraction efficiency will be greater than 70%, recoveries of fortified samples will be between 70% and 120%, and precision, as evaluated by the CV of results across days and analysts, will be less than 20%. For qualitative methods, ideally the LOD or threshold level is set to minimize false positives and false negatives. In some cases, the method will be implemented, even though it did not meet all the criteria in the guidelines. For example, perhaps the extraction efficiency for a particular matrix is lower than 70% but reproducible across samples (<20% CV). If the effect is reproducible and well characterized, it is possible to correct for the limitations. In this case, the limitation should be clarified in the method and the final result may be adjusted to account for the lower efficiency.

Assay performance should be monitored over time and deviations from the validation results should be investigated fully (also see Chapter 9 for common sources of error). The results for the validation study can be used to set acceptance criteria for data during routine analysis. For quantitative methods (ELISA or Western blot), acceptance criteria should incorporate curve performance parameters, acceptable ranges for QC sample results, and acceptable precision of replicate sample wells (FDA, 2001; Viswanathan et al., 2007). For qualitative ELISA assays, a negative and a positive control should be included on each plate. The positive control is generally at the limit of detection, but, depending on application, may be typical of the lowest expressing positive samples. Criteria may include acceptable ranges for ODs of the control samples and acceptable replicate CV values. For LFD, the control line must be clearly visible and the result should be clearly positive (visible test line) or negative (absence of test line). Although a validation study provides the documentation that the method is suitable for use in routine analysis, once the method is implemented, assay monitoring and defined acceptance criteria are critical factors in ensuring the assay continues to perform as expected.

REFERENCES

Codex Procedural Manual 13th Edition, *Considerations of the methods for the detection and identification of foods derived from biotechnology. General approach and criteria for methods*, Codex Alimentarius, Rome, Italy, **2004**, http://www.codexalimentarius.net.

Environmental Protection Agency. Pesticide Registration (PR) Notice 96-1: Tolerance Enforcement Methods—Independent Laboratory Validation by Petitioner, **1996a**, http://www.epa.gov/PR_Notices/pr96-1.html.

Environmental Protection Agency. Residue Chemistry Test Guidelines. OPPTS 860.1340, Residue Analytical Method, **1996b**, http://www.epa.gov/opptsfrs/publications/OPPTS_Harmonized/860_Residue_Chemistry_Test_Guidelines/Series/860-1340.pdf.

Food and Drug Administration. *Guidance for Industry: Bioanalytical Method Validation*, United States Department of Health and Human Services, FDA, Center for Drug Evaluation and Research, Center for Veterinary Medicine, **2001**, http://www.fda.gov/cmv.

Grothaus, G. D., Bandla, M., Currier, T., Giroux, R., Jenkins, G. R., Lipp, M., Shan, G., Stave, J. W., and Pantella, V. Immunoassays as an analytical tool in agricultural biotechnology. *J. AOAC Int.*, **2006**, *89*, 913–928.

Horwitz, W. Protocol for the design, conduct and interpretation of collaborative studies. *Pure Appl. Chem.*, **1988**, *60*, 855–864.

Horwitz, W. Protocol for the design, conduct and interpretation of method performance studies. *Pure Appl. Chem.*, **1995**, *67*, 331–343.

ICH Harmonized Tripartite Guideline. *Validation of analytical procedures: text and methodology Q2(R1)*, **2005**, http://www.ich.org.

Laemmli, U. K. Cleavage of structural proteins during the assembly of the head of bacteriophage T4 *Nature*, **1970**, *227*, 680–685.

Laffont, J. L., Remund, K. M., Wright, D., Simpson, R. D., and Gregoire, S. Testing for adventitious presence of transgenic material in conventional seed or grain lots using quantitative laboratory methods: statistical procedures and their implementation. *Seed Sci. Res.*, **2005**, *15*, 197–204.

Lipton, C. R., Dautlick, J. X., Grothaus, G. D., Hunst, P. L., Magin, K. M., Mihaliak, C. A., Rubio, F. M., and Stave, J. W. Guidelines for the validation and use of immunoassays for determination of introduced proteins in biotechnology enhanced crops and derived food ingredients. *Food Agric. Immunol.*, **2000**, *12*, 156–164.

Official Methods of Analysis, *Appendix D: guidelines for collaborative study procedures to validate characteristics of a method of analysis*. 18th Ed. AOAC INTERNATIONAL, Gaithersburg, MD, **2005**.

Paul, V., Steinke, K., and Meyer, H. H. D. Development and validation of a sensitive enzyme immunoassay for surveillance of Cry1Ab toxin in bovine blood plasma of cows fed Bt-maize (MON810). *Anal. Chim. Acta*, **2008**, *207*, 106–113.

Shan, G., Embrey, S. K., and Schafer, B. W. A highly specific enzyme-linked immunosorbent assay for the detection of Cry1Ac insecticidal crystal protein in transgenic WideStrike cotton. *J. Agric. Food Chem.*, **2007**, *55*, 5974–5979.

Towbin, H., Staehelin, T., and Gordon, J. Electrophoretic transfer of proteins from polyacrylamide gels to nitrocellulose sheets: procedure and some applications. *Proc. Natl. Acad. Sci. USA*, **1979**, *76*, 4350–4354.

United States Department of Agriculture/Grain Inspection, Packers and Stockyards Administration (USDA/GIPSA). *Performance verification of qualitative mycotoxin and biotech rapid test kits*, Directive 9181.2, **2004**, http://archive.gipsa.usda.gov/reference-library/directives/9181-2.pdf.

Viswanathan, C. T., Bansal, S., Booth, B., DeStefano, A. J., Rose, M. J., Sailstad, J., Shah, V. P., Skelly, J. P., Swann, P. G., and Weiner, R. Workshop/conference report—quantitative bioanalytical methods validation and implementation: best practices for chromatographic and ligand binding assays. *AAPS J.*, *9 (1)*, Article 4, **2007**, http://www.aapsj.org.

CHAPTER 7

REFERENCE MATERIALS AND CONSIDERATIONS

Tandace A. Scholdberg
G. Ronald Jenkins

Immunoassays in Agricultural Biotechnology, edited by Guomin Shan
Copyright © 2011 John Wiley & Sons, Inc.

7.1   GENERAL CONSIDERATIONS

Through the advent of modern biotechnology, conventional crop plants are being genetically engineered to express proteins, uniquely different from their parent cultivars that confer insect protection, herbicide tolerance, disease resistance, or combinations of the aforementioned quality characteristics (Ahmed, 2004). Because novel DNA sequences become stably inserted into the plant genome, genetically engineered (GE or GM) crops utilize the internal biomechanics of the cell to express functional proteins with the desired trait characteristics (Uhlen and Moks, 1991). Methods to detect genetically modified traits in plants focus on either novel DNA sequences or resultant GM protein products (Grothaus et al., 2006; Lipp et al., 2005). Indeed, detectable amounts of DNA or GM proteins are commonly found in a diversity of food matrices and feed products that contain GM crops. Enzyme-linked immunosorbent assay (ELISA) and lateral flow devices (LFDs) are the two most commonly used methods to detect the presence of GM proteins in crop plants but have other applications in the field of agricultural biotechnology including product development and seed production (Grothaus et al., 2006; Urbanek-Karlowska et al., 2003; Stave, 2002). Reference materials play a key role in developing, validating, and troubleshooting protein-based methods for detection of biotechnology-derived traits in food and feed products (Anklam and Neumann, 2002; Trapmann et al., 2002). Tables 7.1–7.3 provide details on some of the biotechnology-derived crops that are currently available on the marketplace (AGBIOS, 2008).

Some GM varieties that have approval for commercialization in one country might not have approval for marketing into another (Bertheau and Davison, 2007). Also, some GM varieties have approval for import, but the commodity might require labeling, especially if the GM content is above a legally mandated, specified threshold level (Bertheau and Davison, 2007; EC, 2000). The asynchronous regulations that exist throughout the world create challenges in grain marketing and the need for well characterized reference materials is absolutely essential to establishing compliance with regulatory mandates and providing confidence in analytical measurements for food and feed products that are routinely tested for the presence or absence of GM traits (Holst-Jensen; Bertheau and Davison, 2008). Validated methods that provide both qualitative and quantitative measurements serve as a basis not only to satisfy contractual agreements between buyers and sellers, but also to substantiate economic, political, environmental, and legal decisions as they relate to grain trade issues. Certified reference materials that are also internationally recognized provide the foundation to developing accurate, reliable, reproducible, and cost-effective methods. Technological challenges persist, however, because both DNA and protein detection methods were originally designed and intended to distinguish between plants that express the desired traits and those that do not, and for evaluating products as they moved through the development stream (Grothaus et al., 2006). The original methods used by developers were never intended to provide authenticity to satisfy contractual obligations, for the absence of a gene or its protein product, as they relate to grain marketing or highly processed food products (EC, 2000). Nevertheless, commercially available immunoassays used for detection of GM traits have been developed for a variety of large-scale operations, including testing for compliance and identity

TABLE 7.1 Common Single Stack Traits Found in Maize

OECD identifier or trade name	Maize line (single trait)	Company information	Protein expression derived from introduced genetic elements	
			Pesticide resistance	Herbicide tolerance
YieldGard™	Mon810	Monsanto	Cry1Ab	None
Roundup Ready (MON-00021-9)	GA21	Monsanto	None	CP4 EPSPS[a]
MON-00863-5	Mon863	Monsanto	Cry3Bb1	None
Roundup Ready (MON-00603-6)	NK603	Monsanto	None	CP4 EPSPS
Mon-89034-3	Mon89034	Monsanto	Cry1A.105 and Cry2Ab2	None
NaturGuard™ Knockout™	Bt-176	Syngenta Seeds, Inc.	Cry1Ab	PAT[b]
SYN-BT011-1	Bt-11	Syngenta Seeds, Inc.	Cry1Ab	PAT
Agrisure RW (SYN-IR604-5)	MIR604	Syngenta Seeds, Inc.	mCry3A	PMI[c]
LibertyLink™	T-25	Bayer CropScience (Aventis CropScience)	None	PAT
StarLink	CBH-351	Bayer Crop Science (Aventis CropScience)	Cry9C	PAT (bar gene)
Herculex I	TC-1507, DAS-1507-1	Mycogen (c/o Dow AgroSciences) Pioneer Hybrid Int., Inc. (c/o Dupont)	Cry1F	PAT
Herculex RW	DAS-59122-7	Dow AgroSciences LLC Pioneer Hybrid Int., Inc.	Cry34Ab1/Cry35Ab1	PAT

Source: AGBIOS, 2008.

[a]EPSPS (5-enolpyruvylshikimate-3-phosphate synthase) confers resistance to glyphosate, the active ingredient found in Roundup Ready herbicide.

[b]PAT (phosphinothricin N-acetyltransferase) confers resistance to glufosinate ammonium herbicide.

[c]PMI (mannose-6-phosphate isomerase) protein allows growth on mannose as a carbon source, used as a selectable marker.

TABLE 7.2 Representative Multistacked Traits Found in Maize

			Protein expression derived from introduced genetic elements	
OECD identifier or trade name	Maize line (stacked traits)	Company information	Pesticide resistance	Herbicide tolerance
SYN-Bt011-1, SYN-IR604-5, MON-00021-9	Bt-11 × MIR604 × GA21	Syngenta	Cry1Ab, mCry3A	EPSPS,[a] PAT,[b] PMI[c]
Mon00863-5 × Mon00810-6	Mon863 × Mon810	Monsanto	Cry3Bb1, Cry1Ab	None
DAS-01507-1 × MON00603-6	TC1507 × NK603	Dow Agrosciences LLC, Pioneer Hybrid Int., Inc.	Cry1F	CP4 EPSPS, PAT
DAS-59122-7 × DAS-01507-1 × MON-00603-6	Herculex RW × Herculex I × NK603	Dow AgroSciences	Cry34Ab1/Cry35Ab1, Cry1F	CP4 EPSPS, PAT
SYN-Bt011-1 × SYN-IR604-5	Bt-11 × MIR604	Syngenta Seeds, Inc.	Cry1Ab, mCry3A	PAT, PMI
SYN-Bt011-1, MON00021-9	Bt-11 × GA21	Syngenta Seeds, Inc.	Cry1Ab	EFSPS, PAT
Herculex XTRA (DAS-01507-1, DAS-59122-7)	TC1507 × DAS-59122-7	Dow AgroSciences, Pioneer Hybrid Int., Inc.	Cry34Ab1/Cry35Ab1, Cry1F	PAT

Source: AGBIOS, 2008.

[a]EPSPS (5-enolpyruvylshikimate-3-phosphate synthase) confers resistance to glyphosate, the active ingredient found in Roundup Ready herbicide.

[b]PAT (phosphinothricin *N*-acetyltransferase) confers resistance to glufosinate ammonium herbicide.

[c]PMI (mannose-6-phosphate isomerase) protein allows growth on mannose as a carbon source, used as a selectable marker.

TABLE 7.3 Common Traits Found in Soybean

OECD identifier or trade name	Soybean line (single and stacked traits)	Company information	Protein expression derived from introduced genetic elements	
			Pesticide resistance	Herbicide tolerance
Roundup Ready (MON-04032-6)	GTS 40-32-2	Monsanto	None	CP4 EPSPS[a]
ACS-GE005-3	A2704-12, A2704-21, A5547-35	Aventis Crop Science	None	PAT[b]
ACS-GE006-4	A5547-127	Aventis CropScience, Bayer Crop Science, AgrEvo	None	PAT
DP-356043-5	DP356043	Pioneer Hi-Bred Intl., Inc.	None	GAT4601[c]
				ALS (*gma-hra* gene)[d]

Source: AGBIOS, 2008.

[a]EPSPS (5-enolpyruvylshikimate-3-phosphate synthase) confers resistance to glyphosate, the active ingredient found in Roundup Ready herbicide.

[b]PAT (phosphinothricin *N*-acetyltransferase) confers resistance to glufosinate ammonium herbicide.

[c]GAT (glyphosate *N*-acetyltransferase) confers tolerance to glyphosate by detoxifying the compound. *gat4601* gene is based on the sequences of three *gat* genes from the common soil bacterium *Bacillus licheniformis*.

[d]ALS (acetolactate synthase) enzyme is not affected by the imidazolinone class of inhibiting herbicides. The herbicide tolerant *gm-hra* gene was made by isolating the herbicide-sensitive soybean *gm-als* gene and introducing two specific amino acid changes.

preservation. As an alternative to DNA-based testing, protein-based immunoassay technologies offer one of the simplest and most cost-effective methods for detection of biotechnology-derived traits in grains and oilseeds on raw or semi-processed products (Grothaus et al., 2006; Stave, 2002). However, the ability to detect and analyze a particular protein is not always trivial, unless the protein has a distinctive property (such as that possessed by an enzyme) or a physical property (such as a spectrometric characteristic conferred upon by a nonprotein moiety) (Ahmed, 2004). The relative magnitude of a specified measurement on an intact protein, derived from a source containing an undefined level of a GM trait, forms the basis of comparative procedures by which protein reference materials are used (Trapmann et al., 2000). It is the performance characteristics of reference materials that are the foundation by which protein levels, present in a test sample, are assessed. During the process of cellular translation, proteins fold into unique structural conformations that provide both linear and conformational epitopes. These epitopes act as antigenic determinants that form highly specific complexes with antibody molecules and the basis for LFD and ELISA detection methods. Inherently, proteins degrade when exposed for varying periods of time to high temperatures, pressures, and extremes in pH, as what typically occurs in highly processed food and feed products (Stryer, 1997). For this reason, polyclonal antibodies, for example, a mixed population of antibody molecules recognizing different epitopes on a single protein molecule, provide more versatility for these types of applications, though monoclonal antibodies are used as well. Detection of proteins by LFD or ELISA requires a protein extraction procedure followed by antibody detection of the target protein (Kleiner and Neumann, 1999). Positive results are generally based on a colorimetric reaction. Both positive and negative reference materials are crucial to developing and validating detection methods using both LFD and ELISA technologies (Ahmed, 2004; Stave, 1999).

Production of reference materials is an essential component to fulfill requirements of method validation for both LFD- and ELISA-based technologies (Ahmed, 2004; Ermolli et al., 2006). Generally, reference materials are generated prior to development of a reference method. The purpose of the reference material is to facilitate standardized characterization of biological samples, whatever the type of measurement or method used. During the method development phase, positive reference materials are used to establish accuracy, precision, sensitivity, limit of quantification, and occurrence of false-negative results. Concurrently, negative reference material is required to determine the limit of detection (LOD), specificity, and the occurrence of false-positive results. The behavior of the reference material should resemble as closely as possible to the behavior of the test samples in the assay system used. If more than one reference material is commercially available, then both reference materials should be comparable to one another and provide similar analytical measurements on test samples. When stocks of reference materials become depleted, it is crucial to demonstrate that replacement reference material preserves as closely as possible the same value and performance characteristics as the original reference material. This chapter describes relevance, types, preparation, traceability, metrology, quality, applications, availability, suitability, and performance characteristics of protein-based reference materials as applied to detection and quantification of biotechnology-derived traits in plant and related materials including seed, grain, and products derived from them. Also, a discussion on measurement uncertainty, advantages, limitations, and comparability of commercially available reference materials between protein and DNA-based methods is presented.

7.2 SOURCES AND AVAILABILITY OF PROTEIN REFERENCE MATERIALS

7.2.1 Relevance of Reference Materials for Immunoassays in Agricultural Biotechnology

Protein reference materials are commonly used for routine calibration checks of an apparatus, assessment of a measurement method, and assignment of values to materials (Schimmel et al., 2002; Mihaliak and Berberich, 1995). Thus, development of appropriate and suitable reference materials is a critical component for standardizing measurements among laboratories that use protein-based methods to detect and quantify biotechnology-derived traits in grains. ISO Guide 30:1992 describes the rigorous characterization, general requirements, terms, and definitions used for internationally recognized reference materials and for testing GM traits (ISO, 2004). Reference materials provide the basis for interpreting test results as they relate to the amount of analyte present in a sample and provide a basis for comparability with other laboratories performing similar analyses (JRC, 2008). The timely development of new reference materials and standards is a critically important aspect of ensuring accuracy and reproducibility in analytical measurements as it relates to agricultural biotechnology. Availability of a reference material from which a pure protein extract can be obtained is necessary to perform and evaluate protein-based methods for detecting biotechnology-derived traits (Kleiner and Neumann, 1999).

7.2.2 Types of Reference Materials and Considerations

Commercial reference materials and well-established validated methods have not always been readily available to end users or laboratories that test for GE traits in grains and oilseeds. Special considerations and specific criteria apply to the production and quality evaluation of reference materials, including the inherent variability of biological systems, in addition to variability attributable to biological and immunological assays. Depending upon its application and especially during early developmental stages and safety assessment phases of GM plant products, a reference material may contain varying metrological quantities of analyte. In these instances, an assessment of accuracy for a particular method is generally compared to known values, derived from the best available source of reference material. Thus, relevance of a reference material depends on its use and application, availability, and suitability among other considerations, but must always conform to preset conditions and function in a manner conducive to its specific purpose (Gancberg et al., 2004). Different types of protein-based reference materials and standards with their specific applications can be classified into the following categories (NIST, 2002):

> *Standard Reference Materials*: This material consists of the same matrix as the actual biological material to be tested. If the matrix to be tested is soybean seed, for example, the standardized reference material would contain isogenic seed containing a known proportion of biotechnology-derived seed. Access to standardized reference materials is important during the development, validation, and use of protein-based assays for analysis of transgenic proteins in agricultural commodities. Most analytical laboratories rely on the trait manufacturer to supply this type of reference material.

> *Certified Reference Materials*: An accompanying certificate containing property values that have been certified by a procedure (i.e., accuracy and uncertainty) is provided with this type of reference material. Certified reference material establishes traceability to an accurate realization of the unit in which the property values are expressed and for which each certified value is accompanied by an uncertainty at a stated level of confidence.

Reference Material: Contains one or more property values that are sufficiently homogeneous and well established and are generally used for the calibration of an instrument, the assessment of a measurement method, or the assignment of values to materials. The material consists of both positive and negative controls. A positive control sample will contain the protein of interest (Cry1Ab, Cry9C, Cry1F, etc.), while the negative control sample does not contain the protein of interest, at least not at any detectable level. Plant reference materials generally fall into this category since the expression levels of protein are known to be variable.

National Institute of Standards and Technology (NIST) Reference Material: This type of material, issued by NIST, contains a report of investigation instead of a certificate for the purpose of (1) furthering scientific or technical research, (2) determining the efficacy of a prototype reference material, (3) providing a homogeneous and stable material so that investigators in different laboratories can be ensured that they are analyzing the same material,(4) ensuring availability when a material produced and certified by an organization other than NIST is defined to be in the public interest, or when an alternate means of national distribution does not exist. A NIST reference material usually meets the ISO definition for a reference material and may meet the ISO definition for a certified reference material (depending on the organization that produced it).

Internal Standard: An internal standard is a compound that matches closely, but not completely, to the protein or analyte of interest in the sample. Ideally, the effects of sample preparation, relative to the amount of each species, should be the same for the signal from the internal standard as for the signal from the sample of interest. An internal standard is commonly used to generate a calibration curve, for example, in an ELISA assay, and to compare the analyte signal intensity of the test sample to the signal intensity of the internal standard. Internal standards are generally used during the method development stages of a process, they are not commercially available, and they do not contain property values such as accuracy and uncertainty.

Primary Standard: A primary standard is widely acknowledged or designated as having the highest metrological qualities and whose value is accepted without reference to other standards of the same quantity. Metrology attempts to validate data obtained from testing technologies. Though metrology is the science of measurement, in practical applications, it is the enforcement and validation of predefined standards for precision, accuracy, traceability, and reliability. The process utilizes a traceability approach to assignment of measurement units, unit systems, development of new measurement methods, and realization of measurement standards. Generally, a primary standard needs to be highly pure and stable and should contain low hygroscopic characteristics. Figure 7.1 outlines the general trends in measurement uncertainty associated with the various protein reference materials.

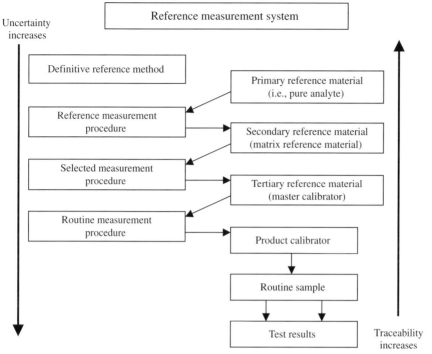

Figure 7.1 Trends in measurement uncertainty associated with the various protein reference materials. (*Source*: Trapmann et al., 2002; Gancberg et al., 2008.)

7.2.3   Forms of Protein-Based Reference Materials and Testing Strategies

The physical form of the reference material determines its suitability for use with any given method (Kleiner and Neumann, 1999). Ground grain, seed, or leaf tissue is often used as reference material when using protein-based methods to detect biotechnology-derived traits. Differences in the particle size distribution between reference materials compared to a routine sample can affect extraction efficiency of the target protein, and thus the method reproducibility due to sampling error (Ahmed, 2004; Stave, 2002). More important, as discussed in detail in Section 7.6, grains may have different levels of target protein compared to either seed or homozygous plants (AGBIOS, 2008). Also, different plant-based reference materials, grown in different regions, might not contain uniform expression levels of target protein. Thus, grain or seeds are sometimes tested using "seed pool" testing strategies (Remund et al., 2001; Laffont et al., 2005). This approach screens for the presence or absence of a particular biotechnology-derived trait on multiple samples (pools), providing a semiquantitative result based on probabilities and applicable Bayesian statistical calculations. Consequently, it is difficult to translate quantitative results obtained with pure protein reference materials into wt% concentration units unless it is performed in a sampling format and to apply Bayesian statistics. A test sample that generates a negative result when using this approach does not necessarily mean that the lot is entirely negative for the trait or protein in question, and statistical calculations should address the uncertainty in the measurement.

7.2.4 Plant-Derived Versus Bacteria-Derived Reference Proteins

Both plant-derived and bacteria-derived reference proteins have applications in agricultural biotechnology (EC 1138/38, 2000). Because recombinant proteins have a single, well-defined sequence and composition, they make ideal candidates for use as references when they can be demonstrated to express and fold in the same manner as the plant proteins from which they were derived. Posttranslational modifications, including proteolytic, phosphorylation, and glycosylation modifications must be known and completely defined, as well as the primary amino acid sequence. In agricultural biotechnology, reference materials using protein detection methods can be derived from recombinant microbes such as *E. coli*, a ground plant matrix (typically leaf or grain), or a processed food fraction. Plant-derived reference materials have limited use for spike and recovery experiments because the level of expression can be variable. The main use of plant-derived reference materials is as positive and negative controls. Expressing proteins in *E. coli* is the easiest, quickest, and cheapest method and is used in ELISA assays for development of standard curves, matrix validation in spike and recovery experiments, and determination of LOQ values (Burns et al., 2006). There are many commercial and noncommercial expression vectors available with different N- and C-terminal tags, as well as many different strains that are optimized for special applications (Uhlen and Moks, 1991). With differences in posttranslational processing of eukaryotic and prokaryotic organisms, the relative immunoreactivity of recombinant proteins derived from *E. coli* versus plant-expressed proteins must be well characterized (Uhlen and Moks, 1991; EC 1138/98, 2000). The physical form of the reference material determines its suitability for use with any given method. In bacteria, proteins sometimes refold into undesirable conformations, bind with other proteins, or degrade more readily due to the fact that posttranslational modifications such as glycosylation and phosphorylation do not occur (Kohno et al., 1991). This result is not to be taken as evidence against the view that the primary structure of a protein determines its conformation, rather it indicates that the protein will fold to its correct conformation under the appropriate conditions and that these conditions are not met in certain types of bacteria (Kohno et al., 1991). Strategies have now been developed to induce recombinant proteins to assume their active conformations (Kohno et al., 1991). The consequence of all these protein posttranslational modifications related to reference material is that many plant proteins cannot be readily represented by one known molecular formula and cannot be assigned a single molecular mass. Nevertheless, prokaryotic systems generally express higher amounts of recombinant protein and are more cost-effective compared to their plant counterpart. For any antibody-based system, designed to bind a unique protein using a specified assay format, a protein folded into a particular conformation may not bind with a high degree of specificity because the epitope may be hidden and therefore not available for antibody recognition. Western blotting is a common method to compare the bacteria-derived protein with the plant-derived protein (Ahmed, 2002). Generally, Western blotting will reveal whether the two proteins are identical in molecular mass (Sambrook and Russell, 2001). The bacteria-derived protein can then be used to evaluate enzyme kinetics and generate standards to determine protein expression in the transgenic plant cultivar. Commercially available ELISA kits for Cry1Ab, Cry1Ac, Cry1F, Cry9c, PAT, and CP4 EPSPS proteins are all based on antibodies generated from microbial-derived reference proteins (Envirologix, 2009).

Proteins expressed in genetically engineered *E. coli* bacteria are often used as standards in ELISA-based assay systems for development of standard curves, for determination of a limit of quantification, and for matrix validation in spike and recovery experiments (Grothaus et al., 2006). In addition, the extraction efficiency from plant matrices highly depends on particle size and extraction conditions. During assay development, protein reference material can be used to help select assay parameters that would minimize interference of the matrix (e.g., nonspecific binding of sample components to antibodies). During validation and use of the assay, reference materials would be extracted and analyzed in parallel with test samples, so that results could be directly compared. Studies need to be conducted on the best procedures for maintaining the integrity and stability of matrix-based, bacteria- and plant-produced protein reference materials. Table 7.4 details the advantages and disadvantages of different types of protein standards (Ahmed, 2004).

7.2.5 Preparing Certified Reference Materials

Certified reference materials are produced, certified, and used in accordance with relevant ISO guidelines (Trapmann et al., 2002; IRMM, 2007). As depicted in Figure 7.2, production of a matrix-based, certified reference material requires the following steps for production: (1) characterization of the base material, (2) decontamination of the coat surface, (3) grinding to a fine particle size, (4) blending of different weight portions, (5) bottling under argon to retain stability, (6) labeling, and (7) quality control checks of the final product (Trapmann et al., 2002). During production and developmental stages of generating certified reference materials for commercial applications, it is relevant to demonstrate compliance with existing internationally recognized standards and guidelines for their preparation. Studies reveal that when it can be demonstrated that a reference material preparation conforms to the highest of standards, precision of analytical measurements improve significantly (Trapmann et al., 2002). However, even the best characterized certified reference materials for detection of GM traits in agricultural biotechnology products must be considered unique in the sense that no direct method exists that can directly measure weight percentage in a test sample. While both DNA- and protein-based methods are commonly used to quantify and detect the presence of GM traits in grains, seeds, and plants, both have shortcomings. Reference materials, for example, may differ from the test sample based on the type of assay being performed. Purified DNA, extracted from plant tissue, may be considered a reference material for PCR-based methods since quantitative PCR in essence counts copies of DNA molecules and compares threshold cycles of test samples to a reference. Conversely, for protein-based methods, it is essential that the reference material remain in the plant matrix to

TABLE 7.4 Considerations for Various Protein Reference Materials

Protein reference material	Advantages	Disadvantages
Bacteria-derived Standards	Easily produced in large quantities	Proteins can become misfolded or degrade quickly
	Short production time	May have problems with binding specificity (antibody-based systems)
	Inexpensive	Need for specialized production facilities/ expertise in molecular biology techniques
	Unlimited supply/availability	Posttranslational processing may result in differences in immunoreactivity
	Broad dynamic concentration range	Indistinguishable among different grain species as well as between approved and unapproved events (e.g., Cry1Ab integrity/instability issues)
	Widely commercialized	
Plant-derived standards (matrix)	Useful in seed pool testing	Limited use for specific applications
	Provides a "true" positive control	Variable extraction efficiencies due to the different particle sizes
	Commutable	Limited availability
	Suitable for both DNA- and protein-based testing	Posttranslational processing may result in differences in immunoreactivity
		Indistinguishable among different grain species as well as between approved and unapproved events (e.g., Cry1Ab)

Source: Ahmed, 2004.

ensure that the proteins remain intact. Certified reference material prepared for detection of Roundup Ready® soybean, for example, has proven to be an invaluable tool using protein detection methods to measure the EPSPS protein derived from *Agrobacterium tumefaciens* strain CP4 (Lipp et al., 2000; Stave, 2002). Studies reveal that matrix-based Roundup Ready soybean reference material and an ELISA method can be used reliably to determine a specified threshold concentration of Roundup Ready soybean in processed food fractions. The study shows that an ELISA method consistently detected Roundup Ready soybeans at 0.3% and was appropriate for quantitative detection of modified protein in processed food fractions (Stave, 2002; Schimmel et al., 2002; Rogan et al., 1999).

When preparing matrix-based reference materials, designed for use with ELISA and LFD methods, because of the complex nature of protein molecules, structural data and degradation rates need to be evaluated. A reference standard needs to behave in a manner that will be comparable with routine test samples (Trapmann et al., 2002). Characterization of the base material and knowledge of its genetic composition are especially important for production of GM matrix-based reference

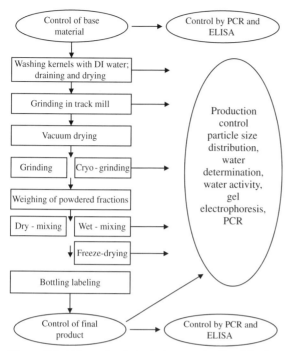

Figure 7.2   Schematic of a typical protocol utilized for the production of matrix-based certified reference material. (*Source*: Trapmann et al., 2002.)

materials. GM and non-GM base materials must be identity preserved, well characterized, derived ideally from isogenic cultivars, and analyzed for purity. The content of homozygous and heterozygous kernels in the base material should be known since this could have an impact on the metrological quantity contained in the internationally recognized reference (Trapmann et al., 2002; Trapmann et al., 2000). In most cases, proteins have to be characterized to ensure "fit for purpose" applications. The relevant parameters have to be defined case by case because of the large variability associated with protein expression and the rate of degradation known to exist in matrices such as processed foods. Homogeneity of the reference material and particle size can directly impact analytical results. For matrix-based reference materials using protein methods, laboratories reported 14–24% higher Mon810 content when using reference material that had an average particle size of 35 µm compared to other laboratories that used an identical source of Mon810 reference material but with an average particle size of 150 µm (Ahmed, 2004; Stave et al., 2000). The particle size distribution of the GM reference flour needs to be consistent, around 40 µm on average, since larger particles increase measurement uncertainty (Trapmann et al., 2000; Holden et al., 2003). During the production of reference material, precautions must be taken to avoid contaminants that could render the reference noncompliant with standards and, thus, the preparation unacceptable for use. It is important, then, that non-GM and GM base materials be prepared and treated in a controlled environment, under laminar airflow, following appropriate quality control guidelines. Knowledge of the particle size distribution, percent water content, and water activity is necessary to evaluate homogeneity and stability of the reference preparation and to demonstrate compliance with specific production control criteria (Trapmann et al., 2002). The status of the protein analyte during production of the reference material and in the final product is usually evaluated by Western blotting and other protein-based detection methods. Even though compliance with stringent quality control guidelines is followed during the preparation of the matrix-based reference material, restrictions and limitations for their application do exist and must not be used for anything other than their "fit for purpose" applications. By controlling parameters such as homogeneity, stability, and traceability, the validity of the reference material preparation can be ensured to function for its intended purpose.

7.3 ROLE OF REFERENCE MATERIALS AND IMMUNOASSAYS FOR TESTING GENETICALLY ENGINEERED TRAITS

All analytical methods have appropriate applications and a number of factors can influence the value of immunoassays for analysis of GM traits in seeds and plants. Immunoassays are used to select specific traits that have optimum levels of protein expression prior to commercialization. Protein expression data are required to gain regulatory approval for new GM crops and quantitative ELISAs are extensively used in this capacity. ELISA test kits contain both positive and negative calibrators supplied by the licensor to the kit manufacturer at the onset of the kit design. Protein-based immunoassays have a wide range of applications as it pertains to agricultural biotechnology (Grothaus et al., 2006). In some instances, qualitative testing is appropriate, while other times quantitative testing is necessary (ISO, 2004). GM grains and processed foodstuffs must demonstrate compliance to regulatory mandates. The application of immunochemistry, using highly reliable reference materials, is a natural extension for a technology that enables life science companies to bring GM traits into the marketplace, provide confidence in analytical measurements, and fulfill contractual agreements between buyers and sellers in grain markets.

7.3.1 Qualitative Analysis

Qualitative analysis as it pertains to the agricultural biotechnology industry is an analytical technique that seeks to identify the presence or absence of novel proteins, expressed in plants, resulting from the advent of biotechnology. Some international standards are exclusively designed for the purpose of qualitative analysis. Asynchronous regulations exist worldwide whereby some GM varieties that have been approved in one marketplace may not have approval in a different one. In the instance where any level of an unapproved GM variety is not allowed into a market (as is the case with StarLink™ maize), appropriate sampling and qualitative testing is sufficient and appropriate to assess the presence or absence of the trait in a lot (Bertheau and Davison, 2008). It is essential at the outset of any approved method of a candidate biological reference standard to state clearly whether the intended use of the material is only for qualitative purposes, since this can have an impact on the applicability of the test. A reference standard is also important to enable an appropriate determination of a LOD (Green, 1996). The LOD provides useful information regarding the sample size to be tested and the confidence that a qualitative analysis delivers (Gancberg et al., 2004). Rapid test protein immunoassays, such as LFDs, have become well established in the biotechnology industry as a relatively simple method for qualitative detection of transgenic traits in plant crops. LFDs allow a rapid, field-based, non-sophisticated laboratory environment to perform qualitative-type testing, whereas plate format immunoassays require more advanced equipment and materials that can be used in both qualitative and quantitative applications.

7.3.2 Quantitative Analysis

Quantitative analysis is the determination of the absolute or relative abundance (often expressed as a concentration) of one, several, or all particular substances present in a sample. The quantitative nature of immunoassays has been well established (Envirologix, 2009). A reference standard for quantitative procedures would require a higher level of data than one for qualitative procedures. Through a series of quantitative procedures inherent in a particular method, specifications of reference standards are generated from available data according to a specified set of guidelines (Green, 1996). A certificate of analysis containing the data is generated detailing all the necessary testing procedures and documentation for a specific reference standard (Pauwels et al., 1998). Protein reference standards must be evaluated for accuracy before they can be used to calibrate a quantitative method and establish equivalence between other types of reference materials. Well-characterized reference standards are critical in product development and method validation, especially as they pertain to quantitative analysis (Gancberg et al., 2004; Codex, 2001; Kupier, 1999). Statistically relevant numbers, generated in an unbiased fashion, and reliable interlaboratory measurements are used to evaluate the performance of reference materials, providing confidence in the quantitative analysis of GM traits in grains and oilseeds.

7.3.3 Antibody Interactions with Plant Proteins

Molecular interactions of antibodies with target proteins in plant tissue are important considerations when developing protein-based reference materials (Stave, 2002; Lipton et al., 2000). An understanding of how much sequence homology is shared by the reference protein and other proteins that might be present in the plant could impact the performance of the assay. The GM trait GA21, for example, contains a modified EPSPS protein derived from *A. tumefaciens* and has a subtle difference of only two amino acids out of the total 445 compared to the native EPSPS protein sequence (AGBIOS, 2008; Thomas et al., 2004). Thus, antibody-based methods provide cross-reactivity between modified EPSPS and native EPSPS proteins, and this type of technology will not discern between the two types. The more knowledge that is available before development begins, the more one is forewarned of possible problems in development of a detection method. In looking at problems associated with the unique nature of the reference protein target, there can be insurmountable difficulties if an identical protein is present elsewhere in the matrix, as is the case with the GA21 trait. However, if the genetically engineered protein contains significantly less than 100% sequence homology to the wild type, then difficulties can be overcome by use of antibodies against unique characteristic peptide sequences.

7.3.4 Traceability of Reference Materials, Standardized Methods, and Analytical Measurements

Traceability to an internationally recognized reference material is necessary to evaluate repeatability data generated by testing laboratories. In addition, analytical measurements must be made with a consistent set of SI units that have an internationally recognized equivalence. Traceability is defined in ISO standards as the "property of the result of a measurement of the value of a standard whereby it can be related to stated references, usually national or international standards, through an unbroken chain of comparisons all having stated uncertainties." Because of the complexity of testing agricultural biotechnology products, there has been a greater emphasis on the precision of results obtained using a specified method, rather than with traceability to a defined standard or SI unit. This has led to the use of "official methods" to fulfill legislative and trading requirements (EC, 2000; ISO, 1999; ISO, 2001). Ideally, analytical measurements should be traceable to a well-defined and internationally recognized reference standard. Internal quality control procedures, well-designed interlaboratory collaborative studies, and accreditation can aid in establishing evidence of traceability to a given standard.

Method specificity and the possibility of generating either false-negative or false-positive results are always relevant concerns to an analyst when applying any analytical method. Both positive and negative reference materials provide a source of control samples, enabling analysts to minimize erroneous results, evaluate possible sources for errors, and establish confidence in analytical measurements. To determine the concentration of an analyte in a sample, the reference material must correlate to known concentrations of the antigen used to produce a linear dose–response curve. The standard curve and the assay response from the samples are used to determine the level of analyte in the test sample (Ahmed, 2002; Lipton et al., 2000). The reference material used to make the standard curve should yield a response that correlates to a metrological concentration of analyte in the sample type and assay conditions specified by the test procedure. It is essential that the standards are of known consistency and account for the effect of sample matrix and sample preparation procedures on antigen reactivity.

When routine methods are calibrated against a single source reference preparation, and compared with results obtained using a specified reference method, bias inherent in the routine method can be identified and an appropriate correction factor applied. The relationship between laboratories that perform different routine methods with a metrological reference preparation compared to a specified reference method forms the basis for relating results of different routine methods. From these comparisons (between a reference method and a routine method), comparability between routinely used methods can be established and correct diagnostic conclusions can be drawn, independent of the test method or the supplier. The value of a reference system consisting of reference materials and reference methods was initially recognized in chemistry of simpler substances and not of substances with the complexity of biological macromolecules such as proteins and DNA. The advances in the development of ELISA and immunochemistry technologies have made it practical to apply this approach to proteins and biological systems as complex as those found in biotechnology-derived traits in grains and oilseeds (Stave et al., 2000).

7.4   QUALITY AND PERFORMANCE CHARACTERISTICS OF PROTEIN-BASED REFERENCE MATERIALS

7.4.1   Quality Characteristics of Protein-Based Reference Materials

For testing of agricultural biotechnology products, as with other diagnostic applications, providing reference materials that demonstrably meet quality assurance criteria is critical to their successful testing applications worldwide. Certified reference materials must be produced under a quality system based on ISO 17025 and ISO Guide 34 principles (ISO, 2001). Participation in comparison studies, when it is possible, will validate measurement capabilities of the certified reference material and show that they conform to acceptable comparability criteria (Bunk, 2007). To this end, there is an international measurement infrastructure that works to promote a worldwide metrology system, supported by international standardization and harmonization activities, by national metrological infrastructure, certification, and accreditation bodies, and by appropriate quality management at the laboratory level, including quality assurance and quality control (Bunk, 2007; Eurachem/CITAC, 2000). In the agricultural biotechnology industry, comparability is the main quality component of measurement results. This term describes the property of measurement results that enables them to be compared independent of the time, place, analyst, and procedure used. It involves the assurance of metrological traceability. Basic practices used by laboratories to assure and demonstrate the quality of their measured results include, but are not limited to, the following: (1) use of standardized methods for sampling and analysis, (2) proper calibration of measuring instruments, (3) routine quality control practices, and (4) participation in interlaboratory comparisons and proficiency tests.

7.4.2   Quality Standards for Reference Materials

According to commonly accepted regulations, the production of reference materials should follow decisive metrological principles and should be traceable to the standard international system when applicable (Gancberg et al., 2008). An arbitrary definition of measurement units could lead to difficulties with noncertified standards and a lack of reproducibility over the long term. Protein reference materials are critical for the validation of externally operated immunochemistry processes and can be obtained from different production sources (Lipton et al., 2000). Some important considerations for reference materials include their stability, purity, and integrity (intact physical structure). A certificate of analysis will generally accompany each reference material (Gancberg et al., 2008; Lipton et al., 2000). The certificate will describe the characteristics of the material, both to the presence of the target material and to the absence of other possibly interfering materials. In addition, a reference material may even be restricted to the method to be used for the analysis of a particular analyte. A certificate of analysis for a GM reference material should address the following factors:

Target Analyte: One must consider whether the reference material is specific for a particular trait and can distinguish one from another. In the case of the Cry1Ab protein, several traits express this protein in the GM plant, albeit at different levels (AGBIOS, 2008). Thus, a reference material of this type would only be useful to screen for the presence or absence of the protein itself and could not discern specifically between different GM varieties that could be present in a test sample.

Adventitious Presence: An important factor that will influence the true concentration of the samples prepared for validation is the level of impurity in the reference material. The starting material used for the preparation of the reference material needs to be characterized for purity with respect to the desired analyte. To do so, a representative subsample must be analyzed for the absence or presence of the analyte in both negative and positive pools. The sample size and the measurement error will determine the confidence level of the final analytical result.

Purity Level: As the highest grade reference material available, the purity of primary standards must be well characterized and documented (Lipton et al., 2000). Most protein reference materials are certified using the most unbiased and precise measurement methods available, often with more than one laboratory being involved in making certification measurements. Certified reference materials are generally used on a national or international level, and they are at the top of the metrological hierarchy of reference materials.

Stability: Proteins naturally degrade or aggregate if they are not properly stored (Stryer, 1997). The stability of the reference material needs to be evaluated under both long-term and short-term storage and test conditions. The short-term stability evaluation of protein reference materials aims to determine whether the value assigned to the target analyte changes in time under the conditions of shipment to the end user. This is accomplished by subjecting the protein reference material to temperature extremes that are expected in the duration of a shipment and then measuring the concentration of target analyte relative to a control sample (maintained at short-term storage conditions). Long-term stability evaluation aims to determine whether the certified values of the analyte remain valid during the shelf life of the certified reference material. There are a number of specific approaches to stability testing, the choice of which often depends on the anticipated stability of the target analyte. It is the experience that influences decisions about the frequency of sampling for long-term stability assessment of protein reference materials used in agricultural biotechnology (WHO, 2005). It must also be demonstrated that protein reference materials remain stable under the conditions of storage and use specified by the method. When LFDs, for example, are exposed to high temperature and humidity, false-positive and/or false-negative results may occur due to the loss of antibody activity. The availability of appropriate and effective reference material can be used to evaluate the performance of a particular method when anomalous or suspect results occur.

7.4.3 Performance Characteristics of Protein Reference Materials in Method Development

When developing an immunoassay for a specific protein antigen, relatively purified preparations of that protein are required for immunizing animals and producing antibodies (Envirologix, 2009). Most novel proteins that have been introduced into transgenic crops are proprietary to the company developing the trait and generally not publicly available. The reference material is used to help select assay parameters that minimize interfering effects of the matrix (e.g., nonspecific binding of sample components to antibodies). Reference materials should be extracted and analyzed in concert with test samples so that the results can be compared directly. Performance-based characteristics include sensitivity, specificity, precision, accuracy, stability, reliability, and fitness for purpose. Kit manufacturers rely on a well-defined reference standard to develop and validate immunoassays that can meet these performance-based characteristics (Envirologix, 2009).

Access to standardized reference materials is critical during the development, validation, and application of immunoassays for the detection of novel proteins in transgenic crops. For ground materials, differences in particle size distribution between reference materials and routine samples may affect extraction efficiency of the GM protein and method reproducibility due to sampling error (Grothaus et al., 2006; Holden et al., 2003; Lipton et al., 2000). To enable objective testing, companies developing biotechnology-derived traits in grains and oilseeds need to provide validated testing methods in addition to providing and commercializing standardized reference materials (Stave, 1999). Uniformity of the reference material is an important performance characteristic that must be thoroughly evaluated during the method development (Trapmann et al., 2002). Uniformity considerations are related to the ability to generate a homogeneous sample. Some tissues may not be amenable to blending. For example, if the plant reference material is a seed that contains high oil content, it could possibly be difficult to combine biotechnology-derived material with trait-free, negative material in such a way as to achieve a homogeneous reference sample.

7.4.4 Value Assignment and Measurement Uncertainty

Several confounders will contribute to uncertainty of measurement and a degree of uncertainty exists with every measured quantity associated with any testing method

(Gancberg et al., 2008; Eurachem/CITAC, 2000). The amount of uncertainty associated with the measured quantity is characterized by a dispersion of values that could reasonably be attributed to the analytical measurement (Remund et al., 2001; Snedecor and Cochran, 1967; Zel et al., 2007). The degree of uncertainty may be, for example, a standard deviation (or a given multiple of it) or the width of a confidence interval (Snedecor and Cochran, 1967). Uncertainty of measurement comprises, in general, many confounders. Some of these confounders may be evaluated from the statistical distribution of the results of a series of measurements and can be characterized by a standard deviation (EC 1138/98, 2000; Eurachem/CITAC, 2000). The other components, which can also be characterized by standard deviations, are evaluated from assumed probability distributions based on experience or other information (Layton and Spiegelhalter, 2008). Where international biological reference standards are to be assigned a value in arbitrary international units (IU), an uncertainty value is generally not given. Measurements on certified reference materials in agricultural biotechnology are normally carried out as part of method validation or revalidation, effectively constituting a calibration of the whole measurement procedure against a traceable reference (Lipton et al., 2000). Because this procedure provides information on the combined effect of many of the potential sources of uncertainty, it provides very good data for the assessment of uncertainty. Information on the variability observed during the course of a collaborative study, to characterize the preparation, is always documented in a collaborative study report and should be made available to end users. In a multimethod collaborative study, differences in fortification estimates of the material using different methods may be necessary. In the absence of a reference method, assumptions about an underlying "true value," or a probability distribution across methods, may not be valid. Summarizing all the components of variability observed in a collaborative study by quoting a single uncertainty value may not be useful information.

When the reference material is only approximately representative of the test materials, additional factors should be considered, including (as appropriate) differences in composition and homogeneity; for example, reference materials are frequently more homogeneous than test samples (Eurachem/CITAC, 2000). Codex Alimentarius Commission has developed guidelines on estimating measurement uncertainty for applications in agricultural biotechnology (Codex Guidelines on Measurement Uncertainty and the draft Guidance Document on "The Use of Analytical Results"). These guidelines require laboratories to estimate the uncertainty of their quantitative measurements. This is particularly important and has consequences for measurements in the sector dealing with foods derived from biotechnology where analytical controls may not be as effective as found in other areas of analysis in the food sector. It is frequently not appreciated that the magnitude of the measurement uncertainty is considerably greater in this analytical sector than would normally be expected.

In the case of protein-based methods, uncertainty may also arise from biological factors. For example, a sample may contain hemizygous grain in which the protein expression levels will vary depending on whether the target sequence was introduced via the male or female parent. In addition, different traits may be introduced into a single plant variety, also known as "stacked traits" (Layton and Spiegelhalter, 2008).

Upon grinding, a sample that contains a single kernel with multiple traits becomes indistinguishable from a sample that contains multiple kernels of only single traits. As the extent of this is unknown, the effect of this cannot be accurately determined, but must be included in the uncertainty of any measurement. In the case of protein-based methods, certainly protein expression level and/or extraction efficiency of proteins vary. In these instances, both DNA- and protein-based methods may be used via a subsampling approach, or on single seeds, where the potential impact of any such biological variation is minimized.

7.4.5   Third-Party Method Verification

Once a standardized method has been developed by a manufacturer, third-party verification needs to be performed to demonstrate compliance of a standard operating procedure, to verify that reference materials are fit for purpose, and to show that the method provides acceptable and appropriate predefined analytical results. The United States Department of Agriculture (USDA) is responsible for grading and assuring the quality of commodity crops and plays a key role in the international marketing of U.S. agricultural products. Grain Inspection, Packers and Stockyards Administration (GIPSA) developed a set of comprehensive guidelines for grain sampling to test for the presence of biotechnology-derived products. Also, GIPSA has been instrumental in initiating collaborative efforts with NIST and test kit manufacturers to assess and validate methods used for detecting the presence of protein products in GM plants and foodstuffs. The USDA–GIPSA program verifies performance statements by manu-facturers of rapid test kits that detect GM events present in grains and oilseeds (USDA, 1995). The manufacturer submits a data package, including a protocol, supporting its claims for a thorough review of the data by a third party. The source reference material is provided by the developer of the GM product and confidentiality of the transferred material is maintained. GIPSA performs an in-house verification procedure and if the declarations are substantiated, GIPSA issues a certificate of performance (COP) to the manufacturer for a period of 3 years. Due to the scale of agricultural production in the United States, typically protein immunoassay methods are most practical for these applications because of their ability to quickly evaluate large numbers of samples. As summarized in Table 7.5, GIPSA has issued certificates of performance on immuno-based assays for StarLink® (Cry9c), Roundup Ready (CP4 EPSPS), Herculex® RW (Cry34Ab/Cry35Ab1), MIR604 (mCry3A), Herculex (Cry1F), and Mon863 (Cry3Bb1) that are found in corn and soybean GM grain products, among others (USDA, 2004).

7.5   HARMONIZING PROTEIN WITH DNA CERTIFIED REFERENCE MATERIALS

Certified Reference Materials can be purchased from the Institute of Reference Materials and Measurements (IRMM) or the American Oil Chemists Society (AOCS) for all U.S. commercialized biotechnology-derived traits currently on the market-place (IRMM, 2007; AOCS, 2009). These certified reference materials have been

TABLE 7.5 GIPSA Verified Rapid Test Kits for the Analysis of Biotechnology Submitted by the Manufacturer for Verification

Company	Test kit	Part number	Event/protein analyte	Test sensitivity	Test format
Agdia, Inc.	Roundup Ready (CP4 EPSPS) ImmunoStrip™ Test	STX 74000	Roundup Ready protein CP4 EPSPS	1 in 1000 corn kernels	Lateral flow strip
Agdia, Inc.	Roundup Ready (CP4 EPSPS) ImmunoStrip™ Test	STX 74000	Roundup Ready protein CP4 EPSPS	1 in 1000 soybeans	Lateral flow strip
Agdia, Inc.	Bt-Cry34Ab1 ImmunoStrip™ Test	STX 04500	Bt-Cry34Ab1/ Herculex RW	1 in 800 corn kernels	Lateral flow strip
Agdia, Inc.	mBt-Cry3A ImmunoStrip™ Test (MIR604)	STX 06711 or STX 06703	mBt-Cry3A (MIR604)	1 in 300 corn kernels	Lateral flow strip
EnviroLogix Inc.	QuickStix ™ Strips for Cry9C (StarLink)	AS 008 BG	StarLink/Cry9C	1 in 800 corn kernels	Lateral flow strip
EnviroLogix Inc.	QuickStix™ Strips for Cry1F (Herculex)	AS 016 BG	Cry1F/Herculex	1 in 200 corn kernels	Lateral flow strip
EnviroLogix Inc.	QuickStix™ Kit for LibertyLink® Rice	AS 013 RB	LibertyLink® (Event LL601RICE expressing the PAT protein)	1 rough rice kernel in 75 rough rice kernels	Lateral flow strip
EnviroLogix Inc.	QuickStix™ Kit for YieldGard® Rootworm Corn Bulk Grain	AS 015 BG	Cry3Bb/rootworm	1 in 200 corn kernels	Lateral flow strip
EnviroLogix Inc.	QuickStix™ Kit for mCry3A (Agrisure™ RW)	AS 037 BG	mCry3A/Agrisure™ RW	6 in 800 corn kernels	Lateral flow strip
EnviroLogix Inc.	QuickStix™ Kit for Cry34 (Herculex™ RW)	AS 037 BG	Cry34/Herculex RW	1 in 200 corn kernels	Lateral flow strip

EnviroLogix Inc.	QuickStix™ Kit for Roundup Ready Bulk Soybeans	AS 0103 BGB or AS065BGBR	Roundup Ready protein CP4 EPSPS	1 in 1000 soybeans	Lateral flow strip
Neogen Corporation	Re√eal® for CP4 (Roundup Ready)	8005	Roundup Ready protein CP4 EPSPS	1 in 800 corn kernels, 1 in 1000 soybeans	Lateral flow strip
Strategic Diagnostics, Inc.	Trait√™ Bt9 Corn Grain Test kit	7000012	StarLink/Cry9C	1 in 800 corn kernels	Lateral flow strip
Strategic Diagnostics, Inc.	Trait√™ LL Bulk Rice Test Kit	7000048	LibertyLink® (Event LL601RICE expressing the PAT protein)	1 rough rice kernel in 50 rough rice kernels	Lateral flow strip
Strategic Diagnostics, Inc.	Trait√™ RUR Bulk Soybeans 5-Minute Test Kit	7000014	Roundup Ready protein CP4 EPSPS	1 in 1000 soybeans	Lateral flow strip
Strategic Diagnostics, Inc.	Trait√™ Cry34Ab1 Test Kit	7000055	Cry34Ab1/Herculex RW	1 in 800 corn kernels	Lateral flow strip

Source: http://archive.gipsa.usda.gov/tech-servsup/metheqp/testkits.pdf.

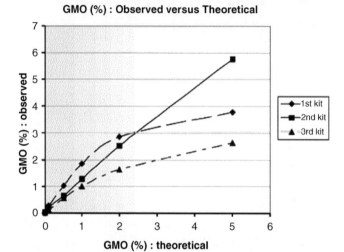

Figure 7.3 Relationship between observed and theoretical values for three kits expressed in % of the GMO content. (*Source*: Ermolli et al., 2006.)

developed and validated almost exclusively for nucleic acid-based testing, however. The correlation between DNA-based methods and immunological-based methods was evaluated in three commercial kits by measuring the amount of Cry1Ab/Cry1Ac in Mon810 IRMM reference materials (Ermolli et al., 2006). Three operators tested the ability of ELISA kits to quantify levels of Cry1Ab/Cry1Ac in IRMM certified reference materials over a 3-day period. As shown in Figure 7.3, the data revealed that different ELISA kits provided discordant quantitative measurements for the Cry1Ab/Cry1Ac protein and that Kit 3 was the most reliable (Ermolli et al., 2006). Only one of the three ELISA kits provided a linear relationship between the theoretical and measured values in the range of 0.0–5.0% Mon810. Provided that immunoassays are going to be used in conjunction with PCR, it is crucial to have protein reference materials that undergo formal validation and meet performance criteria prior to being used as an enforcement tool for regulatory mandates in the grain trade industry (Lipton et al., 2000).

There is no apparent relationship between the DNA content and the mass of a plant or protein expression levels. Identical plants, grown under different environmental conditions, in different stages of growth and maturation, will exhibit different DNA/protein/mass contents. Different parts of the plant differ in genome content (i.e., the endosperm is triploid, while the embryo is diploid) (Springer, 1992; Hirel et al., 2007). The haploid genome equivalent definition of DNA and how it will impact protein methods, particularly as stacked genes become more prevalent, remains challenging. A partial but incomplete solution to the problem of conversion of units of measurement is provided by the use of certified reference materials. The use of a certified reference material that not only provides concordant results between DNA and protein-based methods but is also cost-effective would be ideal. The availability of such a reference material would permit the use of a standard curve to which estimations of the test sample can be applied.

Protein and DNA detection methods can both be applied qualitatively or quantitatively and are essential in the analysis of biotechnology-derived traits in plants. DNA methods potentially offer more distinct targets, for example, the transgene itself, its promoter, and/or border region sequences, than do protein assays. Protein immunoassays are rapid and relatively inexpensive, but are generally not as sensitive as DNA assays. Both have applicability to seeds, meal, and flour, but the methods potentially become more limiting with further processing of these products. For example, lecithin and starch derived from maize generally contain DNA but not proteins, and DNA is only occasionally found in maize-derived oils and syrups (Yoshimura et al., 2005a, 2005b).

The protein content of food products is determined and generally reported on a per weight basis, for example, parts per million (ppm) and parts per billion (ppb), without reliance on an internal standard. The ability to quantify the protein is a function of extraction efficiency and the reproducibility of the analytical method. DNA content cannot be measured on a per weight basis and hence exclusively relies on an endogenous reference gene when estimating the amount of introduced DNA present in a sample (Ahmed, 2004). Such estimates are affected by the number of copies of the introduced gene present, the presence of stacked traits, differential degradation of the DNA, and the amplification efficiency of the qPCR reaction (Van Duijn et al., 2002). DNA extraction methods and sample contamination can impact the reliability of the results (Holden et al., 2003). A variety of studies suggest that there can be a reasonable qualitative agreement between the results of protein and DNA analyses of biotechnology-derived products; however, the quantitative results from such analyses can differ markedly. Measurements may be explicitly expressed as weight or kernels by relative percentage. However, none of the current detection methods (DNA or protein based) are able to measure this directly unless carried out as a multiple subsampling approach on grain. In the case of a DNA-based method used for quantification of foods derived from biotechnology, genome equivalents may typically be measured. Protein methods measure the amount of a specific protein present and may relate that to the desired quantity. Although there may be correlations between kernels/wt% and the amount of DNA or protein, respectively, the very nature of these relationships is influenced by a number of biological factors. The relationship between the two measures is important with respect to international trade and harmonization between the two methods.

7.6 APPLICATIONS AND LIMITATIONS OF PROTEIN-BASED REFERENCE MATERIALS

When evaluating protein-based methods for the detection of biotechnology-derived traits, factors that compromise antigenic determinants on an intact target protein, the potential for antibody cross-reactivity (with both related and unrelated proteins) and the influence of detergents and other agents on antibody–antigen interactions are important considerations. A reliable source of reference material enables developers and life science companies to identify limitations and implement appropriate applications for protein-based technologies that are "fit for purpose" in agricultural biotechnology production, diagnostics, and so on (Dong et al., 2008; Michelini et al., 2008). For example, plant cells express a plethora of proteins, generally at much higher levels compared to GM proteins, and detection of GM proteins using Kjeldahl or near-infrared methods (which cannot discern between GM and non-GM proteins in a sample) does not meet the criteria of "fit for purpose" if the goal is to know the expression levels of a GM trait in a plant, seed, and so on. On the other hand, LFDs- and ELISA-based technologies have proven to be advantageous, appropriate, and "fit for purpose" for screening, compliance, product development, expression level estimation, and many other agricultural applications (Stave et al., 2000).

7.6.1   Limitations Based on Protein Expression Levels

Expression levels of GM proteins in biotechnology-derived crops are reported to range from nanogram to microgram quantities per gram of starting material. Some GM cultivars, with different biological backgrounds, are conventionally inbred so as to maintain a single transformation event. In recent years, providers have crossbred different GM traits to generate hybrids containing multiple GM traits. A number of biological factors contribute to the levels of protein expression for a specified trait, including the number of copies contained in the seed genome (zygosity level), growing conditions, stress levels, environmental factors, and so on.

Detection and measurement of an antibody–antigen interaction may be rendered difficult especially when the transgenic protein is expressed at low levels. This is the case for the Bt-176 trait that expresses low levels of the Cry1Ab protein (Stave, 2002; AGBIOS, 2008; Walschus et al., 2002). Immunoassay methods cannot detect this trait in seed below 1% (Walschus et al., 2002). This example demonstrates the limits imposed by the sensitivity of the technology, but represents an exception rather than the rule with respect to protein expression in GM crops.

To avoid bias when interpreting analytical results, reference materials should not contain other structural analytes that compete with the conjugating or the capturing antibody. In some instances, however, different cultivars of biotechnology-derived crops naturally express different amounts of the same biotechnology-derived protein. In these instances, antibody-based reference materials provide unavoidable cross-reactivity. As shown in Figure 7.4, identical Cry1Ab proteins are expressed at different levels in Mon810, Bt-11, and Bt-176 cultivars of maize. While determination of an absolute quantity of Cry1Ab protein in a sample is feasible, an antibody-based method that can distinguish between the three distinctive cultivars in a ground sample is not. If the objective is to determine wt% content based on expression levels of Cry1Ab in a maize sample, for example, to demonstrate compliance with a regulatory mandate, a high degree of uncertainty would be unavoidable and reveals a limitation of antibody-based detection methods (Stave, 2002). Variable protein expression has been shown to exist within a single biotechnology-derived trait. Mon810 maize plants, for example, have been reported to express levels of Cry1Ab at 7.93–10.34 µg/g in leaves, 0.19–0.39 µg/g in seeds, and 3.65–4.65 µg/g in whole plant (AGBIOS, 2008). Variability of Cry1Ab protein expression in Mon810 has also been reported when growing the cultivar in different

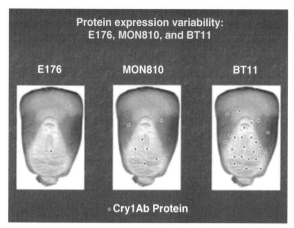

Figure 7.4 Cry1Ab protein expression levels differ in three distinct GM cultivars of BT176 (E176), MON810, and BT11.

regions of the world, and expression levels appear to decline over the duration of the growing season (AGBIOS, 2008).

7.6.2 Stacked Traits Versus Single Traits

Initial varieties of biotechnology-derived corn contained a single transformation of recombinant DNA sequences that conferred one new trait to the plant. In recent years, technology providers have been combining traits through crossbreeding to produce hybrids containing multiple traits (AGBIOS, 2008). An example of a Mon810 × Mon863 stacked trait is depicted in Figure 7.5. These "stacked events" or "stacked traits" pose new and different challenges for grain companies in meeting the regulatory and labeling requirements, or commercial specifications for non-GM corn shipments (Bertheau and Davison, 2008). Stacked traits are usually produced by the

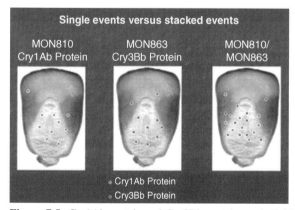

Figure 7.5 Cry1Ab protein and Cry3Bb protein are individually expressed in single traits MON810 and MON863 but are collectively expressed in a Mon810 × Mon863 stacked trait.

natural crossing of two single traits. Indeed, double, triple, and quadruple stacks, conferring a multiplicity of insecticidal properties along with herbicide tolerance, are becoming more prevalent in agricultural industry. Some sources estimate that 63% of corn grown in the United States is stacked with two or more transgenic events (Layton and Spiegelhalter, 2008). Stacked traits are considered novel GM varieties by many regulatory authorities and thus need to be distinguishable from single events. This poses a serious difficulty when developing protein reference materials for stacks because upon grinding a sample, stacked traits cannot be distinguished from a sample that contains a mixture of single traits in any cost-effective manner.

Many regulations or agreements define GM content as wt% so that, for example, one GM corn kernel mixed with 99 non-GM kernels is equivalent to 1% GM maize. If a particular kernel in a sample contained more than one trait, "stacked," then the amount of DNA measured by PCR would be expected to be equal to that of two single event kernels and an erroneous 2% measurement of GM content. The presence of stacked event kernels thus results in an overestimate of measured %GM content compared to the wt% GM corn. As stacked traits become more prevalent, this overestimation is expected to increase, resulting in more barges being rejected for export despite the actual level of GM content being within acceptable levels. Layton and Spiegelhalter 2008, as shown in Table 7.6, examined the current methods used to test and segregate grain along the grain supply chain and compared the results obtained for numerous corn varieties containing different stacked events. LFDs were used to represent testing at the initial screening point and PCR for final export testing. An optical reader was used to read the line intensity on the LFDs and provide an indication of the GM concentration level for each event detected in the grain sample. The optical reader indicated whether the grain would have been accepted or rejected at the receiving location based on different acceptance criteria. The PCR results were compared to the buyer's acceptance criteria. The results demonstrate the projected overestimate of measured %GM and suggest that adjustments need to be made either in regulations or in the screening criteria at the initial receiving location.

7.6.3 Processed Foods

One of the major drawbacks of immunochemical assays is that accuracy and precision can be adversely affected in a complex matrix, such as those found in many processed agricultural food and feed products (Friguet et al., 1984; Anklam et al., 2002). Causes for matrix effect interferences have been attributed to nonspecific interactions with the antibody on nontarget proteins, surfactants (saponins), phenolic compounds, antibody denaturing by fatty acids, and the presence of endogenous phosphatases or enzyme inhibitors (Stave, 2002). One advantage when testing agricultural biotechnology products with protein detection methods is that reference proteins are generally well characterized, available in amounts sufficient for antibody production and assay development, and can be used as an analytical standard. An understanding of how novel proteins respond to denaturing conditions, especially during production and processing stages, is necessary, especially when analytical determinations are going to be made during various stages of the production chain. Posttranslational

TABLE 7.6   An Envirologix QuickStix LFD Reader Used to Provide Semiquantitative Measurements for %GM Content of Single-, Dual-, and Triple-Stacked Events Compared to qPCR Data on Identical Samples

| | PCR results | | | Buyer acceptance | Strip screening | | |
wt%	Single	Dual	Triple	<5%	>2%	>1%	+/−
<0.1				Accept	Accept	Accept	Accept
0.5	0.5			Accept	Accept	Accept	Redirect
0.5	0.8			Accept	Accept	Accept	Redirect
0.5		1.4		Accept	Accept	Accept	Redirect
0.5		1.0		Accept	Redirect	Redirect	Redirect
0.5		1.0		Accept	Accept	Redirect	Redirect
0.5		1.1		Accept	Accept	Accept	Redirect
0.5			2.1	Accept	Accept	Accept	Redirect
0.5			1.6	Accept	Redirect	Redirect	Redirect
2.0	2.1			Accept	Accept	Redirect	Redirect
2.0	2.6			Accept	Redirect	Redirect	Redirect
2.0		4.8		Accept	Redirect	Redirect	Redirect
2.0		3.1		Accept	Redirect	Redirect	Redirect
2.0		4.0		Accept	Redirect	Redirect	Redirect
2.0		5.2		Reject	Redirect	Redirect	Redirect
2.0			6.1	Reject	Redirect	Redirect	Redirect
2.0			7.7	Reject	Redirect	Redirect	Redirect

Based on specified buyer's acceptance criteria, in some instances, qualitative test results would require redirecting a barge that would otherwise be accepted based on semiquantitative and qPCR test results. *Source*: Layton and Spiegelhalter, 2008.

modifications, thermal breakdown, chemical and enzymatic treatments all give rise to either conformational changes or peptide fragments of the original protein that an antibody used in diagnostics may or may not recognize (Stave, 1999). The subcommittee on Food Definition of the AOAC International Task Force on Methods for Nutrition Labeling proposed a "food matrix organizational system" (Wolf, 1993) to systematically judge the applicability of collaboratively studied methods over a range of food matrices, including processed foods. This system describes a food matrix by its location in one of the 3×3 sectors in a triangle, with each point of the triangle defined as representing 100% and the opposite side representing 0% of the normalized contents of each of the three major components of fat, protein, and carbohydrate. Foods falling within the same sector would be chemically similar and thus behave in a similar analytical manner. This same scheme can be used to select one or two food matrices representing each sector for development of a series of reference materials representing different processed foods. However, given the nature of antibody/protein relationships, no assumptions can be made regarding the applicability of a protein assay to a processed food. Protein methods are almost exclusively designed for minimally processed plant materials and reference materials are generally not recommended for testing against highly processed food products unless a particular method can demonstrate that it is indeed fit for purpose (Van Duijn et al., 2002). Degradation of proteins associated with thermal treatment or simply by poor antibody affinity of a commercially available source of antibodies limits the applicability and use of protein-based detection methods (Ahmed, 2004; Beier and Stankar, 1996). Currently, there are minimal data related to the application of immunoassays to complex food matrices, as found in processed food and feed products. Developing standards or reference materials for partially degraded/denatured proteins is under consideration by many research groups.

7.7 CHALLENGES AND FUTURE NEEDS

Since the fundamental goal of the agricultural biotechnology industry is to generate test results on identical samples that are comparable, regardless of which laboratory produces the result or whether it was a DNA- or protein-based test, developing validated methods in conjunction with reference materials that meet appropriate analytical specificities and have calibration traceability will be the focus of research projects over the next several years (Stave, 1999). As biotechnology advances, methods that provide accurate and reliable measurements for detecting the presence or absence of biotechnology-derived traits in plants will become increasingly more challenging. Experts from various disciplines, including life science organizations, test kit manufacturers, laboratory testing organizations, and governments worldwide, are working independently and collaboratively to identify the need and the best approach for development of reference materials to be used in testing of GM crops. The consensus from these organizations includes the following:

- Protein and DNA reference materials are needed to develop and validate reliable methods, to harmonize testing on a global basis, and to demonstrate compliance with regulatory mandates. The ideal reference material will satisfy both DNA and protein testing protocols. With the proliferation of GM crops in the marketplace, and limited resources, priorities must be established with respect to the development of reference materials for GM crop plants.

- Matrix-based reference materials have potential use for DNA- and protein-based testing. Alternative/nonmatrix protein and DNA reference materials, for example, bacteria-produced protein and/or plasmid DNA, may be useful as control standards, but matrix-based reference materials are necessary for the development of valid methods and globally consistent testing measurements.

- A globally recognized convention for reporting test results must be defined and accepted on a global basis. The grain industry and food processors recommend wt% as the appropriate unit for DNA-based measurements, while ppb is used for protein-based measurements. Reference materials must be prepared at levels consistent with regulations.

- Government organizations need to take an active role in ensuring the development of reference materials for GM crops. Organizations specifically identified include NIST, USDA–GIPSA, IRMM, AOCS, and the National Research Council (NRC).

- Development of reference materials will require cooperation among the life science organizations, governments, grain and food industry, test developers and manufacturers, and testing organizations. Also noteworthy, it is becoming increasingly difficult to obtain event-free, negative control materials; that is, that do not contain trace amounts of transgenic seed/grain.

- Research is a critical component in developing reference materials. For example, developing reference materials that satisfy both DNA and protein testing needs will require investigating the relationship among a number of factors including protein expression, genetics (zygosity, copy number, seed versus grain, etc.), environmental influences, and so on. In addition, research needs to be conducted with respect to testing parameters including protein extraction, matrix interference, stability, and instrumentation. This topic is discussed in more detail in Chapters 6, 9 and 10.

To develop protein-based certified reference materials for the agricultural biotechnology industry, the following criteria must be considered:

- Although biotechnology-derived crops are being developed in other countries, the vast majority of events have been developed and commercialized in the United States. The USDA–GIPSA, NIST, and IRMM have established confidentiality and/or material transfer agreements with U.S. life science organizations. Therefore, it is reasonable that these organizations work cooperatively in the establishment and coordination of reference materials.

- Establish or revise confidentiality and/or material transfer agreements between governments and life science organizations. The life science organizations are the most appropriate source for both nontransgenic and transgenic materials to be used in the preparation of reference materials. It may be necessary to establish new or revise existing confidentiality and/or material transfer agreements to allow the sharing of new information and needed materials.

- Develop formal relationships between grain importing and exporting countries to achieve cooperation and harmonize conventions with respect to measurement units, and the development of reference materials.

- Involve international scientific organizations in this process. Scientific organizations include CODEX Alimentarius Commission, the International Organization of Standards (ISO), Committee on European Norms (CEN), AOAC International, American Association of Cereal Chemists (AACC), American Oil Chemists Society (AOCS), and the Asian-Pacific Economic Consortium (APEC).

The following suggestions are offered with respect to creating reference standards that accommodate the variability associated with protein expression:

- Identify a minimal expression level and use this value as a worst case scenario.

- Develop a pooled average/representative standard as necessary (i.e., this could be related to single event, or multiple events expressing the same protein).
- Define a protein concentration value as equivalent to a wt% (theoretical definition that is universally applied; that is, 1.0 μg/g concentration of protein is equivalent to 0.5%).
- For different traits that express the same GM protein, identify a representative value assigned to protein concentration based on market share and expression level, that is, for Cry1Ab determine this assigned value based on the relative market shares of Mon810 and Bt-11, ignoring any contribution from Bt-176.
- Assign a protein concentration that is consistent with the DNA test value.

A challenge facing the agricultural biotechnology industry is the ability to quantitatively assess protein expression levels in GM traits, fulfill contractual obligations, and demonstrate compliance with government enforced regulatory mandates. Developing validated methods with the use of well-characterized reference materials may in the future facilitate the analysis of crops that have a significantly altered content. Both ELISA and PCR should be regarded as complementary techniques rather than exclusive of each other. In critical cases, one technique may be used for screening and the other for confirmation. The production of matrix reference materials certified for protein and DNA analysis is essential to fulfill a critical role in method validation and enforcement of regulatory mandates using ELISA-based methods. Proprietary proteins need to be made available to testing laboratories to ensure standardized test performances.

In many instances, immunoassays are viewed as a complementary tool along with PCR as a robust, specific, sensitive, and cost-effective analytical method to detect genetically engineered proteins in GM plants. It is not too difficult a task to develop immunoassays of the desired sensitivity from antibodies generated with modern molecular and biotechnology methods, and there are many methods described in the literature for obtaining immunoassay sensitivities down to extreme levels such as 600 molecules, if so desired (Harris et al., 1979). Indeed, extreme sensitivity can be associated with new and difficult analytical problems related to contamination, particularly where (as with food and agricultural matrices) the analytical samples are available in bulk. The rate-limiting steps in immunoassay development and applications are the required availability of appropriate antibodies and reference materials. The difficulty of producing antibodies to particular sequences from a protein, and the inability to generate on demand antibodies, capable of reacting normally at extremes of pH or in high concentrations of salt or solvent, provides challenges to kit manufacturers. Applications using recombinant antibody technologies provide considerable benefits as it becomes possible to more easily select antibodies with rare properties and manipulate the properties of antibodies already available. As our understanding of food materials and food processing increases, it will become possible to realize changes proteins undergo, as well as other analytes that get subjected to thermal and other denaturing forces.

A final consideration of immunoassays is the development of a system that can provide greater automation and throughput and could include biosensor technology

(Kalogianni et al., 2006; Feriotto et al., 2003). Because future generations of GM products may possess two or more introduced or altered proteins, an appropriate approach to determining whether the material is above or below the labeling threshold will need to be established. With the anticipated introduction of food crops containing multiple introduced proteins, access to reference materials and the development of antibody reagents might challenge the ability of analysts to develop and apply relevant immunoassays to detect biotechnology-derived products. An area where advances are currently being made in the agricultural biotechnology arena is with rapid test kits by combining antibody methods with instrumental techniques. In addition to the "hyphenated" methods such as immunoassay and mass spectrometry, considerable advances are now being made in real-time observations of antibody binding to target molecules—the biosensors that have been sought for so long. Instruments using optical methods for detection of molecular interactions are now being used for analysis of real samples (Elliott et al., 1998). While there is still much work to be done on development of newer technologies, the need for reliable, cost-effective, and stable reference materials provides much optimism for furthering progress and providing confidence in analytical measurements when using protein-based methods for detecting bioengineered traits in grains and oilseeds.

REFERENCES

AGBIOS. Information and Database on GM-Approved Products, **2008**.

Ahmed, F. E. Detection of genetically modified organisms in foods. *Trends Biotechnol.*, **2002**, *20*, 215–223.

Ahmed, F. E. *Testing of Genetically Modified Organisms in Foods*, Vol. *1*, The Haworth Press, Inc., New York, 2004, p. 324.

Anklam, E. and Neumann, D. A. Method development in relation to regulatory requirements for detection of GMOs in the food chain. *J. AOAC Int.*, **2002**, *85*(3), 754–761.

Anklam, E., Heinze, P., Pijnenburg, H., and Van Den Eede, G. Analytical methods for detection and determination of genetically modified organisms in agricultural crops and plant-derived food products. *Eur. Food Res. Technol.*, **2002**, *214*, 3–26.

AOCS, **2009**, available from http://www.aocs.org/index.cfm.

Beier, R. C. and Stankar, L. H. *Immunoassays for Residue Analysis: Food Safety*, American Chemical Society, Washington, DC, 1996.

Bertheau, Y. and Davison, J. Implementation of GMO labelling legislation: technical issues and corresponding outcomes of the European research project Co-Extra. In *1st Global Conference on GMO Analysis*, European Commission, Joint Research Center, Como, Italy, **2007**.

Bertheau, Y. and Davison, Y. The theory and practice of European traceability regulations for genetically modified food and feed. In *International Traceability Symposium (RDA)*, INRA, Institut. Nat. de la Recherche Agronomique, Versailles, France, 2008, www.biosafety-info.net/file_dir/217634888375cb71c7.pdf.

Bunk, D. M. Reference materials and reference measurement procedures: an overview from a national metrology institute. *Clin. Biochem. Rev.*, **2007**, *4*, 131–137.

Burns, M., Corbisier, P., Wiseman, G., Valdivia, H., McDonald, P., Bowler, P., Ohara, K., Schimmel, H., Charels, D., Damant, A., and Harris, N. Comparison of plasmid and genomic DNA calibrants for the quantification of genetically modified ingredients. *Eur. Food Res. Technol.*, **2006**, *224*, 249–258.

Codex Committee for In-House Method, Validation. In 23rd Session on Methods of Analysis and Sampling, **2001**.

Dong W., Shen, K., Kim, B., Kleter G. A., Marvin H. J., Guo, R., Liang, W., and Zhang D. GMDD: a database of GMO detection methods. *BMC Bioinf.*, **2008**, *9*, 260–268.

EC 1138/98, A.C.R.E.N., Commission Regulation (EC), **2000**, pp. 13–14.

EC Regulation, Amending Council Regulation (EC) No 1138/98 concerning the compulsory indication on the labeling of certain foodstuffs produced from genetically modified organisms of particulars other than those provided for in directive 79/112/EEC. *Off. J. Eur. Commun.*, **2000**, 4–7.

Elliott, C. T., Baxter, G. A., Hewitt, S. A., Arts, C. J., van Baak, M., Hellenas, K. E., and Johannson A. Use of biosensors for rapid drug residue analysis without sample deconjugation or clean-up: a possible way forward. *Analyst*, **1998**, *123*, 2469–2473.

Envirologix, **2009**, http://envirologix.com/artman/publish/index.shtml.

Ermolli, M., Fantozzi, A., Marinin, M., Scotti, D., Balla, B., Hoffman, S., Querci, M., Paoletti, C., and Van den Eede, G. Food safety: screening tests used to detect and quantify GMO proteins. *Accredit. Qual. Assur.*, **2006**, *11*, 55–57.

Eurachem/CITAC. Quantifying uncertainty in measurement. In Ellison, S. L. R., Rosslein, M., and Williams, A. (Eds.), *Guide 4*, 2000.

Feriotto G., Gardenghi, S., Bianchi, N., and Gambari, R. Quantitation of Bt-176 maize genomic sequences by surface plasmon resonance-based biospecific interaction analysis of multiplex polymerase chain reaction (PCR). *J. Food Agric. Chem.*, **2003**, *51*, 4640–4646.

Friguet, B., Djavadi-Ohaniance, L., and Goldberg, M. E. Some monocolonal antibodies raised with a native protein bind preferentially to the denatured antigen. *Mol. Immunol.*, **1984**, *21*, 673–677.

Gancberg, D., Corbisier, P., Schimmel, H., and Emons, H.Guidance document on the use of reference materials in genetic testing. © European Communities 2008. Available from http://www.eurogentest.org/web/files/public/unit1/reference_materials/Guidance%20doc%20for%20Use%20RMsGT_report_complete.pdfISO 21572, Foodstuffs—Methods for the detection of genetically modified organisms and derived products—Protein based methods, **2004**.

Green, M. J. A practical guide to analytical method validation. *J. Anal. Chem.*, **1996**, *68*, 305A–309A.

Grothaus, D. G., Bandla, B., Currier, T., Giroux G. R., Jenkins, G. R., Lipp, M., Shan G., Stave, J. W., and Pantella, V. Immunoassay as an analytical tool in agricultural biotechnology. *J. AOAC Int.*, **2006**, *89*, 913–928.

Harris, C. C., Yolken, R. H., Krokan, H., and Hsu, I. C. Ultrasensitive enzymatic radioimmunoassay: application to detection of cholera toxin and rotavirus. *Proc. Natl. Acad. Sci. USA*, **1979**, *76*, 5336–5339.

Hirel B., Le Gouis, J., Bernard, M., Perez, P., Falque, M., Quetier, F., Joets, J., Montalent, P., Rogowsky, P., Murigneux, A., and Charcosset A. Genomics and plant breeding: maize and wheat. In Morot-Gaudry, J. F., Lea, P.,and Briat J. F. (Eds.), *Functional Plant Genomics*, Science Publishers, Shrub Oak, NY, 2007, p. 708.

Holden, M. J., Blasic, J. R., Jr., Bussjaeger, L., Kao, C., Shokere, L. A., Kendall, D. C., Freese, L., and Jenkins, G. R. Evaluation of extraction methodologies for corn kernel (*Zea mays*) DNA for detection of trace amounts of biotechnology-derived DNA. *J. Agric. Food Chem.*, **2003**, *51*, 2468–2474.

Holst-Jensen, A. *What is the future of GMO detection? A freely speaking scientists opinion.* Co-Extra Research Live, 2006, http://www.biosafety-info.net/file_dir/233934888312b53edd.pdf.

IRMM, **2007**, available from http://www.irmm.jrc.be/html/reference_materials_catalogue/index.htm.

ISO 1369. *Cereals, pulses and milled products—sampling of static batches*, **1999**.

ISO 21570. *Detection of genetically modified organisms and derived products—quantitative nucleic acid based methods*, **2001**.

ISO 21572. *Foodstuffs—Methods for the detection of genetically modified organisms and derived products—Protein based methods*, **2004**.

JRC, **2008**, available from http://gmo crl.jrc.ec.europa.eu/statusofdoss.htm.

Kalogianni, D. P., Koraki, T., Christopoulos, T. K., and Ioannou, P. C. Nanoparticle-based DNA biosensor for visual detection of genetically modified organisms. *Biosens. Bioelectron.*, **2006**, *21*, 1069–1076.

Kleiner, J. and Neumann, D. A. *Summary Report of a Workshop on Detection Methods for Novel Food Derived from Genetically Modified Organisms*, ILSI European Novel Food Task Force in collaboration with ILSI International Food Biotechnology Committee, Brussels, Belgium, ILSI Press, Washington, DC, 1999.

Kohno, T., Carmichael, D. F., Sommer, A., and Thompson, R. C. Refolding of recombinant proteins. In Goeddal, D. E. (Ed.), *Gene Expression Technology*, Methods in Enzymology, Vol. *185*, Academic Press, Inc., San Diego, CA, 1991.

Kupier, H. A. Detection methods for novel foods derived from genetically modified organisms. *Food Cont.*, **1999**, *19*, 339–349.

Laffont, J. L., Remund, K. M., Wright, D., Simpson, R. D., and Gregoire, S. Testing for adventitious presence of transgenic material in conventional seed or grain lots using quantitative laboratory methods: statistical procedures and their implementation. *Seed Sci. Res.*, **2005**, *15*, 197–204.

Layton D. T. and Spiegelhalter, F. How grain companies are managing challenges posed by "stacked raits" in meeting the global regulatory commercial requirements for non-GMO corn shipments. In First Global Conference on GMO Analysis, Villa Erba, Como, Italy, **2008**.

Lipp, M., Anklam, E., and Stave, J. W. Validation of an immunoassay for detection and quantification of a genetically modified soybean in food and food fractions using reference materials. *J. AOAC Int.*, **2000**, *83*, 919–927.

Lipp M., Shillito, R., Giroux R., Spiegelhalter F., Charlton S., Pinero D., and Song P. Polymerase chain reaction technology as analytical tool in agricultural biotechnology. *J. AOAC Int.*, **2005**, *88*, 136–155.

Lipton, C. R., Dautlick, J. X., Grothaus, G. D., Hunst, P. L., Magin, K. M., Mihaliak, C. A., Rubio, F. M., and Stave, J. W. Guidelines for the validation and use of immunoassays for determination of introduced proteins and biotechnology enhanced crops and derived food ingredients. *Food Agric. Immunol.*, **2000**, *12*, 153–164.

Michelini E., Simoni, P., Cevenini, L., Mezzanotte, L., and Roda, A. New trends in bioanalytical tools for the detection of genetically modified organisms: an update. *Anal. Bioanal. Chem.*, **2008**, *3*, 355–367.

Mihaliak, J. and Berberich, S. A. Guidelines to the validation and use of immunochemical methods for generating data in support of pesticide registration. In Nelson, J. O., Karu, A. E., and Wong, B. (Eds.), *Immunoanalysis of Agrochemicals*. American Chemical Society, Washington, DC, 1995, pp. 56–69.

NIST, 2002, available from http://ts.nist.gov/MeasurementServices/ReferenceMaterials/DEFINITIONS.cfm.

Pauwels, J., Lamberty, A., and Schimmel, H. Quantification of the expected shelf-life of certified reference materials. *Fresenius J. Anal. Chem.*, **1998**, *361*, 395–361.

Remund, K., Dixon, D. A., Wright, D. L., and Holden, L. R. Statistical considerations in seed purity testing for transgenic traits. *Seed Sci. Res.*, **2001**, *11*, 101–119.

Rogan, G. J., Dudin, Y. A., Lee, T. C., Magin, K. M., Astwood, J. D., Bhakta, N. S., Leach, J. N., Sanders, P. R., and Fuchs, R. L. Immunodiagnostic methods for detection of 5-enolpyruvylshikimate-3-phosphate synthase in Rounup Ready soybeans. *Food Cont.*, **1999**, *10*, 407–414.

Sambrook, J. and Russell, D. W. *Molecular Cloning: A Laboratory Manual*, 3rd edition, Cold Springs Harbor Laboratory Press, Cold Springs Harbor, NY, 2001.

Schimmel, H., Corbisier, P., Klein, C., Phillip, W., and Trappman, S. New reference materials for DNA and protein detection. In 12th Conference on Euroanalysis, Dortmund, Denmark, **2002**.

Snedecor, G. W. and Cochran, W. G. *Statistical Methods*, 6th edition, Iowa State University Press, Ames, IA, 1967.

Springer, P. S. *Genomic Organization of Zea mays and its close relatives in biology*. Ph.D. Thesis, Purdue University, West Lafayette, **1992**, p. 190.

Stave, J. W. Detection of new or modified proteins in novel foods derived from GMO-future needs. *Food Cont.*, **1999**, *45*, 497–501.

Stave, J. W. Protein immunoassay methods for detection of biotech crops: applications, limitations, and practical considerations. *J. AOAC Int.*, **2002**, *85*, 780–786.

Stave, J. W., Magin, K., Schimmel, H., Lawruk, T. S., Wheling, P., and Bridges, A. R. AACC collaborative study of protein method for detection of genetically modified corn. *Cereal Foods World*, **2000**, *45*, 497–501.

Stryer, L.(Ed.). *Biochemistry*, W. H. Freeman and Company, New York, 1997.

Thomas, W. E., Pline-Srnic, W. A., Thomas, J. F., Edmisten, K. L., Wells, R., and Wilcut, J. L. Glyphosate negatively affects pollen viability but not pollination and seed set in glyphosate-resistant corn. *Weed Sci.*, **2004**, *52*, 725–734.

Trapmann, S., Schimmel, H., Le Guern, L., Zeleny, R., Prokisch, J., Robouch, P., Kramer, G. N., and Pauwels, J. Dry mixing techniques for the production of genetically modified maize reference materials. *Eighth International Symposium on Biological and Environmental Reference Materials*, Bethesda, MD, **2000**.

Trapmann S., Schimmel, H., Kramer, G. N., Van den Eede, G., and Pauwels, J. Production of certified reference materials for the detection of genetically modified organisms. *J. AOAC Int.*, **2002**, *85*, 775–779.

Uhlen, M. and Moks, T.In Goeddel, D. V. (Ed.), *Gene Expression Technology*, Methods in Enzymology, Vol. *185*, Academic Press, Inc., San Diego, CA, 1991.

Urbanek-Karlowska B., Sawilska-Rautenstrauch, D., Jedra M., and Badowski P. Detection of genetic modification in maize and maize products by ELISA-test. *Rocz. Panstw. Zakl. Hig.*, **2003**, *54*, 345–353.

USDA, Grain Inspection Packers and Stockyards Administration. Grain Inspection Handbook; Book 1, **1995**.

USDA, Grain Inspection, Packers and Stockyards Administration, U.S. Department of Agriculture, **2004**, available from http://www.gipsa.usda.gov/GIPSA/webapp?area=home&subject=landing&topic= landing.

Van Duijn, G., Van Biert, R., Bleeker-Marcelis, H., Van Boeijen, I., Adan, A. J., Jhakrie, S., and Hessing, M. Detection of genetically modified organisms in foods by protein- and DNA-based techniques: bridging the methods. *J. AOAC Int.*, **2002**, *85*, 787–791.

Walschus U., Witt, S., and Wittmann, C. Development of monoclonal antibodies against Cry1Ab protein from *Bacillus thuringiensis* and their application in an ELISA for detection of transgenic Bt-maize. *Food Agric. Immunol.*, **2002**, *14*, 231–240.

WHO. *Modern food biotechnology, human health and development: an evidence-based study*. World Health Organization, Food Safety Department, **2005**, available from http://www.who.int/foodsafety/ publications/biotech/biotech_en.pdf.

Wolf, W. Methods of analysis for nutrition labeling. *J. AOAC Int.*, **1993**, 115–120.

Yoshimura, T., Kuribara, H., Takashi, K., Seiko, Y., Futo, S., Watanabe, S., Nobutaro, A., Tayoshi, L., Akiyama, H., Tamio, M., Shigehiro, N., and Hino, A. Comparative studies of the quantification of genetically modified organisms in foods processed from maize and soy using trial producing. *J. Agric. Food Chem.*, **2005a**, *53*, 2060–2069.

Yoshimura, T., Kuribara, H., Matsuoka, T., Kodama, T., Takashi, L., Mayu, W., Takahiro, A., Akiyama, H., Tamio, M., Satoshi, F., and Hino, A. Applicability of the quantification of genetically modified organisms to foods processed from maize and soy. *J. Agric. Food Chem.*, **2005b**, *53*, 2052–2059.

Zel, J., Gruden, K., Cankar, K., Stebih, D., and Blejec, A. Calculation of measurement uncertainty in quantitative analysis of genetically modified organisms using intermediate precision—a practical approach. *J. AOAC Int.*, **2007**, *90*, 582–586.

CHAPTER 8

AUTOMATION OF IMMUNOASSAYS

Michele Yarnall

8.1   INTRODUCTION TO IMMUNOASSAY AUTOMATION

Since there has been an emphasis on screening large number of patient samples with less trained technologists, the evolution of the automated immunoassay has advanced tremendously in recent years, especially in the pharmaceutical and clinical diagnostic fields (Bock, 2000; Price, 1998). Nonetheless, with the introduction of the first genetically engineered crop about 15 years ago, the incentive for automated technology in the agriculture detection field has not been very great. However, the steadily increasing annual growth of genetically engineered crops and necessity for constant monitoring of these crops through every phase of the product life cycle (development, supply chain, and termination of product) have created a need for methods to screen large numbers of samples.

Many automated systems currently used in the pharmaceutical and clinical diagnostic fields are not suited for testing plant extracts due to the particulate nature of

Immunoassays in Agricultural Biotechnology, edited by Guomin Shan
Copyright © 2011 John Wiley & Sons, Inc.

the extract. A summary of the automated immunoassays systems being used in the pharmaceutical or clinical diagnostic industry is provided elsewhere (Wild, 2005). Many of these existing systems are designed to minimize the amount of serum required, but small sample size is not a limitation when working with plant extracts. Furthermore, even though the throughput of these systems can be up to several hundred samples per day, this number may still be insufficient for testing the number of samples collected from the field during peak growing seasons. In addition, a great deal of in-house development may be required to adapt the manual immunoassay method to the various fully automated platforms. This would be required for each unique protein to be measured. Time involved for learning a new process and setting it up in the lab to be fully functional could be significant. Costs of consumables and instrument service must also be considered when evaluating the overall economics of a full automation system.

An alternative to a fully automated platform is the automation of the individual steps within the process. One advantage of step automation is that most of the equipment needed may already be existing in the lab; therefore, a lab may not incur a huge expense when converting manual immunoassays to a semiautomated platform. Current manual ELISA methods can be used with minimal development, optimization, and validation. Also, converting to a step automation platform is rather a more rapid process than setting up a fully automated platform.

Deciding upon a completely automated system versus automation of the different steps will depend on several factors. These include cost of instrument(s), cost of consumables, available space for instrument(s), and the ability to convert existing ELISAs to a new format. In order to develop realistic and practical automation, various technologies for each process in immunoassay analysis must be evaluated. Mechanical manipulations should be simplified, and a general format that can be adapted to a number of different traits should be chosen. Methods to manage reagents and reaction mixtures precisely and avoid risks such as cross-contamination must be employed.

The majority of field samples to be tested require information only on whether or not the trait is present. Samples carrying the trait will continue with the breeding process and those without the trait will be discarded. Because of the large number of samples to be screened, this chapter describes some practical and flexible ways to automate lab procedures specifically for the qualitative testing of plant extracts. It illustrates how different instruments can be used to automate each process and how to minimize cross-contamination, while keeping costs at reasonable levels. These processes can be adapted for quantitative ELISA testing as well. Currently available systems as well as alternative technologies potentially useful for agriculture biotechnology in the future are also discussed.

8.2 INCREASING THROUGHPUT BY AUTOMATION

Automation is any process that minimizes human activity in performing the task (i.e., tedious manual methods). The ELISA method of analyzing samples involves many steps—plate coating, sample and reagent additions, plate washing, detection, and analysis. Automation of any of these steps will accelerate the entire process, especially if automation allows the processing of 384-well plates. Converting a 96-well plate to a 384-well plate can significantly increase throughput. For example, a person who can typically run 10 96-well plates per day can process the equivalent of 40 plates per day just by switching to the 384-well format.

It is fairly easy to convert a preexisting ELISA to a 384-well format by simply decreasing the volume of reagents. The advantage is that performance of the preexisting assay (precision, limit of detection, and accuracy) is already known and in most cases, converting to a 384-well format does not significantly change the assay performance (Massé et al., 2005). Since newer instruments for washing or reading plates are capable of processing both 96- and 384-well formats, the only additional instrument needed to automate the 384-well ELISA would be a liquid dispenser and/or an instrument capable of transferring samples from a 96-well format to a 384-well format. Details of how this can be accomplished are described.

8.3   TISSUE COLLECTION

The first step in increasing throughput of sample testing is to devise a method of sample collection that allows automated processing downstream. Tissue collected from the field and submitted for ELISA analysis can be brought to the lab in many forms (e.g., in plastic bags and individual tubes), and organizing the samples for extraction and subsequent ELISA analysis can be time-consuming. In order to simplify this process for downstream processing, leaf tissue should be sampled in a 96-well block format. Many different manufacturers offer deep well blocks suitable for sampling. A normal paper punch can be used to sample plant tissue, which can then be placed within the 96-well block (Figure 8.1). It is important to clean the paper punch between samples to minimize possible cross-contamination. Also, the amount of tissue to be punched into each well will depend on the assay sensitivity as previously determined by the lab.

An alternative method is to use 1.2 mL tubes that fit within a larger block (2.2 mL wells). The tissue-filled tube can then be placed in its corresponding position within a 96-well block. This method of sampling is more time-efficient than sampling directly into the block and prevents sampling errors in the field (Figure 8.2).

8.4   SAMPLE PROCESSING

Once the tissue has been sampled in the field, it must be prepared for extraction. The samples must be lyophilized or frozen in order to achieve sufficient tissue grinding. Both methods work equally well, but lyophilized tissue samples are easier to deliver to the testing lab because they do not require dry ice. Alternatively, for field locations without specialized lyophilization equipment, food dehydrators can be used to dry the tissue before shipping.

Figure 8.1 An example of a 96-well block for tissue collection is shown. A leaf disk created from a normal 1/4″ sized paper punch fits easily within the wells.

8.4.1 Tissue Grinding

To grind the tissue, stainless steel or glass balls are added to each tube. A device manufactured to add 96 balls at one time is needed for maximum throughput. The size and number of the balls added will depend on the tissue type and the amount of tissue collected. Blocks containing tissue and steel or glass balls are shaken on a grinding instrument such as the Kleco grinder (Figure 8.3a and b) or other commercially available 96-well grinders (Spex CertiPrep Geno/Grinder 2000 or Quiagen Mixer Mill MM300). Since tissue ground to a fine powder results in better extraction and thus better assay results, select the conditions that result in the finest grind of the tissue. One minute of shaking is usually sufficient to grind either frozen or lyophilized tissue to a fine powder. Following tissue grinding, blocks can be spun briefly in a centrifuge to remove tissue clinging to the lid.

8.4.2 Protein Extraction

Protein extraction is a simple procedure of adding buffer to the blocks and briefly shaking the contents. Liquid dispensing devices, as shown in Figure 8.4, can be used to

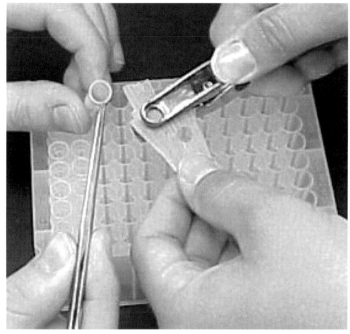

Figure 8.2 A paper punch is used for sample collection and sample is placed into a 1.2 mL tube. Tubes containing the leaf disk are inserted in the correct position in a 2 mL sampling block.

automate the addition of extraction buffer. When deciding on a liquid dispensing device, several features should be examined. One should consider volume(s) to be dispensed, whether the equipment can dispense into a 96- or 384-well plate and/or a deep well block, and the availability of stacking options. After adding extraction buffer, shaking of the block can be done manually or with the same instrument used for initial grinding of the tissue. It is important to use a lid for the block that fits tightly in order to prevent carryover between wells during the extraction process. If samples are to be tested the following day, no centrifugation of the blocks is required. The plant particulate matter will settle to the bottom of the block and will not interfere with subsequent steps.

8.5   SAMPLE TRANSFER

The biggest bottleneck in automating the process of testing plant samples in an ELISA format is the transfer of plant extract contained in four 96-well blocks to one 384-well microtiter plate. Pipetting samples from a 96-well block and dispensing them into the corresponding wells of a 384-well plate is very difficult and tedious since it requires strict focus to avoid errors or contamination. Automation of this step can be achieved with a liquid handling instrument such as the Becton Dickenson Fx (Beckman

(a)

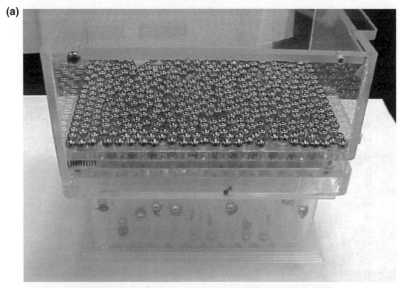

(b)

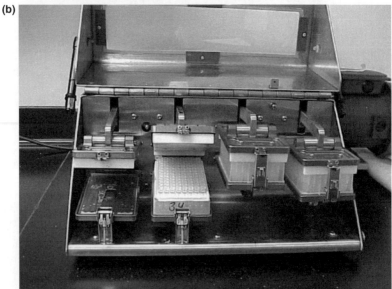

Figure 8.3 (a) A device is used to add stainless steel balls to all 96-wells simultaneously. (b) Shaking devices such as the Kleco can grind four blocks at the same time.

Coulter, Fullerton, CA), as shown in Figure 8.5. Many other liquid handling instruments can be formatted to perform this step of the process. Some of these instruments include the Tomtec Quadra 3 (Hamden, CT), the Tecan Freedom EVOlyzer or other Tecan platforms (Switzerland), the Eppendorf epMotion (Westbury, NY), and Hamilton platforms such as the MICROLAB STAR or MICROLAB

(a)

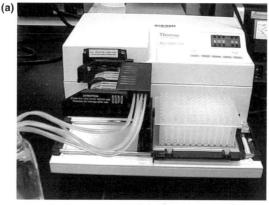

(b)

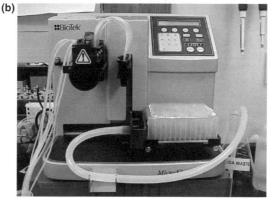

(c)

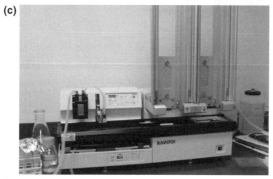

Figure 8.4 The Thermo Multidrop (a), BioTek MicroFlo Select (b), and WellMate Microplate Dispenser (c) are examples of different liquid dispensing instruments that can accurately pipette small volumes into 384-well plates as well as larger volumes into deep well blocks.

NIMBUS (Reno, NV). Qualities to consider when choosing the appropriate liquid handling instrument include the size of the instrument (large, medium, or compact footprint), type of dispensing head (8-, 96-, or 384-well), stacking options, flexibility, and programming. One inexpensive device that recently has become available is the

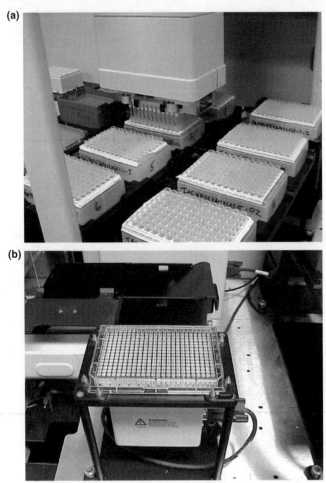

Figure 8.5 Many liquid handling machines are capable of transferring samples from a 96-well format (a) to a 384-well format (b).

Rainin Liquidator 96 (Woburn, MA). This is a manual benchtop system that can transfer 96-well samples into a 384-well plate.

8.6 REAGENT ADDITIONS

Plate coating and other reagent additions can be performed with dispensing equipment that can accommodate a 384-well ELISA plate. Simple and relatively inexpensive dispensing instruments make coating plates and adding reagents a simple process. Three types of dispensers (Thermo Multidrop, Thermo Fisher Scientific, Milford, MA; BioTek MicroFlo Select, Bio-Tek Instruments, Inc., Winooski, VT; and WellMate Microplate Dispenser, Thermo Fisher Scientific) are illustrated in Figures 8.4a–c, respectively. To minimize cross-contamination, the removable tubing cassettes can easily be exchanged for each reagent addition step (i.e., one tubing cassette for detection antibody, and one for substrate addition, etc.).

8.7   WASHING AND READING STEPS

Many plate washers and plate readers on the market accommodate both 96- and 384-well plates (Figure 8.6a and b). Again, most of the equipment that already exist in an immunoassay laboratory can be utilized to increase throughput, simply by converting to a 384-well format. Several kinds of plate washers and readers have stacking ability, allowing increased walkaway time during these processes.

8.8   ANALYSIS

Most immunoassay analyses for field samples require qualitative results (i.e., either the trait is present or it is absent), making analysis for trait detection simpler than quantitative protein analysis. Only a cutoff value needs to be determined to identify which samples are positive and which are negative. However, software is needed to convert the raw data from the plate reader into an easy-to-read format that people in the field can easily interpret. Most plate readers have recommended integrated software that analyzes the data and is then reported. Alternatively, macros can be written in Microsoft Excel to simplify the analysis process. The macro can import the raw data from the plate reader directly into the submission form completed by the sample submitter. With macros in place, ELISA results can be directly linked with each specific plant sampled by simply clicking a button or two.

8.9   ACHIEVING FULL AUTOMATION BY CONNECTING THE AUTOMATED STEPS

Semiautomation as described above can be converted to a completely automated system by connecting dispensers, washer, and plate reader with a robotic arm for plate movement. By having an open system, the operator can assemble the equipment according to the requirements of the different ELISA procedures. Most of the large robotic workstations have this capability. After samples are transferred to a 384-well plate, the instrument can move these plates to a plate washer, can dispense subsequent reagent additions, and finally place the plate in a plate reader for analysis. Stackers attached to the instrument and software for coordinating all the steps with the incubation times allow a number of plates to be processed at once.

8.10   DUPLEXING

Another way to increase throughput is to develop assays that can measure more than one trait at a time. This will save time on the number of samples that must be extracted.

(a)

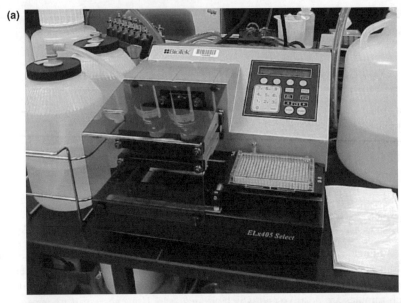

(b)

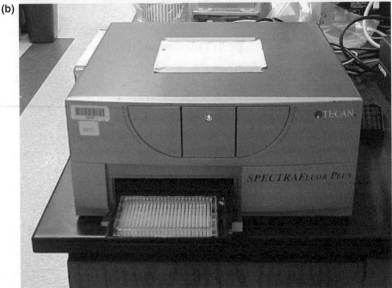

Figure 8.6 (a) The Bio-Tek ELx405 instrument is an example of a plate washer that can process either 96- or 384-well plates. (b) The Tecan SpectraFluor Plus is capable of reading 384-well plates.

The simplest way is to create duplex assays using existing ELISAs. This technique is a well-established procedure in immunofluorescence studies (Hunter et al., 1979) as well as in ELISA (Mason and Sammons, 1978; Blake et al., 1982; Farshy et al., 1984). With this technique, one of the assays utilizes a horseradish peroxidase (HRP)-labeled detection antibody and the other assay uses an alkaline phosphatase (AP)-labeled detection antibody. After the final incubation with the detection antibodies, the substrate for AP is added first and a reading of the plate is taken. Then, the plate is washed and the substrate for HRP is added. By combining the two assays into one process, two sets of data are generated with one set of consumables (plates, extraction blocks, pipette tips, etc.), saving both time and money. This duplex assay format will work only if no cross-reactivity exists between the different antibodies or between the different traits being measured.

8.11 FUTURE TECHNOLOGIES

Innovative new technologies for protein detection, such as microarrays, flow cytometry, and biosensors, are being developed and utilized in the pharmaceutical and/or clinical diagnostic industries. ELISA-based formats have evolved into miniaturized, highly multiplexed, and microarray formats (Chiem and Harrison, 1998; Ekins and Chu, 1991, 1999; Gushin et al., 1997; Rowe et al., 1999; Silzel et al., 1998; Mendoza et al., 1999; Liu et al., 2003; Massé et al., 2005). High-throughput screening through miniaturizing the ELISA format into 384-well microtiter plates has recently made its way to the agricultural biotechnology field (Brett et al., 1999; Roda et al., 2006). Microarrays, flow cytometry, and biosensors are being explored and these technologies will eventually be useful for screening traits in plants.

Technology that can measure several traits within a single sample will be the most useful feature of future technologies. With many traits being stacked within a plant, new multiplexing technology is needed to increase analysis efficiency. The FlowMetrix system from Luminex is an example of this type of multiplexing technology. This system can analyze several different analytes from a single sample using a flow cytometer and digital signal processing of a range of color dyes. Various types of tissues, including plant extracts, can be used with this system (Bergervoet et al., 2007). Antibodies from existing ELISA formats can be used, but extensive optimization is needed to ensure that the different antibodies used will not interfere with each other.

Multiplexing can also be accomplished with microarrays. Antibodies can be spotted on the microarray surface, allowing simultaneous detection of as few as 4 to greater than 20 different analytes (Mendoza et al.1999; Ekins and Chu, 1999; Varnum et al., 2004).

Microarrays that are being used for proteomics have been shown to be useful for detecting and identifying proteins in complex mixtures (Stoll et al., 2003; Templin et al., 2002; Letarte et al., 2005; Gulmann et al., 2006). Therefore, microarrays should be applicable to testing plant extracts. In addition, microarrays should be able to utilize the same antibodies that are functional in the ELISA format. Reverse arrays spot the sample to be tested on the array surface. The sample can be spotted as many times as necessary to accommodate the number of different antigens to be tested. Each spot can be probed with a different antibody, allowing the detection of several different antigens from one sample. If the antigen in the sample is not at a high enough concentration to be detected in the reverse array format, the array can be set up in a sandwich format similar to the ELISA (Varnum et al., 2004; Nielsen and Geierstanger, 2004). An antibody is coated on the array surface, followed by the addition of the sample and then detection with a labeled antibody. The advantages of the microarray format are its flexibility to test for different antigens, the ability for existing antibodies used in the manual ELISA to be adapted to this format, and the capability of testing many samples at one time. More detail regarding multiplexing technology will be discussed in Chapter 16.

Biosensors are another type of technology that could have applications for measuring traits in plant. A biosensor is an integrated receptor–transducer device. One end contains the recognition element or bioreceptor (i.e., antibodies, peptides, nucleic acids, and aptamers) that is in contact to a transducer. When the bioreceptor is activated, the transducer converts the signal into an electrical response that can be measured (Turner, 1989; Marco and Barcelo, 1996; Strehlitz et al., 2008). In the future, biosensors may allow the analysis to be performed directly in the field, but they must first overcome a number of obstacles (Rogers and Mascini, 2007). Currently, development costs are high, shelf life is limited, and the assay format is complex for potentially portable biosensor systems.

8.12 SUMMARY

When automating any laboratory test, flexibility of the system, operation complexity, and system throughput must be considered. The automated assay performance characteristics and its applications are of equal importance and also need to be examined. The right type of automation will ultimately depend on the needs of the lab.

Step automation offers a high degree of flexibility since washing, liquid handling, and detection steps are all done separately. Different instruments, many of which already exist in the laboratory, can be used. This saves on costs since no new instrument needs to be purchased. The degree of operation complexity associated with step automation is minimal since lab personnel will be familiar with the existing equipment; this allows a relatively quick transition from the manually run ELISA to the step automation process. All liquid handling and washing steps can be automated, saving time, preventing possible human error during sample transfer, and simplifying each step. The step automation described still allows walkaway time for the scientist. This time can be used to prepare extracts for the next run or to analyze and report the data from previous runs. Using this system and converting from a 96-well to a 384-well plate format, the throughput can easily be increased to 5000–10,000 samples tested per day for one trait. If a duplex ELISA format is used, the amount of data generated doubles. Integrating all the steps using an open robotic system will increase the walkaway time and may lead to increased throughput. Still newer technologies for protein detection that can increase throughput by multiplexing will be of importance for agricultural biotechnology in the future. By exploiting these technologies, screening for traits in plants will become easier and more efficient.

REFERENCES

Bergervoet, J. H. W., Peters, J., van der Wolf, J. M., Haasnoot, W., Bienenmann, M. E., du Pre, J. G., Wessels, R., and Meulenberg, E. *Multiplex Detection in the Food Production Chain.* Plenet Xmap, Amsterdam, The Netherlands, 2007, 2–3.

Blake, C., Al-bassam, M. N., Gould, B. J., Marks, B., Bridges, J. W., and Riley, C. Simultaneous enzyme immunoassay of two thyroid hormones. *Clin. Chem.*, **1982**, *28*, 1469–1473.

Bock, J. L. The new era of automated immunoassay. *Am. J. Clin. Pathol.*, **2000**, *113*, 628–646.

Brett, G. M., Chambers, S. J., Huang, L., and Morgan, M. R. A. Design and development of immunoassays for detection of proteins. *Food Control*, **1999**, *10*, 401–406.

Chiem, N. H. and Harrison, D. F. Microchip systems for immunoassay: an integrated immunoreactor with electrophoretic separation for serum theophylline determination. *Clin. Chem.*, **1998**, *44*, 591–598.

Ekins, R. and Chu, F. W. Multianalyte microspot immunoassay: microanalytical "compact disk" of the future. *Clin. Chem.*, **1991**, *37*, 955–967.

Ekins, R. P. and Chu, F. W. Microarrays: their origins and applications. *Trends Biotechnol.*, **1999**, *17*, 217–218.

Farshy, C. E., Hunter, E. F., Larsen, S. A., and Cerny, E. H. Double-conjugate enzyme-linked immunosorbent assay for immunoglobulins G and M against *Treponema pallidum*. *J. Clin. Pathol.*, **1984**, *20*, 1109–1113.

Gulmann, C., Sheehan, K. M., Kay, E. W., Liotta, L. A., and Petricoin, E. F., III. Array-based proteomics: mapping of protein circuitries for diagnostics, prognostics, and therapy guidance in cancer. *J. Pathol.*, **2006**, *208*, 595–606.

Gushin, D., Yershow, G., Zaslavsky, A., Gemmell, A., Shick, V., and Proudnikov, D. Manual manufacturing of oligonucleotide, DNA, and protein microchips. *Anal. Biochem.*, **1997**, *250*, 203–211.

Hunter, E. F., McKinney, R. M., Maddison, S. E., and Cruce, D. D. Double-staining procedure for the fluorescent treponemal antibody-absorption (FTA-ABS) test. *Br. J. Vener. Dis.*, **1979**, *55*, 105–108.

Letarte, M., Voulgaraki, D., Hatherley, D., Foster-Cuevas, M., Saunders, N. J., and Barclay, A. N. Analysis of leukocyte membrane protein interactions using protein microarrays. *BMC Biochem.*, **2005**, *6*, 2–14.

Liu, S., Boyer-Chatenet, L., Lu, H., and Jiang, S. Rapid and automated fluorescence-linked immunosorbent assay for high-throughput screening of HIV-1 fusion inhibitors targeting gp41. *J. Biomol. Screen.*, **2003**, *8*, 685–693.

Marco, M.-P. and Barcelo, D. Environmental applications of analytical biosensors. *Meas. Sci. Technol.*, **1996**, *7*, 1547–1572.

Mason, D. Y. and Sammons, R. Alkaline phosphatase and peroxidase for double immunoenzymatic labelling of cellular constituents. *J. Clin. Pathol.*, **1978**, *31*, 454–460.

Massé, G., Guiral, S. Fortin, L.-J., Cauchon, E., Ethier, D., Guay, J., and Brideau, C. An automated multistep high-throughput screening assay for the identification of lead inhibitors of the inducible enzyme mPGES-1. *J. Biomol. Screen.*, **2005**, *10*, 599–605.

Mendoza, L. G., McQuary, P., Mongan, A., Gangadharan, R., Brignac, S., and Eggers, M. High throughput microarray-based enzyme-linked immunosorbent assay (ELISA). *BioTechniques*, **1999**, *27*, 778–788.

Nielsen, U. B. and Geierstanger, B. H. Multiplexed sandwich assays in microarray format. *J. Immunol. Methods*, **2004**, *290*, 107–120.

Price, C. P. The evolution of immunoassay as seen through the journal *Clinical Chemistry*. *Clin. Chem.*, **1998**, *4*, 2071–2074.

Roda, A., Mirasoli, M., Guardigli, M., Michelini, E., Simoni, P., and Magliulo, M. Development and validation of a sensitive and fast chemiluminescent enzyme immunoassay for the detection of genetically modified maize. *Anal. Bioanal. Chem.*, **2006**, *384*, 1269–1275.

Rogers, K. R. and Mascini, M. *Biosensors for analytical monitoring*, **2007**, available at http://www.epa.gov/heasd/edrb/biochem/intro.htm.

Rowe, C. A., Tender, L. M., Feldstein, M. J., Golden, J. P., Scruggs, S. B., MacCraith, B. D., Cras, J. J., and Ligler, F. S. Array biosensor for simultaneous identification of bacterial, viral, and protein analytes. *Anal. Chem.*, **1999**, *71*, 3846–3852.

Silzel, J. W., Cercek, B., Dodson, C., Tsay, T., and Obremski, R. J. Mass-sensing, multianalyte microarray immunoassay with imaging detection. *Clin. Chem.*, **1998**, *44*, 2036–2043.

Stoll, D., Templin, M. F., Schrenk, M., Traub, P. C., Vohringer, C. F., and Joos, T. O. Protein microarray technology. *Front. Biosci.*, **2003**, *7*, 13–32.

Strehlitz, B., Nikolaus, N., and Stoltenburg, R. Protein detection with aptamer biosensors. *Sensors*, **2008**, *8*, 4296–4307.

Templin, M. F., Stoll, D., Schrenk, M., Traub, P. C., Vohringer, C. F., and Joos, T. O. Protein microarray technology. *Trends Biotechnol.*, **2002**, *20*, 160–166.

Turner, A. P. F. Current trends in biosensor research and development. *Sens. Actuators*, **1989**, *17*, 433–450.

Varnum, S. M., Woodbury, R. L., and Zangar, R. A protein microarray ELISA for screening biological fluids. In Fung, E. (Ed.), *Methods in Molecular Biology*, Vol. *264*, Protein Arrays, Humana Press Inc., Totowa, NJ, 2004, pp. 161–172.

Wild, D. (Ed.) *The Immunoassay Handbook*, 3rd edition, Elsevier, 2005.

CHAPTER 9

DATA INTERPRETATION AND SOURCES OF ERROR

Rod A. Herman
Guomin Shan

9.1   INTRODUCTION

While several different immunoassay formats are commonly applied to agricultural biotechnology, including Western blot analysis, dot blot analysis, and lateral flow devices (LFDs), this chapter will mainly focus on the interpretation and sources of error for quantitative sandwich enzyme-linked immunosorbent assay (ELISA). However, many of the principles described in this chapter are applicable to other immunoassay formats and even to nonantibody-based quantitative analytical assays.

As described in Chapter 4, sandwich ELISA typically employs one antibody coated on a solid surface (typically a 96-well microtiter plate) and another antibody, labeled with a reporter, in solution. The antigen, in this case a transgenic protein, is sandwiched between these two antibody layers, and then excess solution-phase antibody is washed away. The reporter labeled on the solution-phase antibody is typically an enzyme that catalyzes the reaction of a chemical substrate. The cleaved substrate develops a color as the enzymatic reaction progresses. Measuring the optical density (OD) of the solution allows the extent of the reaction to be estimated providing an index of the bound liquid-phase antibody and thus the protein antigen that is sandwiched between the antibody layers. Development of a calibration curve allows this index to be quantified relative to a purified standard.

Immunoassays in Agricultural Biotechnology, edited by Guomin Shan
Copyright © 2011 John Wiley & Sons, Inc.

Various assay characteristics can influence the interpretation and potential sources of error for a particular ELISA. We will treat these characteristics sequentially in this chapter, but often mention several factors together when the interaction of these factors is important for interpreting assay results and identifying potential errors. We will also repeat some of the assay validation discussion covered in Chapter 6 to emphasize the importance of this process in interpreting results.

9.2   DATA INTERPRETATION

9.2.1   Assay Validation

Assay validation, as described in more detail in Chapter 6, is critically important in the interpretation of ELISA data. The accuracy and precision of the particular assay will affect the confidence that can be placed on results. The quality of the calibration curve will likewise be very important for the interpolation of samples containing unknown quantities of transgenic proteins.

9.2.1.1   Accuracy   Paramount to the interpretation of the quantitative ELISA analysis is the accuracy of the assay. Accuracy is the relevance of assay results to the use that results will be put. More simply, it is a measure of how well assay results reflect the concentration of the correct analyte in the relevant structural conformation. Accuracy is distinguished from precision, which is a measure of reproducibility, in that it reflects the fitness of the assay for its intended purpose (Figure 9.1).

One aspect of accuracy is how well the assay distinguishes the form of the transgenic protein that is of interest from other forms of the protein and other components of the matrices that may contain the protein. This is especially important for detecting proteins because proteins as a class are typically safe (and when eaten nutritious), so only the active conformation of a transgenic protein is

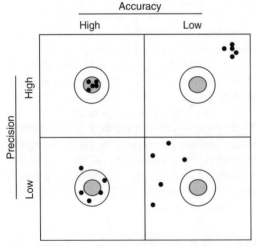

Figure 9.1   Schematic illustrating the difference between accuracy and precision.

Stylized protein structure:

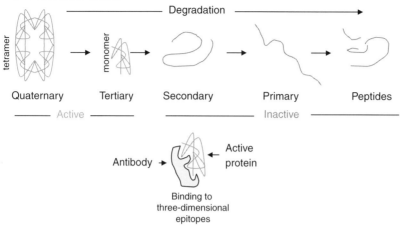

Figure 9.2 Stylized depiction of structural organization of proteins and antibody binding to protein epitope.

typically considered to have potential for adverse effects and thus is of interest in a monitoring situation.

Proteins have a primary or linear amino acid structure. Functional proteins most often also have secondary (partially folded) and tertiary (fully folded/native) structures. In addition, some proteins require association of multiple protein molecules or other molecules (quaternary structure) to display functionality (Figure 9.2). Thus, if one wishes to measure the concentration of functional protein in a sample, the assay must be able to distinguish the functional form from denatured or fragmented protein species that may also be present in a sample. Developing antibodies using active protein as the analyte often allows selection of antibodies that recognize three-dimensional epitopes specific to the active form of a protein (Figure 9.2). However, this must be investigated in the validation process. If a quantitative activity assay is available for the protein, the specificity of an assay can be investigated by looking for a correlation between ELISA results and activity results and by checking that denatured protein is not detected by the assay. Western blot analysis may also be employed to check the antibodies used in the ELISA for reactivity to inactive protein fragments, if they are present. This type of analysis may help determine the specificity of these antibodies but is not a replacement for comparing activity results with ELISA results.

An example of this approach can be illustrated for insecticidal crystal proteins originally isolated from the soil bacterium *Bacillus thuringiensis*. Several of these proteins have been expressed in transgenic crops to control insect pests. Laboratory bioassays can be conducted to compare the potency of plant material with that of the purified standard, and these results can be compared with ELISA results to see if a good correlation exists (Figure 9.3). Similarly, transgenic enzymes from tissue and standard preparations can be tested in kinetic experiments, and these results compared with ELISA results. Investigations of this type help ensure that an ELISA is selectively measuring functional protein.

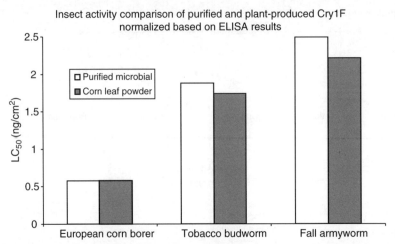

Figure 9.3 Plot comparing the activity of plant-produced and microbe-produced Cry1F insecticidal protein. (*Source*: Evans, 1998.)

In addition to evaluating the selectivity of the ELISA for active conformations of the transgenic protein, it is important in the validation process to ensure that similar proteins do not cross-react in the assay. This can be accomplished by simply checking the reactivity of proteins with known similar structure (amino acid sequence) in the ELISA. In addition to evaluating the cross-reactivity of similar proteins, it is important to check the matrices of interest for their ability to influence results. This influence could be through binding to antibodies, binding to the protein of interest, interference with OD readings through light absorption at the desired frequency, or catalyzation of the reporter enzymatic reaction. More detail regarding matrix effect will be discussed later. Such effects can be evaluated by testing matrix samples that have been fortified with know quantities of a purified standard and determining the maximum matrix concentration that will not materially affect results. If significant cross-reactivity or matrix effects are seen, then these effects will confound ELISA data and interpretation of results will be difficult or impossible. In some cases, samples can be diluted sufficiently to minimize matrix effects while maintaining acceptable levels of sensitivity to the target transgenic protein.

Also, key to the accuracy of ELISA is the quality of the purified standard used to develop the calibration curve. Simply, the protein standard must possess the same immunoreactivity as the protein present in samples containing unknown quantities of the protein (typically the same native structure). This is almost always achieved by investigating the equivalence of the protein standard to that of the protein in the matrix. Biochemical and activity measures are often used to investigate this equivalence. Parallelism between curves generated for dilutions of the purified standard and matrices containing the transgenic protein also helps confirm the quality of the standard. The correct quantification of the purified standard is likewise of paramount importance to the accuracy of quantitative ELISA. Activity assays, Western blot or sodium dodecyl sulfate polyacrylamide gel electrophoresis (SDS-PAGE), mass spectrometry, dye-based protein assays, and amino acid analysis are all possible techniques for quantifying the concentration of the purified protein.

Finally, the extraction efficiency of the transgenic protein from relevant matrices must be checked during the validation process to ensure that ELISA results are accurate. Spike recovery experiments and repeated extraction studies are normally completed to evaluate this property. However, results from these studies may only determine the percent recovery of soluble protein from the matrices. In some cases, activity assays comparing matrix extract to purified protein may help determine whether biological availability is predicted by solubility. If extraction efficiency is poor or highly variable, misinterpretation of ELISA results will likely occur.

9.2.1.2 *Precision*

Precision is a measure of the repeatability of an assay. Precision influences the accuracy of an assay, but good precision does not guarantee accuracy. That is, one can be very precise at measuring the wrong thing. However, if an assay is imprecise, the accuracy of individual estimates will necessarily be poor (Figure 9.1).

Precision within a single 96-well ELISA plate involves several levels of repeatability. Single samples can be extracted and aliquoted across a single plate to understand variability among individual plate wells and pipetting steps and to investigate the homogeneity of a single extract. If minimal variability is seen in these experiments, then multiple samples of the matrix can be extracted and tested in a single plate to determine the homogeneity of the matrix sample and the repeatability of the extraction process. Precision can often be improved by using larger matrix samples that will both decrease weighing errors (as a percentage of the total) and partially compensate for imperfect matrix homogeneity. Increased replication will also necessarily improve precision. In addition, multiple dilutions of the matrix can be tested to evaluate potential matrix effects or lack of parallelism between the matrix samples and the calibration curve (to be discussed in more detail later). Matrix effects or lack of parallelism will be evident in systematic changes in estimates of concentration as dilutions change (Figure 9.4). Finally, the robustness of the ELISA can be evaluated across plates, kits, days, analysts, and laboratories to investigate repeatability. Reasonable precision is needed to allow for ELISA results to be useful in characterizing the concentrations of transgenic proteins in complex matrices.

9.2.1.3 *Calibration Curves*

The quality of calibration curves is very important for the interpretation of quantitative ELISA results. The unknown quantity of transgenic protein in samples is estimated by interpolating OD results using a calibration curve developed using a purified protein standard. Various aspects of characterization of protein standards have previously been covered and the importance of this characterization cannot be overstated. Here, we will review aspects of determining the quantitative range for the assay and fitting the curve. A more integrated and detailed discussion of this topic is presented later.

One tangential factor to be aware of when developing calibration curves is the potential for perpetuating pipetting errors in the serial dilution of purified standards for use in calibration curves. Here, we refer to perpetuation of an error made in an

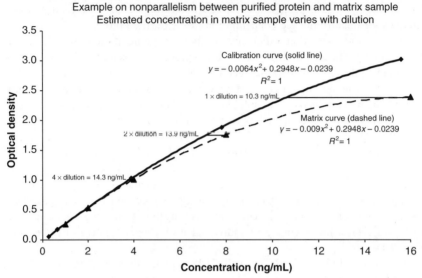

Figure 9.4 Hypothetical comparison of calibration curve and matrix curve illustrating impact of nonparallelism on interpolated values.

initial dilution in the subsequent dilutions. If such an error is very large, it may be easily detected using historical data and appropriate quality control procedures. However, if a sizable error (improper pipetting) occurs only in preparing the highest concentration on the calibration curve but OD results fall within the variability seen in historical data sets, then it may go undetected and result in biased results (Figure 9.5).

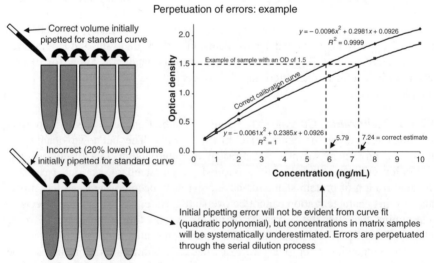

Figure 9.5 Hypothetical evaluation of the effect of a single pipetting error in the first step of a serial dilution used to prepare a calibration curve.

This situation is distinguished from propagation of error associated with acceptable pipette variability that can be compounded with error in subsequent serial dilutions (Higgins et al., 1998). This latter concept is often overlooked when assessing variability at different concentrations for the purpose of weighting regressions that are fit to calibration curves. Often, only measurement error among subsamples from a single serial dilution is considered in this process that can result in improper weighting. Weighting should be based on the variability among points on curves generated from multiple calibration curves assayed on the same plate. These curves should be prepared from independent samples of the standard protein.

The quantitative range for an assay is bracketed by the lower limit of quantitation (LLOQ) and the upper level of quantitation (ULOQ). The LLOQ and ULOQ levels are determined during the assay validation process. The limits of the quantitative range are set based on the variation in estimated concentration values across replications at various concentrations in conjunction with a consideration of acceptable levels of error. In general, the quantitative range falls on the steepest part of the calibration curve where small changes in concentration are expected to result in relatively large changes in OD. Thus, small errors in measuring OD will result in only minor errors in the prediction of concentration. As described below, the quantitative range is typically in the middle part of the concentration–OD curve, and concentrations of the standard across this region are typically included in each ELISA plate. If estimates of concentration in matrix samples are made by interpolation to flat portions of the calibration curve, unacceptable levels of error are often expected since small errors in measuring OD will result in large errors in the interpolated estimation of protein concentration in matrix samples.

Since the unknown concentration of an analyte is predicted by interpolation to a calibration curve, the accuracy of results will depend on the fit of a model to the limited number of purified protein concentrations included in each plate. While mechanistic models have been explored for fitting ELISA curves, empirical models are most often used (Diamandis and Christopoulos, 1996; Little, 2004). Model selection is based on goodness of fit and a minimization of systematic departures from the model (residuals). Common models include four- and five-parameter logistic models and power models (quadratic and cubic polynomials; Figure 9.6). Logistic models may fit data over a wide concentration range, while power models are typically only useful over the quantitative range of the assay (Little, 2004; Findlay and Dillard, 2007; Herman et al., 2008).

Four common models for fitting calibration curves

Quadratic (three parameters) OD = $a + b^*$Conc $+ c^*$Conc^2

Cubic (four parameters)   OD = $a + b^*$Conc $+ c^*$Conc$^2 + d^*$Conc^3

Four-parameter logistic   OD = $(a - d)/[1 + ($Conc$/c)^b] + d$

Five-parameter logistic   OD = $(a - d)/[1 + ($Conc$/c)b]^m + d$

Figure 9.6 Common polynomial and logistic equations used for fitting calibration curves.

Overfitting models with many parameters (relative to the number of concentrations) must be balanced with potential systematic errors that may be present with simpler models. For quantitative ELISA, optimal fit over the quantitative range should be the priority. Choosing a model with poor fit or systematic errors will result in an increased error or biased results. Overfitting a model may allow a good fit to pathological data sets. That is, errors in the preparation of the calibration samples may produce atypical results and these results may be well fit by a model that has too much flexibility in shape (too many parameters relative to the number of data points being fit). Model selection is discussed in detail later.

Once a satisfactory model for the calibration curve is selected, the parallelism between the protein standard curve and dilutions of matrix samples containing transgenic protein should be evaluated. This is typically accomplished by comparing the interpolated values for the matrix samples at multiple dilutions and looking for systematic patterns in predicted concentration values. Different dilutions of matrix samples should result in similar predictions of concentration if the curves are parallel. In addition, formal statistical methods for comparing the regressions can be employed (slopes, asymptotes, inflexion points, etc.), but results from such tests must be interpreted with caution since they can be overly sensitive to biologically insignificant differences (Plikaytis et al., 1994). If the curves are not parallel, then resulting predictions of concentration will be biased depending on the dilution used in the assay and the actual concentration of the transgenic protein in the matrix sample (Figure 9.4). A more detailed discussion of calibration curves can be found in Section 9.2.3.

In summary, the interpretation of quantitative ELISA results and the error associated with these results highly depends on a high-quality assay validation protocol. The accuracy/relevance of the assay to the purpose with which results will be applied is of the utmost importance. The precision/reproducibility of the assay is a prerequisite to obtaining accurate results with individual samples. The key to the usefulness of ELISA is the proper choice of purified protein standards and the concentrations of these standards for use in construction of calibration curves. Fitting calibration curves with appropriate models is equally critical in obtaining accurate and precise results.

9.2.2 Quality Control/Monitoring

Minimizing error during the routine use of quantitative ELISA requires rigorous quality control and monitoring processes. These processes include monitoring of both the protein standard used for constructing calibration curves and the performance of individual ELISA plates and kits. Furthermore, the stability of transgenic protein in matrix samples should be evaluated.

The purified protein standards used for constructing calibration curves must be stable to prevent errors in interpolating the concentration in matrix samples over time. Recertification of these standards should be scheduled on a regular basis and include both purity assessments and comparisons with historical immunoreactivity data. This will help prevent degraded protein standards with altered immunoreactivity from being used in the construction of calibration curves, with resulting errors in estimation of concentrations of transgenic protein in matrix samples (overestimation).

As a database of standard curves is accumulated, data should be evaluated for trends in error (residuals) as well as for potential systematic errors from the chosen model. A graphical representation of residuals can be an effective technique for comparing different models for fit to calibration curves (Herman et al., 2008). Monitoring calculated model parameters over time can also be useful for detecting drift in kit performance over time (Findlay and Dillard, 2007). Simply looking at the OD of certain concentrations over time (typically the lowest and the highest concentration in quantitative range) is also a good way to detect anomalies or trends in kit performance or stability of the protein standard.

Finally, including quality control samples from a homogeneous batch of matrix containing transgenic protein is a highly effective way to monitor kit performance. Use of these quality control samples at a few concentrations over the quantitative range of the assay can be a useful way to monitor the accuracy of the kit, potential unexpected matrix effects, and parallelism between matrix samples and the standard curve.

Continually monitoring ELISA kit performance is required to maintain quality control. Without this quality control step, assay problems may go undetected and results may be misleading.

9.2.3 Considerations in Developing Calibration Curves

9.2.3.1 Purified Protein Standard The first step in developing a calibration curve for ELISA is the production of a suitable protein standard. Key to this development step is isolating a high-purity protein preparation that shares key features with the relevant conformation of the transgenic protein present in matrices. Standards for ELISAs designed to estimate transgenic proteins expressed in crops are typically isolated from microbial sources. Protein standards are typically in low concentrations in crops, making it impractical to isolate the needed quantities from the crops themselves. However, the genes encoding these proteins can be engineered into microbes and overexpressed under laboratory conditions. This allows protein standards with the same amino acid sequence as those expressed in transgenic plants to be isolated in sufficient quantities to be used as ELISA standards.

ELISAs designed to quantify transgenic proteins expressed in crops are often undertaken to estimate potential exposure to the proteins. The quantity of active conformations of a transgenic protein is often of most interest in assessing exposure, since any potential negative effects of exposure to a protein are almost certainly associated with protein function (Delaney et al., 2008). One exception to this is allergenic potential; however, allergenic potential is assessed using a weight of evidence approach based on the original source of the gene, physiochemical properties of the protein, and bioinformatic analyses (amino acid homology). No lower safety limit for exposure to allergens has yet been determined, and so exposure estimates are an unknown value in predicting allergenic reactions (Bindslev-Jensen et al., 2002). Proteins are generally considered a nutrient and few proteins are toxic. When proteins are toxic, the mode of toxicity is associated with the function of the protein. Thus, inactive confirmations of the protein represent negligible risk. For this reason, it is often desirable to develop ELISAs that selectively detect active protein conformations relevant to estimating exposure and potential risk.

Proteins have several layers of complexity to their structure. The first level of structure is the linear amino acid sequence. Secondary protein structure refers to simple folding of the protein, while tertiary structure refers to complete folding into a native conformation. Some proteins also associate in fixed ways with other proteins of the same structure or with different molecules to form quaternary conformations (Figure 9.2). Protein function is most frequently associated with correct tertiary or quaternary structure (native conformation). These native conformations typically have three-dimensional epitopes that are not present in inactive protein forms. This often allows choice of antibodies specific to active transgenic protein (Figure 9.2).

Confirming that purified protein standards have the appropriate structure that is homologous to the protein as expressed in transgenic plants is a key first step in developing a relevant ELISA for transgenic crops. Several tools are often employed to evaluate this equivalence. Using small quantities of the transgenic protein isolated from plants, comparisons of linear amino acid sequence can be made using N- and C-terminal sequencing and matrix-assisted laser desorption/ionization time-of-flight mass spectrometry (MALDI-TOF-MS) peptide mass fingerprinting.

Measuring the activity of the protein standard is also important for assays designed to specifically detect active protein conformations. This can be done through use of activity assays if available (biological or biochemical). A good correlation between activity and ELISA results for proteins expressed in crops and present in the purified standard provides reasonable assurance that conformational equivalence is present.

It should be noted that some crops may contain nutritious storage proteins in the future for which no biological or enzymatic activity is known. However, it is unclear what risk (beyond potential allergenicity) these proteins might present, and consequently how knowledge of exposure will inform a risk assessment. From a practical standpoint, stability of storage proteins in plants will likely depend on tertiary or quaternary structure and native Western blot analysis may be useful for discriminating different protein forms. Equivalence between crop-produced and microbe-produced transgenic protein might be possible by comparing ELISA results with densitometer readings from Western blots developed with polyclonal antibodies.

A final check on the quality of the protein standard is to compare dilutions of the standard with that of matrix samples containing the transgenic protein for parallelism. This is a measure of the shape/slope of the two curves produced from these dilution series.

9.2.3.2 *Parallelism*

Parallelism is a comparative measure of the behavior/shape of concentration–OD curves between microbe-produced and plant-produced transgenic proteins. Parallelism may be affected by both antibody choice and protein form. Antibody choice is a complex process and will not be discussed in detail in this chapter. If concentration–OD curves are parallel, estimating the concentration of transgenic protein for different dilutions of the matrix sample will not systematically affect interpolated results. If systematic trends in predicted concentration are present for different dilutions, the equivalence of the protein standard to plant-produced protein, the choice of standard buffer and extraction buffer (solubility), and antibody choice should be investigated as potential causes. Formal statistical approaches for comparing regression curves are available, but the simple approach described previously is usually sufficient and even preferable for detecting meaningful departures from parallelism (Plikaytis et al., 1994). The comparison should cover the quantitative range of the assay (Figure 9.4).

9.2.3.3 *Quantitative Range*

The quantitative range of an ELISA refers to the range of concentrations across which precise and accurate estimates of concentration can be made. The required precision is a subjective assessment but should be related to the use with which results will be applied. Precision is a measure of repeatability and error. For ELISA, the precision of quantitative measurements is associated with how concentration and OD are related across the concentration range. To minimize error, small errors in estimating OD will need to have minimal effect on interpolated concentration estimates. This relationship exists for the steepest portion of the concentration–OD curve. In addition, the LLOQ needs to result in OD measurements that are sufficiently above background such that the signal to noise ratio is high enough to prevent excessive variability (false positives (FP)). Once the required level of precision is determined, the LLOQ and the ULOQ can be established by repeatedly estimating the concentrations in samples near these limits and estimating variability. If variability at any given concentration is unacceptable, narrower limits will be required.

9.2.3.4 *Choice of Standard Protein Concentrations*

A quantitative ELISA requires that a minimum number of concentrations of a protein standard are included to accurately reflect the concentration–OD relationship across the quantitative range. Efficiency concerns within ELISA kit plates dictate that the space allocated to determining the standard curve is kept to a minimum. The number of concentrations required to estimate the calibration curve and their distribution across the curve is somewhat dependent on the model chosen to fit the data. While power models (e.g., quadratic or cubic polynomials) can be adequately fit by including both concentrations at the extremes of the quantitative range and additional concentrations equally spaced across the quantitative range (Rozet et al., 2007), logistic models (four or five parameters) require additional concentrations at a low and a high concentration outside the quantitative range and a logarithmic serial dilution scheme is typically recommended (DeSilva et al., 2003; Findlay and Dillard, 2007). The extreme concentrations are required to accurately estimate lower and upper asymptotes of these sigmoidal models (Figure 9.7 inset). In practice, both approaches have been employed, but under a constraint of limited space for multiple concentrations, a fundamental question exists as to whether the quantitative range is best modeled by distributing these concentrations across the quantitative range or by replacing some of the concentrations within the quantitative range with those at the concentration extremes to allow estimation of asymptotes for employment of logistic models (Figure 9.7). Generally, five to eight concentrations are used to define the standard curve (Findlay and Dillard, 2007). Our research suggests that modeling the

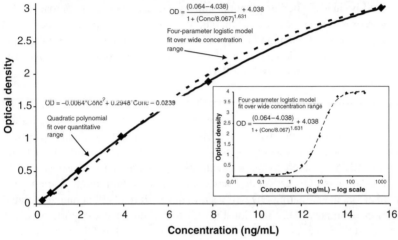

Figure 9.7 Calibration curves fit over the quantitative range with a four-parameter logistic and quadratic polynomial equation showing typical bias associated with the logistic model.

quantitative range using polynomial equations is preferable in many cases (Herman et al., 2008).

9.2.3.5 Calibration Curve Model Choice

Various models have been employed to fit ELISA calibration curves. The ELISA curve typically appears sigmoidal across a wide concentration range and is often well fit by a four- or five-parameter logistic model (Figure 9.7 inset). ELISA calibration curves are also well fit by power models (e.g., quadratic of cubic polynomials) over the quantitative assay range (Figure 9.7). The practical benefits of each approach may vary across different ELISA kits. It is however intuitively appealing to maximize the characterization of the calibration curve across the quantitative range by placing a maximum of concentrations within this range. This will preclude the preferential use of logistic models because estimation of asymptotes will often be inaccurate in the absence of a very low and a very high concentration (Herman et al., 2008). It should be noted that a linear regression of the logarithmically transformed concentration and OD has sometimes been suggested, but this approach has been found to be greatly inferior to the use of power or logistic models (Findlay and Dillard, 2007; Herman et al., 2008). An additional factor to consider in modeling calibration curves using power models is using of a zero concentration when fitting these models. It is advisable not to include zero concentration data when fitting power models because these data are not well fit and can result in systematic departures from the model for the concentrations at or near the LLOQ (Herman et al., 2008).

A final consideration in fitting models to ELISA calibration curve is the need for weighting different concentrations to deal with heterogeneity of variance (Diamandis and Christopoulos, 1996; Warwick, 1996). In simple terms, this involves giving greater weight to concentrations where less variability is seen. When the variance structure of the data is known, weighting will improve the accuracy of the resulting model fit. However, precise execution of ELISA procedures, combined with a good model choice, may make the contribution of weighting negligible in terms of improving accuracy.

9.3 SOURCES OF ERROR

Physical interaction between antibody and antigen is the foundation of immunoassay. Quantitative and qualitative measurements of target analyte in an unknown sample or matrix by immunoassay are mainly governed by the affinity and specificity of antigen–antibody reaction for the analyte of interest. Any biological and chemical components, sample processing, or assay operation interfering with the nature of antibody–antigen binding will compromise the accuracy of immunoassay. There are two main error sources in immunoassay processes: biological source and analytical or instrumental source. Biological factors routinely causing measurement error in an immunoassay analysis include antigen protein forms, cross-reactivity, sample treatment or extraction, matrix effect, variability of protein expression level in plants, and stability of reference antigen and other reagents. Analytical factors include liquid handling, signal reading, quality of test kits, and human errors caused during analysis. In addition, assay error may be affected by other sources such as assay format, buffer choice, nature of sample processing, and interpretation of results.

9.3.1 Biological Source

9.3.1.1 Reference Antigen   For a quantitative ELISA, the reference antigen plays a critical role for accurate and consistent analysis. Unlike small-molecule chemicals, macromolecules such as proteins are less stable and normally susceptible to degradation during storage. Therefore, storing protein under proper conditions is critical. Lyophilized protein is usually more stable and suitable for long-term storage compared to aqueous solutions. A major cause of error for lyophilized protein standards is the absorption of moisture, which results in a change of protein purity (mass based) or protein aggregation. It is recommended to store the protein in a dry, sealed container with desiccant at $-80°C$ if possible.

Compared to powder protein, protein solutions are more problematic. Proteins may degrade or aggregate during storage if the buffer condition, temperature, or protein concentration is not appropriate. Many proteins function well in near-neutral pH buffers such as phosphate buffered saline (PBS). However, some proteins may require specific conditions such as high or low pH and high salt levels. Otherwise, the protein solubility may be affected significantly. Some proteins may form oligomers in certain buffers, which may significantly increase or decrease immunoreactivity. As a reference standard in an immunoassay, the protein needs to be thoroughly characterized and studied.

Protein concentration may impact stability. Low-concentration protein solutions (e.g., $<0.5\,mg/mL$) are more prone to inactivation and loss due to the hydrophobic binding to the storage container. A relatively high concentration such as 1.0 mg/mL or higher is recommended. To improve long-term storage stability of solubilized protein, the common practice is to add a sacrificial protein such as purified bovine serum albumin (BSA) at 0.1–0.5% level to protect target protein from degradation and loss.

Temperature is vital for storage of protein solutions. Generally, proteins should be stored as cold as possible in sterilized glass or polypropylene containers. Storage at ambient temperature usually causes protein degradation and inactivation due to microbial growth. For short-term storage (up to a few weeks), many proteins may be stored in buffer solution at 2–8°C. For long-term storage (>1 month), proteins are commonly stored at −20 to −80°C or in liquid nitrogen. Some laboratories prepare single-use aliquots (e.g., 50 µL) of the protein in clean plastic containers for storage. Because freeze–thaw cycles decrease protein stability, the best practice for frozen storage is to dispense and prepare single-use aliquots. Once thawed, the protein solution should not be refrozen. Alternatively, addition of cryoprotectants (e.g., 50% glycerol) will prevent solutions from freezing at −20°C, thus enabling repeated use from a single stock without thawing.

To increase protein shelf life, another common practice is to include additives in protein solutions. The following are some of reagents used for this purpose:

- *Cryoprotectants*: Glycerol or ethylene glycol is normally added to protein solution at final concentration of 25–50% to stabilize proteins by preventing the formation of ice crystals at −20°C that can damage protein structure.

- *Antimicrobial Agents*: Sodium azide (NaN_3) at a final concentration of 0.02–0.05% (w/v) or thimerosal at a final concentration of 0.01% (w/v) is used to inhibit microbial growth in protein solution.

- *Metal Chelators*: Ethylenediaminetetraacetic acid (EDTA) at a final concentration of 1–5 mM is used to avoid metal-induced oxidation of –SH groups and help maintain the protein in a reduced state.

- *Reducing Agents*: Dithiothreitol (DTT) and 2-mercaptoethanol (2-ME) at final concentrations of 1–5 mM are also used to help maintain the protein in the reduced state by preventing oxidation of cysteines.

In some cases, the reference antigen may be derived from microbial sources and may be slightly different from plant protein in its amino acid sequence or with respect to glycosylation, phosphorylation, and/or other posttranslational modifications of the target protein. Thus, the equivalence between reference antigen and target protein regarding their immunoreactive response in the assay needs to be thoroughly evaluated and optimized during method validation.

Finally, methods used for certifying reference protein purity or concentration are a very common source of error. If the reference protein is pure enough, the preferred method is to conduct a quantitative amino acid analysis. In some cases, it is almost impossible to achieve high purity. SDS-PAGE and densitometry are usually applied for quantification of such protein preparations. Different amounts of reference protein, such as BSA, and target protein are loaded onto the same gel and a calibration curve is established with the BSA for quantification of the reference antigen. If a "gold standard" exists, ELISA can serve as an analytical tool to quantify the reference protein. To ensure the validity of a reference concentration, at least two methods are needed to certify a reference antigen.

9.3.1.2 *Protein Stability in Samples*

Plant tissues contain a large quantity of enzymes (e.g., proteases) and/or other reactive compounds (e.g., quinones and polyphenols) that may interact with or degrade the target protein during the time of storage or sample processing. One of the most common issues is the degradation of protein in plant tissues. In certain cases, we have observed a 30–50% drop in protein concentration measured by ELISA in 24 h when fresh samples were kept at 4°C. It is recommended that fresh tissues and sample extracts be analyzed immediately unless stability under the appropriate storage conditions has been demonstrated. If fresh samples need to be stored for a short time (days) to accommodate shipping or waiting for further processing, the best practice is to freeze the samples immediately using dry ice or liquid nitrogen and store at −20 or −80°C thereafter.

For longer term storage, plant tissues may be lyophilized and stored at −80 or −20°C. Proteins in grain, seed, and processed fractions may be stable for long periods under conditions of controlled temperature and humidity. To ensure that data are accurate, storage stability of protein in the bulk samples should be thoroughly investigated during method validation.

9.3.1.3 *Matrix Effects*

A major challenge for the analysis of genetically engineered (GE or GM) proteins in plant matrices is the effect of the complex and variable mixture of proteins, enzymes, carbohydrates, chlorophylls, plant secondary products, and metals. The effect of these components ("matrix effects") may seriously compromise assay performance. In addition, reagents used in immunoassay may contribute to matrix effects. In general, interferences from plant matrix can be categorized into three types (Gegenheimer, 1990):

1. Plant proteins may cross-react in the immunoassay. Rabbits and goats are common hosts for polyclonal antibody production, and plant materials may be part of their diet and it is possible that their sera contain antibodies that bind to plant proteins. To minimize such cross-reactivity, a practical approach is to use at least one monoclonal antibody in the assay system. In some cases, plant proteins may have similar epitopes as the target protein and thus cause false positives. This interference can be eliminated by antibody pair selection during assay development and matrix-specific method validation.

2. Enzymes present in plants such as proteases or peroxidases may cause protein degradation or nonspecific bindings. Plant tissues contain a variety of proteases across all four groups of proteases: the serine proteases, the cysteine proteases, the aspartic proteases, and the metalloproteases (Ryan and Walker-Simmons, 1981; Dalling, 1986). Seeds or grains are particularly rich in proteases, which are activated to process storage proteins during germination and embryonic development. Other plant parts, such as flower, meristem, young boll (cotton), or new leaf, normally contain high concentrations of proteases. Proteases in plant extracts may directly interact with a target analyte and quickly degrade the protein in assay solution, thus giving an underestimate of the true concentration.

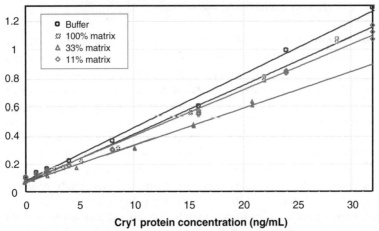

Figure 9.8 Matrix effect of corn pollen extracts on Cry1 ELISA.

In some cases, the proteases in a plant extract may cause little effect in ELISA while the extract is undiluted; however, a greater interference may be measured when the extract is further diluted with buffer (Figure 9.8). The reason for such unusual phenomenon is that proteases may be inhibited by coextracted chemicals in the solution. Once diluted, the proteases are free to function and result in a significant suppression of ELISA readings (Figure 9.8). Use of protease inhibitors during preparation of plant extracts is a common practice to eliminate protease interference (Gray, 1982; Beynon and Bond, 1994; Zollner, 1993). Detailed information of protease inhibitors is provided in Table 9.1. These inhibitors can be used alone if a specific class of proteases is identified as a major source of interference. In many cases, it is not likely or not necessary to know the specific information regarding the proteases in the test matrix. Protease inhibitor cocktails are a logical choice for plant extracts and a variety of ready-to-use cocktails are commercially available. Some cocktails are listed in Table 9.2 for reference.

Peroxidases are often present in plant tissue, especially in roots, and may be a potential source of interference in ELISA using horseradish peroxidase (HRP) as the labeling enzyme. This effect can be eliminated by choosing proper antibodies and using specific blocking techniques during plate coating. One simple solution to eliminate such effects is to use a different labeling enzyme in the assay such as alkaline phosphatase.

3. Plant extracts contain abundant alkaloids and the polyphenolic compounds including flavonoids and tannins (Loomis, 1974). These compounds may adversely affect protein extraction, antibody–antigen binding, and enzyme activities. The interaction between these compounds and proteins includes four types of reactions (Loomis, 1974): (a) Hydrogen bonding: isolated phenolic hydroxyl groups may form strong hydrogen bonds with the oxygen or nitrogen atoms of proteins. (b) Oxidation of phenols to form highly reactive quinones:

TABLE 9.1 List of Protease Inhibitors

Inhibitors	Description	Recommended working concentration[a]	Reference
Serine protease inhibitor			
Phenylmethylsulfonyl fluoride (PMSF)	Irreversibly inhibits serine proteases by sulfonylation of the serine residue in the active site of the protease. Inhibits also papain (reversible by DTT treatment) and acetylcholinesterase. Does not inhibit metallo-, aspartic- and most cysteine proteases	1 mM	Prouty and Goldberg (1972)
Benzamidine HCl or p-aminobenzamidine-2HCl	A more potent and expensive inhibitor of thrombin and trypsin	1 mM	Ensinck et al. (1972)
Aprotinin	Basic single-chain polypeptide that inhibits numerous serine proteases by binding to the active site of the enzyme, forming tight complexes. It inhibits above all plasmin, kallikrein, trypsin, chymotrypsin, and urokinase, but not carboxypeptidase A and B, papain, pepsin, subtilisin, thrombin, and factor X. Since aprotinin contains three disulfide bonds, it might be inactivated by the high levels of reductants used for plant extraction	1–2 μg/mL	Zyznar (1981)
4-(2-Aminoethyl) benzenesulfonyl fluoride (AEBSF-HCl)	Irreversible inhibitor of thrombin and other serine proteases. Inhibits by acylation of the active site of the enzyme. Much less toxic than PMSF and diisopropyl fluorophosphate (DFP)	1 mM	Marwardt et al. (1973)
Trypsin inhibitor from soybean	Monomeric protein. Reversible serine protease inhibitor. Inhibits trypsin, factor Xa, plasmin and plasma kallikrein, but not tissue kallikrein	Equimolar	Kassell (1970)
Cysteine (thio) protease inhibitors			
Antipain	Reversible inhibitor of serine and cysteine proteases. Inhibits papain and trypsin specificity more than leupeptin. Plasmin is inhibited only slightly. Also involved in inhibition of RNA synthesis	1 μg/mL	Miyata et al. (1988)

(Continued)

205

TABLE 9.1 (*Continued*)

Inhibitors	Description	Recommended working concentration[a]	Reference
Leupeptin	Tripeptide aldehyde. Reversible competitive inhibitor of serine and cysteine proteases. Inhibits also phospholipase D and C activation in rat hepatocytes	1 µg/mL	Umezawa (1976)
N-Ethylmaleimide	Binds stoichiometrically to SH groups. This covalent inhibitor generally is not very effective in plant leaf extracts	Equimolar	Riordan and Vallee (1972)
L-trans-Epoxysuccinyl-leucylamido(4-guanidino)butane (E-64)	Noncompetitive irreversible inhibitor of papain and other cysteine proteases. Forms a thioether bond with the sulfhydryl group in the active center of the enzyme	Equimolar	Barrett et al. (1982)
Aspartate (acidic) protease inhibitors			
Pepstatin A	Pentapeptide derivative. Reversible inhibitor of aspartic proteases, for example, pepsin, cathepsin D, chymosin, and renin	0.1 mg/mL	Umezawa (1976)
Diazoacetylnorleucine methyl ester (DAN)	In the presence of $Cu(II)$, this is a covalent inhibitor. Cupric acetate is usually co-used at equimolar concentration	5 mM	Gray (1982)
Metalloprotease inhibitors			
Ethylene glycol-bis (β-aminoethyl ether)-N, N,N',N'-tetraacetic acid (EGTA)	An efficient chelators of divalent metal cations other than Mg^{2+}. Inhibits metalloproteases	10 mM	Mortensen and Novak (1992)
1,10-Phenanthroline	Metalloprotease inhibitor; chelates iron, zinc and other divalent metals	5 mM	Sigman et al. (1991)
Bestatin-HCl	Metalloprotease inhibitor with multipharmacological functions. Inhibits cell surface aminopeptidases and leucine aminopeptidase. Inhibitor of leukotriene A4 hydrolase and of enkephalin degradation in cell preparations	10–50 µM	Wilkes and Prescott (1985)
EDTA-Na$_2$	Reversible inhibitor of metalloproteases	1–10 mM	Janas (1994)

[a]Recommended working concentration is based on concentration from literature or commonly used in the laboratory. http://www.sigmaaldrich.com/sigma/general%20information/protinhibselectionguide.pdf.

TABLE 9.2 List of Commercially Available Protease Inhibitor Cocktails

Cocktails	Description	Recommended working concentration	Reference
Sigma P9588	Designed for plant matrix. Inhibits serine, cysteine, aspartic and metalloproteases, as well as aminopeptidases. Containing AEBSF, bestatin, E-64, leupeptin, pepstatin A, and 1,10-phenanthroline	1 mL is good for 100 mL of extract from 30 g of plant tissue	http://www.piercenet.com/Objects/View.cfm?type=ProductFamily&ID=02040806
Sigma P2714	Designed for general use. Inhibits serine, cysteine, aspartic and metalloproteases. Contains AEBSF, aprotinin, bestatin, EDTA, E-64, and leupeptin	1 mg powder is good for 100 mL of extract	http://www.piercenet.com/files/1988as4.pdf
Sigma S8820	Designed for general use. Inhibits serine, cysteine, aspartic and metalloproteases. Contains AEBSF, aprotinin, bestatin, EDTA, E-64, leupeptin	Soluble tablet	http://www.emdbiosciences.com/html/CBC/inhibitor_cocktails.htm
Pierce 78430	Designed for general use. Inhibits serine, cysteine, aspartic and metalloproteases. Contains AEBSF, aprotinin, bestatin, EDTA, E-64, leupeptin, and pepstatin A	Use in 1:100 dilution	https://www.roche-applied-science.com/sis/complete/protease.jsp
Pierce 78425	Designed for general use. The same as Pierce 78430 without EDTA.	Use in 1:100 dilution	
Calbiochem 539133	Designed for plant matrix. Inhibits serine, cysteine, aspartic and metalloproteases, as well as aminopeptidases. Contains AEBSF, bestatin, E-64, leupeptin, pepstatin A, and *o*-phenanthroline	One milliliter is recommended for 30 g of various plant tissues	
Roche Complete Protease Inhibitor Cocktail 11697498001	Designed for bacterial, mammalian, yeast, and plant matrix. Inhibits serine, cysteine and metalloproteases, aminopeptidases	One tablet is recommended for 50 mL extracts	

207

the resulting quinones can covalently react with free functional groups of protein such as hydroxyls, thiols, and primary amines and further polymerize proteins through covalent linkages. This is the principal cause of the browning reactions in plant matrices. (c) Ionic interactions: under basic conditions, phenolic hydroxyl groups may form salt linkage with the amino acid residues of proteins. (d) Hydrophobic interactions: the aromatic rings of polyphenolic compounds are essentially lipophilic and can physically interact with hydrophobic region of proteins.

Preventing damage from phenolics and quinones, or removing them from the extracts, is necessary to achieve an accurate and consistent transgenic protein measurement both quantitatively or qualitatively. A general practice is to add adsorbents or protective agents to the extracts that can compete with target proteins in reacting with the phenols and quinones and also prevent oxidation of the phenols. Polymers are frequently added to the extraction buffer to serve as phenol adsorbents or quinone scavengers. Polyvinylpyrrolidone (PVP) is an effective polymer for binding phenolic compounds by forming strong hydrogen (H) bonded complexes, and this tactic has been used by many laboratories (Loomis and Battaile, 1966; Koenig, 1981; Devaraja et al., 2005; Shan et al., 2007). One to five percent soluble PVP in buffer is usually used for plant matrices depending on the plant material–buffer ratio. In a 0.1 g dry weight tissue to 10 mL buffer setting, 1–2% PVP is sufficient (Shan et al., 2007). Bovine serum albumin is another widely used and effective additive in plant extraction (Laties and Treffry, 1969; Raison and Lyons, 1970; Schrader et al., 1974). BSA has high capacity to react with plant phenolic compounds. As a protein, it is expected to bind phenolic compounds by H-bonding to its oxygen or nitrogen atoms. BSA is also known for its capacity to bind lipids by hydrophobic forces due to its high content of hydrophobic amino acids (Spahr and Edsall, 1981). Moreover, BSA is rich in lysine and cysteine and appears to be an effective quinone scavenger by covalent condensation of quinones with ε-amino groups of lysine. The concentration of BSA used ranges from 0.1% up to 2% (Throneberry, 1961) with common practice at 0.5–1%.

Another approach is to add antioxidants or phenol oxidase inhibitors to prevent oxidation of phenols. Ascorbic acid and 1,4-dithiothreitol (DTT) are the most common antioxidants or reducing agents for this purpose (Lieberman and Biale, 1956; Lieberman, 1960; Shesworth et al., 1998). At a concentration of 5–10 mM, both reagents effectively protect the target protein in plant extracts. Sulfites such as metabisulfite, bisulfite, and sulfite are a group of reducing agents used in plant extracts (Loomis, 1974) that irreversibly inhibit phenol oxidases (Anderson and Rowan, 1967). One thing to be mindful of when using sulfite is its potential side effect. Certain disulfide bonds of proteins may be cleaved by sulfite (Means and Feeney, 1971).

Similarly, soil is a very challenging matrix for ELISA system and more detail on this matrix is provided in Chapter 12.

9.3.1.4 Protein Extraction Efficiency
Extraction of protein from plant tissues is a key step to achieve satisfactory protein analysis. Protein extraction is usually thoroughly investigated during assay development and method validation (Chapters 6

and 10). An ideal extraction includes three characteristics: high efficiency, consistent extraction rate, and no alteration in protein conformational/functional structure. It is important to ensure that the extracted protein is still intact and compatible with the analytical method. Any protein denaturation, degradation, or conformational change will likely cause substantial error in measurement. Some proteins are susceptible to heat, and extreme extraction conditions may partially inactivate the target protein. This effect needs to be considered while choosing extraction conditions. Some reagents such as sodium dodecyl sulfate (SDS) are common in Western blot analyses; however, SDS may severely affect the ELISA results since it modifies protein tertiary structure by destroying hydrogen bonds. Extraction efficiency depends on sample type, sample particle size, extraction buffer, extraction time, and buffer–sample ratios. In a quantitative analysis, sample particle size could be a significant factor for data variation. Grothaus et al. (2006) reported an effect of particle size on Cry9C protein extraction efficiency from corn processed products. Flour (fine powder), meal (medium size), and grits (coarse size) were prepared from the same lot of grain and extracted by an identical procedure. Less than 50% of Cry9C protein was extracted in grits compared to that extracted from flour. To avoid such error, it may be necessary to further process the coarse samples to obtain more uniformed particle sizes. An extraction condition for one plant tissue may not be applicable to another. Testing Cry1Ac protein in cotton tissues, Shan et al. (2007) found that buffer to tissue ratios directly affected the extraction efficiency. Under the same extraction procedure for pollen samples, only 26% of Cry1Ac was extracted at a ratio of 60 μL buffer per milligram of sample, and a ratio of 150 is needed to achieve a satisfactory extraction efficiency (94%). However, a ratio of 75 to 1 mg is enough for other tissues such as cotton bolls, seed, and whole plant. Therefore, it is important to thoroughly investigate each parameter and define the optimum extraction condition for a specific sample type. Even under optimum conditions, it is unlikely that all of the target proteins will be extracted from the sample. Usually, 70% or greater extraction efficiency is considered acceptable (Lipton et al., 2000). However, lower extraction efficiency (e.g., 60%) may be still acceptable if it is consistent. In the latter situation, results can be converted using the extraction factor. Extraction efficiency is not as important as consistency of results, since results can be corrected based on a low but consistent extraction efficiency.

9.3.1.5 *Protein Expression Level*
Protein expression levels in genetically modified (GM) plants vary significantly depending on the germplasm (variety), individual plant, plant growth stage, type of tissue, geographic location of the field of planting, and environmental conditions. The same plant tissues can express very different levels of GM protein depending on these factors. According to published results of DAS-59122-7 corn at http://www.aphis.usda.gov/brs/aphisdocs/, the leaf tissue at R4 growth stage has the highest Cry34Ab1 protein expression, which is about four times higher than that in V9 stage leaf tissue (Table 9.3). At the same growth stage, protein expression levels in different parts of a plant may be significant. For example, in a study conducted with the DAS-59122-7 corn event at the R1 stage, the mean Cry35Ab1 protein expression in leaf tissue was 52.2 ng/mg dry weight, while the pollen tissue contained as little as 0.02 ng of Cry35Ab1 protein per milligram of

TABLE 9.3 Comparison of Cry34Ab1 Protein Expression in DAS-59122-7 Maize Leaf at Different Growth Stages

Growth stage	Mean (ng/mg dry weight)	S.D.	Range (ng/mg dry weight)
V9 leaf	49.5	7.79	37–81.4
R1 Leaf	80.6	12.4	59.1–103
R4 Leaf	220	37.5	143–302
R6 Leaf	163	83.6	4.26–296

All data were from six different sites in Chile in 2002–2003 seasons. http://www.aphis.usda.gov/brs/aphisdocs/ 03_35301p.pdf (accessed on February 26, 2008).

dry weight (Table 9.4). It is not uncommon that protein expression levels of the same variety planted in United States are significantly different from that planted in other geographies (Table 9.5). The difference in protein expression levels at the same growth stage, within the same tissue, could vary one to two orders of magnitude across different planting sites within the same country (Table 9.3). In protein quantitative analysis, such variability is a major factor for data discrepancy. To obtain an accurate and consistent measurement of target protein in plant tissues, these potential sources of variation need to be thoroughly considered. Detailed sampling information is very important for data analysis and troubleshooting. Another sampling related source of

TABLE 9.4 Comparison of Cry35Ab1 Protein Expression in Different DAS-59122-7 Maize Tissues (at R1 Growth Stage)

Tissue	Mean (ng/mg dry weight)	S.D.	Range (ng/mg dry weight)
R1 leaf	52.2	12.9	29.2–80.8
R1 root	5.08	1.57	2.49–8.85
R1 pollen	0.02	0.04	0–0.15
R1 stalk	10	2.26	5.64–14.2
R1 whole plant	21.3	3.54	9.02–18.1

All data were from six different sites in Chile in 2002–2003 seasons. http://www.aphis.usda.gov/brs/aphisdocs/ 03_35301p.pdf (accessed on February 26, 2008).

TABLE 9.5 Comparison of CP4 EPSPS Expression in MON 40-3-2 Soybean Leaf (at Second Month Growth Stage) from Different Geographic Locations

Location	Mean CP4 EPSPS (ng/mg fresh weight)	S.D.	Range (ng/mg dry weight)
United States	264	293	46–490
Puerto Rico	657	138	523–798

The U.S. data were from two sites planted in 1992 season. The Puerto Rico data was from one site planted in 1992 season. http://www.aphis.usda.gov/brs/aphisdocs/93_25801p.pdf (accessed on February 26, 2008).

variability is the choice of fresh sample or lyophilized sample. Some laboratories prefer to use fresh sample (e.g., leaf) and the data are described as amount of target protein per unit area. Due to the difference in thickness and moisture content in different regions of the same leaf, higher variability is expected for analyzing fresh leaf samples. In a comparison study conducted in our laboratory using cotton leaf samples expressing Cry1 Bt protein, the fresh leaf group resulted in a high CV of 49%, while lyophilized sample group produced a much lower variability with a CV of 15% (unpublished data, Dow AgroSciences). Lyophilized samples are usually more uniform or homogeneous, and small samples (e.g., 15 mg) are sufficient for a quantitative analysis. On the other hand, a larger sample is recommended for fresh leaf sample such as 1–5 g.

In certain immunoassay formats, including LFDs, high expression levels may cause false-negative (FN) results due to a "hook effect" (Selby, 1999). The "hook effect" is defined by the production of artificially low results from samples that have extraordinarily high concentrations of target protein, far exceeding the concentration of the upper standard in the assay and thus results in a characteristic shape of the analytical dilution curve. It has been described by Rodbard et al. (1978) that the hook effect is most commonly found in LFD assays that have an analyte concentration range over several orders of magnitude.

In a simultaneous one-step assay, the capture antibody, analyte, and detection antibody are incubated together. The huge excess of free analyte would saturate specific and less specific binding sites on both antibodies and thereby limit the formation of the sandwich. This causes low-affinity attachment of the detection antibody to the solid phase. In the presence of huge excesses of analyte, the protein may weakly bind to secondary binding sites (less specific binding) of the capture antibody and directly to the solid phase support. Although the protein may bind to the detection antibody to form an antigen–antibody complex, it may eventually be desorbed during the washing step, which again prevents the formation of the sandwich on the plate (Selby, 1999). Ryall et al. (1982) used a two-site model and simulated the events leading to a hook effect. Based on calculations of chemical equilibrium, the cause of the hook effect is governed by three factors: the quantity of analyte immobilized to the capture antibody in the first stage, the extent of competing side reactions in the second stage, and the extent of complexing immobilized analyte by detection antibody in the second stage. The authors concluded that the hook effect was much more likely to occur in the absence of an adequate secondary antibody concentration.

Fenando and Wilson (1992a), (1992b) studied reagent-limited assays and both single-step and two-step reagent excess assays and concluded that in reagent-limited assays the hook effect was the result of cooperative interactions between mixtures of antibody molecules forming a linear or circular aggregate. In a two-step sandwich assay, the hook effect is caused by desorption of bound analyte due to a conformational change after reaction of the labeled antibody with several epitopes of the adsorbed analyte.

However, kit manufacturers usually have optimized and validated the assay with high expressing varieties to insure that this does not occur, and thus following the user's guide during testing is required.

9.3.2 Analytical/Instrumental Source

9.3.2.1 Lateral Flow Devices
In general, a LFD gives a qualitative result: positive or negative. Other than a test line(s), an LFD also contains a control line, which should be present in any assay. If this line is absent, it indicates that the device is not functional and the test result is invalid. LFDs are usually simple, robust, and easy to perform; however, there are a few sources of error that can cause either false-positive or false-negative results.

LFDs are typically manufactured under conditions of low humidity and packaged with desiccant to keep them dry. It is important to keep the LFD kit in an appropriate storage condition recommended by manufacturer for a long shelf life. Humidity can shorten a LFD's shelf life, which may cause false positives or false negatives due to aggregation of the colloidal gold and/or loss of antibody activity (Grothaus et al., 2006).

One common problem an analyst may have in using a LFD is impaired liquid flow, which may result in either false-positive or false-negative readings. The major causes of impaired liquid flow include the extraction ratio of buffer volume to sample (μL/mg sample) being too low or a sample being too finely ground. In addition, oily components in the sample may impede fluid flow. Fine particles can clog the membrane and block the gold conjugate from entering the strip. Partially clogged membranes may cause nonspecific binding at the test line by preventing sufficient washing past the test line, which may produce a faint false-positive reading. On the other hand, insufficient analyte–gold complex reaching the test line may result in a false-negative reading or difficulty in interpreting results. It is necessary to check the sample extract prior to assay and follow the instruction procedure carefully.

Improper handling of LFDs by the operator is a common cause of errors. LFD kits usually have a sample volume limit and a maximum depth that the strip can be inserted in the sample. If the strip is immersed too deep into the sample extract, the conjugate pad may be placed under the liquid, which may cause gold conjugate to be dissolved and released into the liquid. As a result, the test line will be absent or very faint. If the LFD assay is too sensitive or the level of target protein in testing matrix is extremely high, false positives may become more common due to contamination. Keeping the testing area organized and clean is essential to avoid such errors.

Young plant tissues generally contain high concentration of chlorophyll that may bind nonspecifically to the test line of LFD resulting in a light green line. An inexperienced analyst may interpret such a test as positive. Usually, the manufacturers have optimized and validated the devices with target plant matrix and the green line signal is minimized or eliminated. If the LFD is going to be applied to a new type of plant matrix, it is necessary to perform some preliminary testing or optimization with the matrix. Analysts may refer to Chapter 5 for more detail and specific information. In general, a LFD is a simple and robust analytical tool for all levels of operators. If the analyst reads the instruction carefully and follows the procedure step by step, an accurate and reliable result is easily achievable.

9.3.2.2 ELISA Plate Assay
Common sources of error in ELISA plate assays are discussed in the literature (Butler, 1991; Tijssen, 1985; Grothaus et al., 2006;

Law, 1996) and readers may use these references for more detail. Basic trouble-shooting skills are important for any analyst to successfully perform daily analyses. Additional discussion of common error sources and basic troubleshooting tips are summarized (Wu, 2000; Tulip Group, 2008; Abcam technical support, 2008) in Table 9.6.

One common goal for an immunoassay laboratory or ELISA kit manufacturer is to develop a robust assay or a user-friendly kit in which all the reagents are ready to use. In high-throughput laboratories, automation is the choice for routine analysis. However, for whatever level of robustness or automation is achieved, one must accept that problems will occur since human error will be part of the assay. Although ELISA is a simple and relatively easy method, basic training is necessary for a person new to this technology to avoid the most common errors, such as not following the product instructions, improper pipetting of samples and reagents, and improper or inconsistent washing of microplates. It is desirable to have experienced laboratory personnel perform the assay, since one of the key attributes of a good analyst is the ability to troubleshoot and quickly find a solution when a problem occurs. Here, we discuss some common problems in routine analysis to aid analysts in solving problems in a timely fashion.

One commonly encountered problem in plate format ELISA is random high OD wells on the plate called "hot spots." This usually indicates nonspecific binding of enzyme in the wells. This can occur for a number of reasons, such as incomplete washing (due to a malfunctioning plate washer), buffer contamination, or reagent aggregation. The washing step is the most common cause for "hot spots." If an automatic washing system is used in the laboratory, needle clogging by salts or mold is very common. Plate washers need to be cleaned with detergent and bleached regularly. To diagnose whether or not the washer is the source, a quick approach is to repeat the assay and wash the plate by hand or to rotate the plate 180° during washing steps. In some cases, higher detergent concentration is used in the washing buffer and this may cause severe foaming during the process and extra liquid along with conjugate may remain in the well even after tapping the plate on paper towels. A simple practice is to gently rinse the plate with distilled water once before drying the plate. Another possible cause is that the assay buffer or washing buffer is contaminated or the container contains mold. Analysts need to check the buffer container and make sure no visible particles are observed in the solution. If it is suspected that the buffer may be the cause, a new buffer should be prepared in a clean container and the assay repeated. Finally, aggregation of the enzyme conjugate can also lead to hot spots. In this case, other symptoms such as lower overall ODs in the assay may be present. A good practice is to pass the antibody conjugate through a 0.22 μm filter to remove any aggregates that may have been formed during storage.

Another common problem in ELISA is the edge effect, in which the edge wells of a plate produce high OD readings. This may be caused during the plate coating and blocking process. More details on this issue are discussed in Chapter 4. Humidity and temperature in the laboratory play important roles in ELISA reactions. In a routine analytical laboratory, the edge effect often occurs when the plate is used before it warms to room temperature. This can usually be avoided by bringing the kit to ambient temperature for at least 30 min prior to initiating the assay. In addition, dry

TABLE 9.6 ELISA Troubleshooting

Problem	Possible causes	Corrective steps
Positive results in negative control wells	Contamination of negative control wells	Washing carefully, do not allow wells to overflow
	Contamination of negative control samples and reagents	Use fresh reagents and pipette carefully
	Insufficient washing	Check washer and make sure they are working properly. Perform routine maintenance
	Nonspecific binding due to too much antibody used	Check the recommended amount of antibody. Try using less antibody if needed
High background across the plate	Conjugate too concentrated or incubated too long	Check dilution of conjugate; use it at the recommended dilution
	Substrate solution or stop solution is not fresh	Use fresh substrate solution. Visually check substrate and stop solutions and make sure they are in good condition
	Waited too long before reading the plate	Read the plate within 30 min
	Contaminants from laboratory glassware	Ensure reagents are fresh and prepared in clean container.
	Incubation temperature too high	Ensure that the incubation temperature is correct. The normal range is 22–25°C
	Nonspecific binding of antibody	Make sure a blocking step is included and a suitable blocking buffer is used
	Insufficient washing	Make sure wells are completely filled. While washing ensure residual reagents is removed from well
Low positive control value or low absorbance	Incubation time too short	Record time of incubation. Ensure recommended incubation time is applied
	Incubation temperature too low	Check the temperature in work area or the incubator
	Substrate solutions not fresh or combined incorrectly	Ensure substrate solution is not expired
	Reagents not fresh or prepared in wrong buffer (e.g., pH)	Ensure reagents are prepared correctly and not expired
	Insufficient antibody	Check that the recommended antibody being used

Symptom	Possible cause	Solution
Inconsistent absorbance across the plate	Added sample volume too low	Ensure pipette tips are fitted correctly and tightly. Ensure pipette is calibrated
	Stop solution not added	Stop solution increases the color intensity and stabilizes the final color
	Washing too vigorous	Adjust the pressure in washing system
	Pipeting inconsistent	Ensure pipettes are working correctly and are calibrated
	Antibody/conjugate not well mixed	Ensure all reagents and samples are mixed well before transferring onto the plate; visually check the reagent solution whether any particle exist. If needed, filter the solution before use (0.2 μm)
	Plates stacked during incubation	Avoid stacking; stacking of plates affect even distribution of temperature across the plate
Substrate solution is blue	Substrate solution is contaminated	Use fresh substrate solution
Color develops too slow	Incubation temperature too low	Check the temperature in work area or the incubator
	Conjugate solution concentration is too low	Ensure the conjugate stock is in date and stored properly. Ensure it is used at the correct concentration.
Color develops too fast	Contaminated with enzyme	Make sure all reservoirs and containers are clean
No color even after 30 min incubation with substrate	Substrate is not working	Use or prepare a fresh substrate solution
	Conjugate solution is not working	Test an aliquot of conjugate solution with substrate:(1) If it turns blue (HRP) or yellow (AP), either a wrong conjugate reagent is used or the conjugate has degraded. Use correct conjugate reagent or contact manufacturer2) If it shows no color, the conjugate is no good any more. Use new conjugate reagent
Stop solution is yellow	Contamination	Use fresh stop solution

Source: Adapted from Wu 2000 and Abcam technical support 2008.

laboratory conditions or incubation of ELISA plates close to air conditioning or heating ducts may result in uneven evaporation of liquid from the wells, especially on the periphery of the plate. The best practice is to cover the plate during incubation.

Many immunoassay laboratories likely have experienced problems with sudden low ODs or loss of binding. For sandwich ELISA, this may be caused by decomposition of conjugate, denaturing of capture antibody, or degradation of protein standard.

Numerous laboratories routinely use automated plate washers and plate readers for ELISA. Automated plate washers may be problematic if they are not properly maintained. In particular, washers may not perform well when washing plates with samples containing particulate matter that may block the washing probes. This results in inconsistent washing that can have a dramatic impact on many aspects of method performance including background, accuracy, and precision. To avoid the malfunction of a washer, it is necessary to schedule regular cleanings with bleach and mechanical checks every 3 months or as needed. As a daily practice, other than routine rinses including priming before and after use each day, we recommend rinsing with distilled water after each use. Water left in the system will prevent salt crystallization during the day. In some laboratories, automated liquid handling systems are used for higher throughput analysis. The major error in automated systems is inaccuracy of liquid transfer due to bubbles in the system. Therefore, it is important to ensure that there are no air bubbles in the tubes and syringes before use. With proper maintenance and calibration, plate readers, plate washers, and liquid handling equipment contribute very little error to ELISA methods.

REFERENCES

Abcam technical support, *Troubleshooting tips—ELISA*, available at http://www.abcam.com/index.html? pageconfig=resource&rid=11390 (accessed on September 4, **2008**).

Anderson, J. W. and Rowan, K. S. Extraction of soluble leaf enzymes with thiols and other reducing agents. *Phytochemistry.* **1967**, *6*, 1047–1056.

Barrett, A. J., Kembhavi, A. A., Brown, M. A., Kirschke, H., Knight, C. G., Tamai, M., and Hanada, K. L-trans-Epoxysuccinyl-leucylamido(4-guanidino)butane (E-64) and its analogues as inhibitors of cysteine proteinases including cathepsins B, H and L. *Biochem. J.*, **1982**, *201*, 189–198.

Beynon, R. J.and Bond, J. S. (Eds.). *Proteolytic Enzymes: A Practical Approach*, 1994, pp. 241–249.

Bindslev-Jensen, C., Briggs, D., and Osterballe, M. Can we determine a threshold level for allergenic foods by statistical analysis of published data in the literature? *Allergy*, **2002**, 741–746.

Butler, J. E. *Immunochemistry of Solid-Phase Immunoassay*, CRC Press, Boca Raton, FL, 1991.

Dalling, M. J. (Ed.). *Plant Proteolytic Enzymes*, Vols *1 and 2*, CRC Press, Boca Raton, FL, 1986.

Delaney, B., Atwood, J. D., Cunny, H., Conn, R. E., Hetouet-Guicheney, C., MacIntosh, S., Meyer, L. S., Privalle, L., Gao, Y., Mattsson, J., and Levine, M. Evaluation of protein toxicity in the context of agricultural biotechnology. *Food Chem. Toxicol.*, **2008**, *46*, S71–S97.

DeSilva, B., Smith, W., Weiner, R., Kelley, M., Smolec, J., Lee, B., Khan, M., Tacey, R., Hill, H., and Celniker, A. Recommendations for the bioanalytical method validation of ligand-binding assays to support pharmacokinetic assessments of macromolecules. *Pharma. Res.* **2003**, *20*, 1885–1900.

Devaraja, N. K., Savithri, H. S., and Muniyappa, V. Purification of Tomato leaf curl Bangalore virus and production of polyclonal antibodies. *Curr. Sci.*, **2005**, *89*, 181–183.

Diamandis, E. P. and Christopoulos, T. K. Curve fitting of immunoassay data. In Diamandis, E. P.and Christopoulos, T. K. (Eds.), *Immunoassay*, Academic Press, New York, 1996, pp. 42–50.

Ensinck, J. W., Shepard, C., Dudl, R. J., and Williams, R. H. Use of benzamidine as a proteolytic inhibitor in the radioimmunoassay of glucagon in plasma. *J. Clin. Endocrinol. Metab.*, **1972**, *35*, 463–467.

Evans, S. L. Equivalency of microbial and maize expressed Cry1F protein; characterization of test substances for biochemical and toxicological studies. *Submitted to United States Environmental Protection Agency*, MRID# 44714803, 1998.

Fenando, S. A. and Wilson, G. S. Studies of the hook effect in the one step sandwich immunoassay. *J. Immunol. Methods*, **1992a**, *151*, 47–66.

Fenando, S. A. and Wilson, G. S. Multiple epitope interactions in the two-step sandwich immunoassay. *J. Immunol. Methods*, **1992b**, *151*, 67–86.

Findlay, J. W. A. and Dillard, R. F. Appropriate calibration curve fitting in ligand binding assays. *AAPS J.*, **2007**, *9*, E260–E266.

Gegenheimer, P. Preparation of extracts from plants. In Weissbach, A. and Weisbach, H. (Eds.), *Methods in Enzymology*, Vol. 182, Academic Press, New York, 1990, pp. 174–193.

Gray, J. C. Use of proteolytic inhibitors during isolation of plastid proteins. In Edelman, M., Hallick, R. B., and Chua, N. H. (Eds.), *Methods in Chloroplast Molecular Biology*, Elsevier, Amsterdam, 1982, pp. 1093–1102.

Grothaus, G. D., Bandla, M., Currier, T., Giroux, R., Jenkins, G. R., Lipp, M., Shan, G., Stave, J. W., and Pantella, V. Immunoassay as an analytical tool in agricultural biotechnology. *J. AOAC Int.*, **2006**, *89*, 913–928.

Herman, R. A., Scherer, P. N., and Shan, G. Evaluation of logistic and polynomial models for fitting sandwich-ELISA calibration curves. *J. Immunol. Methods*, **2008**, *339*, 245–258.

Higgins, K. M., Davidian, M., Chew, G., and Burge, H. The effect of serial dilution error on calibration inference in immunoassay. *Biometrics*, **1998**, *54*, 19–32.

Janas, R. M., Marks, D. L., and Larusso, N. F. Purification and partial characterization of a heat-resistant, cytosolic neuropeptidase from rat liver. *Biochem. Biophys. Res. Commun.*, **1994**, *198*, 574–581.

Kassell, B. Naturally occurring inhibitors of proteolytic enzymes. *Methods Enzymol.*, **1970**, *19*, 839–905.

Koenig, R. Indirect ELISA methods for the broad specificity detection of plant virus. *J. Gen. Virol.*, **1981**, *55*, 53–62.

Laties, G. G. and Treffry, T. Reversible changes in conformation of mitochondria of constant volume. *Tissue Cell*, **1969**, *1*, 575.

Law, B. Assay problems and troubleshooting. *Immunoassay: A Practical Guide*, Taylor & Francis, Bristol, PA, 1996, pp. 204–212.

Lieberman, M. Oxidative activity of cytoplasmic particles of apples: electron transfer chain. *Plant Physiol.*, **1960**, *35*, 796–801.

Lieberman, M. and Biale, J. B. Oxidative phosphorylation by sweet potato mitochondria and its inhibition by polyphenols. *Plant Physiol.*, **1956**, *31*, 420–424.

Lipton, C. R., Dautlick, J. X., Grothaus, G. D., Hunst, P. L., Magin, K. M., Mihaliak, C. A., Rubio, F. M., and Stave, J. W. Guidelines for the validation and use of immunoassays for determination of introduced proteins in biotechnology enhanced crops and derived food ingredients. *Food Agric. Immunol.*, **2000**, *12*, 156–164.

Little, J. A. Comparison of curve fitting models for ligand binding assays. *Chromatogr. Suppl.*, **2004**, *59*, S177–S181.

Loomis W. D. Overcoming problems of phenolics and quinones in the isolation of plant enzymes and organelles. In Fleischer S. and Packer L. (Eds.), *Methods in Enzymology*, Vol. 31, Academic Press, 1974, pp. 528–539.

Loomis, W. D. and Battaile, J. Plant phenolic compounds and the isolation of plant enzymes. *Phytochemistry*, **1966**, *5*, 423–438.

Marwardt, F., Hoffman, J., and Korbs, E. The influence of synthetic thrombin inhibitors on the thrombin-antithrombin reaction. *Thromb. Res.*, **1973**, *2*, 343–348.

Means, G. E. and Feeney, R. E. *Chemical Modification of Proteins.* Holden-Day, San Francisco, CA, 1971, p. 152.

Miyata, S., Shimazaki, T., Okamoto, Y., Motegi, N., Kitagawa, M., and Kihara, H. K. Inhibition Of RNA synthesis in embryo of *Xenopus Zaeuis* by protease inhibitor. *J. Exp. Zool.*, **1988**, *246*, 150–155.

Mortensen, A. M. and Novak, R. F. Dynamic changes in the distribution of the calcium-activated neutral protease in human red blood cells following cellular insult and altered Ca^{2+} homeostasis. *Toxicol. Appl. Pharmacol.*, **1992**, *117*, 180–188.

Plikaytis, B. D., Holder, P. F., Pias, L. B., Maslanka, S. E., Gheesling, L. L., and Carlone, G. M. Determination of parallelism and nonparallelism in bioassay dilution curves. *J. Clin. Microbiol.*, **1994**, *32*, 2441–2447.

Prouty, W. F. and Goldberg, A. L. Effects of protease inhibitors on protein breakdown in *Escherichia coli*. *J. Biol. Chem.*, **1972**, *247*, 3341–3352.

Raison, J. K. and Lyons, J. M. The influence of mitochondrial concentration and storage on the respiratory control of isolated plant mitochondria. *Plant Physiol.* **1970**, *45*, 382–385.

Riordan, J. F. and Vallee, B. L. Reagent for the modification of sulfhydryl groups in proteins. *Methods Enzymol.*, **1972**, *25*, 449–456.

Rodbard, D., Feldman, Y., Jaffe, M. L., and Miles, L. E. Kinetics of two-site immunoradiometric (sandwich) assays-II. Studies on the nature of the "high dose hook effect." *Immunochemistry*, **1978**, *15*, 77–82.

Rozet, E., Ceccato, A., Hubert, C., Ziemons, E., Oprean, R., Rudaz, S., Boulanger, B., and Hubert, P. Analysis of recent pharmaceutical regulatory documents on analytical method validation. *J. Chromatogr. A*, **2007**, *1158*, 111–125.

Ryall, R. G., Story, C. J., and Turner, D. R. Reappraisal of the causes of the "hook effect" in two site immunoradiometric assays. *Anal. Biochem.*, **1982**, *127*, 308–315.

Ryan, C. A. and Walker-Simmons, M. Plant proteases. In Marcus, A. (Ed.), *The Biochemistry of Plants*, Vol. 6, Academic Press, New York, 1981.

Schrader, L. E., Catalog, D. A., Peterson, D. M., and Vogelzang, R. D. Nitrate reductase and glucose-6-phosphate dehydrogenase activities as influenced by leaf age and addition of protein to extraction media. *Physiol. Plantarum*, **1974**, *32*, 337–341.

Selby, C. Interference in immunoassay. *Ann. Chin. Biochem.*, **1999**, *36*, 704–721.

Shan, G., Embrey, S. K., and Schafer, B. W. A highly specific Enzyme-linked immunosorbent assay for the detection of Cry1Ac insecticidal crystal protein in transgenic WideStrike cotton. *J. Agric. Food Chem.*, **2007**, *55*, 5974–5979.

Chesworth, J. M., Stuchbury, T., and Scaife, J. R. *An Introduction to Agricultural Biochemistry*, Chapman and Hall, 1998.

Sigman, D. S., Kuwabara, M. D., Chen, C. H., and Bruice, T. W. Nuclease activity of 1,10-phenanthroline-copper in study of protein–DNA interactions. *Methods Enzymol.*, **1991**, *208*, 414–433.

Spahr, P. F. and Edsall, J. T. Amino acid composition of human and bovine serum mercaptalbumins *J. Biol. Chem.* **1964**, *239*, 850–854.

Throneberry, G. O. Isolation of metabolically-active subcellular particles from etiolated cotton seedling hypocotyls using bovine serum albumin in preparative medium. *Plant Physiol.*, **1961**, *36*, 302.

Tijssen, P. *Practice and Theory of Enzyme Immunoassays*, Elsevier, Amsterdam, The Netherlands, 1985.

Tulip Group. *ELISA Troubleshooting Aspects*, available at http://www.tulipgroup.com/Common/html/ELISATech.pdf (accessed September 4, **2008**).

Umezawa, H. Structures and activities of protease inhibitors of microbial origin. *Methods Enzymol.*, **1976**, *45*, 678–695.

Warwick, M. J. Standardisation of immunoassay. In Law, B. (Ed.), *Immunoassay: A Practical Guide*, Taylor and Francis, Bristol, PA, 1996, pp. 150–170.

Wilkes, S. H. and Prescott, J. M. The slow, tight binding of bestatin and amastatin to aminopeptidases. *J. Biol. Chem.*, **1985**, *260*, 13154–13162.

Wu, J. T. *Quantitative Immunoassay: A Practical Guide for Assay Establishment, Troubleshooting, and Clinical Application*, AACC Press, 2000.

Zollner, H. (Ed.). *Handbook of Enzyme Inhibitors*, 2nd edition, VCH, Weinheim, 1993.

Zyznar, E. S. A rationale for the application of trasylol as a protease inhibitor in radioimmunoassay. *Life Sci.*, **1981**, *28*, 1861–1866.

CHAPTER 10

IMMUNOASSAY APPLICATIONS IN TRAIT DISCOVERY, PRODUCT DEVELOPMENT, AND REGISTRATION

Beryl Packer
Andre Silvanovich
G. David Grothaus

10.1 INTRODUCTION

During the development and commercialization of biotechnology-derived crops (GE or GM crops), immunoassays are used over an extensive range of applications from initial discovery during research and development, through proof of concept, international product registration and reregistration, and long after product launch until a GM crop is discontinued. Agricultural biotechnology (Ag Biotech) companies use results from immunoassays to influence key decisions ranging from those made in early product discovery to those made in production, sales, and support of GM crops.

Immunoassays in Agricultural Biotechnology, edited by Guomin Shan
Copyright © 2011 John Wiley & Sons, Inc.

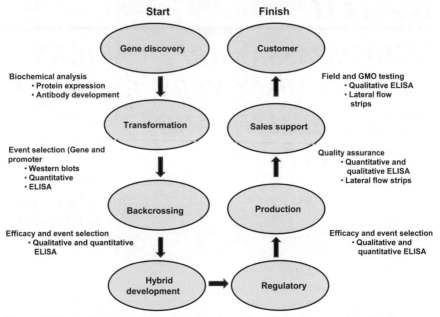

Figure 10.1 Applications of immunoassays in product development and registration.

The commercialization of GM crops commenced well over a decade ago with the launch of Roundup Ready® soybean containing CP4 EPSPS synthase that confers tolerance to glyphosate (Padgette et al., 1996) and insect-resistant YieldGard® corn (MON810) containing Cry1Ab protein. Immunoassays were critical to the development of these initial biotechnology-derived crops and have remained essential to the development and support of subsequent GM products.

Figure 10.1 shows some of the key phases of product development, registration, and commercialization of biotechnology-derived crops and some of the different immunochemical techniques that are typically applied during each phase. Quantitative, semiquantitative, and qualitative immunochemical methods all play an important role at various stages in the product development process.

10.2 DISCOVERY, TRANSFORMATION, AND EVENT SELECTION

The discovery stage is characterized by activities involving trait and gene identification and gene cloning, followed by transformation of a model organism or crop plant of interest. Through the evaluation of agronomic and/or phenotypic characteristics of the transformed plants, the transformation events that produce a desirable trait are identified. For example, if the product concept is to increase yield, then transformation events demonstrating yield within a defined range or above a certain level relative to the parental control that had not undergone transformation, or the negative isoline that had gone through the transformation process but did not contain the trait, would be selected for further analysis and characterization. Events showing commercial promise are advanced into a proof of concept stage where immunodiagnostic tools are developed and used for product characterization.

During the discovery phase, immunodiagnostic methods and reagents may not always be necessary, but they are in some instances invaluable. For instance, if a particular construct or a set of constructs yields no distinctive phenotype, immunoassays are used to determine whether or not the protein of interest is being produced. The absence of expression of the protein of interest might lead to an evaluation of the constructs used for transformation, whereas if the protein of interest is being produced, the absence of indicative phenotype might be the product of improper protein localization at the tissue or subcellular level, or the lack of function, both of which might require further investigation.

In the proof of concept stage of product development, immunoblotting, ELISA, immunoprecipitation, and immunohistochemical localization methods are used to address diverse questions regarding basic product characterization, mode of action, and event selection.

Immunoblotting can be used to determine the relative or absolute quantity of the protein of interest, to determine protein stability and whether the protein undergoes posttranslational degradation/modification, and to determine temporal and spatial protein production. Temporal gene expression profiles provide an important characterization of gene function, whereas spatial gene expression profiles demonstrate the expression pattern or location of protein expression in the transformed plant itself. Both are important in protein characterization and the assessment of protein and plant safety because it is important to know not only whether the gene of interest is being expressed, but also at what levels it is expressed in the various plant tissues. This information is thoroughly evaluated when assessing potential allergenicity and toxicity of food and feed derived from GM plants.

- Immunoprecipitation may be used to determine if the protein of interest is part of a complex. If it is so, it can be used to help identify other members of the protein complex by permitting the rapid enrichment of complexes for subsequent analysis by quadrupole/time-of-flight/matrix-assisted laser desorption/ionization (QTOF-MALDI) or sodium dodecyl sulfate polyacrylamide gel electrophoresis (SDS-PAGE) and N-terminal sequencing.

- ELISA analysis, either qualitative or quantitative, can be employed as a means of comparison of protein levels between transformation events, or absolute quantitation of protein levels. More details regarding these technologies are provided in Chapters 4 and 5.

- Immunohistochemical staining is employed to determine intracellular protein localization. This can be important in determining mode of action, and with the advent of next-generation GM crops containing novel traits that have not been previously characterized, this immunological remains an important tool in the arsenal of immunochemical test methods.

During the proof of concept stage, immunoreagents and test methods are invaluable as they facilitate the early characterization of transgenic events and provide a foundation for regulatory studies needed for further development and eventual commercialization. The following detailed descriptions of these immunoassay methods provide a framework for understanding their applications during the course of GM crop product development.

10.2.1 Immunoblotting/Western Blots

Immunoblotting became a widely used method using antibodies to characterize proteins after methods were developed to transfer proteins that had been separated electrophoretically on a gel to an immobilized matrix (Gershoni and Palade, 1983). Immunoblotting can be used to determine the relative or absolute quantity of the protein of interest, to determine protein stability and whether the protein undergoes posttranslational degradation/modification, and to determine temporal and spatial protein expression through the analysis of specific tissues or tissue collection time points, respectively. Immunoblotting is also an important tool in determining mechanism of action (MOA) of a particular introduced protein, where certain proteins may only be expressed or efficacious under specific conditions such as drought, high or low temperature, or varied lengths of light exposure.

Several aspects of immunoblotting make it an ideal tool for use during the proof of concept stage. First, immunoblotting is relatively insensitive to tissue extraction conditions because sample extracts are subjected to SDS-PAGE and electroblotting to a nitrocellulose or polyvinylidene difluoride (PVDF) membrane prior to protein detection. As a consequence, harsh extraction buffers of nonphysiological pH containing detergents or reducing agents may be used. The harsh buffers, sample preparation conditions used for SDS-PAGE, and the PAGE have the added advantage of effectively eliminating protein complexes and tertiary structures that could mask antibody (Ab) binding to epitopes. Second, immunoblotting yields molecular weight information for detected proteins. Through molecular weight estimates and the banding pattern of the protein of interest, it is possible to assess whether a protein is intact, has undergone processing such as removal of a targeting peptide, is unstable and rapidly degraded (Doran, 2006), or has undergone certain posttranslational modifications such as N-linked glycosylation (Sonnewald et al., 1990). Proteins that undergo posttranslational removal of a targeting peptide will by definition have a predictably lower molecular weight. Proteins that are subject to degradation often display "chatter," where a band of the expected molecular weight and a series of bands of a lesser molecular weight are observed. Glycosylated proteins tend to display a diffuse banding pattern higher than predicted molecular weight. Finally, immunoblotting is able to yield qualitative, and under certain circumstances quantitative, information on the abundance of the protein of interest. For instance, it is possible to evaluate levels of a protein of interest in multiple transgenic events by comparing the band intensities within an immunoblot of comparably prepared and diluted sample extracts (Ares et al., 1998). For absolute quantitation of a protein of interest, blotting can be performed using test samples and a series of lanes containing serial dilutions of the protein of interest. Once the blot is developed, a densitometric scan can be performed on the band(s) produced by the serially diluted protein of interest, and a standard curve can be established. Band intensities for the protein of interest in test samples can then be interpolated from the standard curve to get a quantitative estimate of the levels of the protein of interest (Duncan et al., 2006).

It should be noted that quantitative immunoblotting is subject to two significant limitations in its applications. The first limitation is matrix effects. Depending upon the complexity of the sample extract loaded onto a gel and the identity of the protein of interest, the per unit mass of signal produced by a plant-produced protein may be higher or lower than the signal produced by the purified protein standard spiked into a plant extract matrix or the purified protein in the absence of matrix. The second limitation is that the usable linear range of the protein standards determines the ability to quantitatively evaluate samples using an immunoblot. Typically, the linear range is approximately one order of magnitude (although a number of product vendors have suggested that two orders of magnitude are possible) for blots developed using either chromogenic or chemiluminescent substrates.

Although immunoblotting has several excellent attributes, it is a labor-intensive activity not well suited to automation, as it involves a complicated gel manipulation step prior to the electrotransfer to PVDF or nitrocellulose. In addition, sample throughput is limited by gel size and well number with a practical upper limit of 20 wells per gel.

Useful references for Western blot or immunoblotting protocols can be found in Towbin's publications (Towbin and Gordon, 1984; Towbin et al., 1979). The following is a general protocol that would typically be used for a Western blot:

1. Proteins are separated by size using SDS-PAGE. A nitrocellulose membrane is placed on the gel and electrophoresis is used to transfer the protein (polypeptide) bands onto the membrane. The gel should face the negatively charged cathode while the nitrocellulose membrane should face the positively charged anode. In this configuration, proteins that have an overall negative charge due to hydrophobic interaction with SDS will, in the presence of an applied electrical potential, travel toward and bind to the membrane. As a result, the nitrocellulose membrane will have a faithful transfer of protein that parallels the pattern of the source gel.

2. A variety of parameters such as transfer buffer pH, transfer voltage and current, and transfer temperature are critical for successful transfer in protein blotting. Likewise, simple mechanical manipulations such as removing all air bubbles from the gel–nitrocellulose sandwich prior to the transfer helps the application of voltage and assures a uniform transfer of proteins and an accurate representation of the gel electrophoresis.

3. Following electroblotting, the nitrocellulose membrane is blocked using a protein such as BSA or a nonionic detergent prior to incubation with a primary antibody specific for the target protein. The purpose of blocking is to fill any unoccupied protein binding sites on the nitrocellulose membrane such that antibodies specific for the protein of interest do not uniformly bind to the entire membrane, but rather only to the region of the blot containing the immobilized protein of interest. For example, if a plant is transformed to contain CP4 EPSPS

synthase, then the primary antibody needs to specifically bind to the CP4 EPSPS protein after it is electrophoretically transferred to the nitrocellulose, forming a protein–antibody complex. In the absence of blocking, indiscriminate binding of the anti-CP4 EPSPS antibody would yield a blot displaying substantial background signal such that it might obscure the signal that reflects binding of an antibody to CP4 EPSPS. Likewise, the specificity of the antibody is critical to assuring that only the target protein, and not the many thousands of endogenous proteins, is detected.

4. Following washes with an appropriate buffer, the nitrocellulose membrane containing the target protein–antibody complex is then incubated with a secondary antibody that is conjugated to an enzyme used for visualization of the target protein. This secondary antibody is an antibody–enzyme conjugate such as HRP-anti-IgG, where HRP is horseradish peroxidase enzyme conjugated to an antibody that reacts specifically with the antibody already bound to the target protein.

5. The nitrocellulose membrane now contains the HRP-anti-immunoglobulin bound to the primary antibody that in turn is bound to the immobilized target protein. This membrane is then incubated with a buffer containing substrate for the HRP enzyme to produce a visual color wherever there is a protein–primary antibody–secondary antibody–enzyme complex, that is, wherever the target protein is located on the nitrocellulose. With good technique, appropriate methodology, and well-characterized antibody reagents, Western blotting artifacts such as high background, band distortions, and inconsistent transfer can be eliminated.

10.2.2 Immunoprecipitation

If a transgene encoded protein is contained in a protein complex *in planta*, the identification of partner proteins may be of considerable importance in elucidating a mechanism of action that yields the desired trait. Immunoprecipitation or immunoaffinity purification of the protein of interest will frequently result in sufficient enrichment of protein complexes, if they exist, to facilitate identification of partner proteins. Protein complexes are typically captured using antibodies that are bound to immobilized protein A (Goding, 1978) or protein G. Tissue extracts prepared in a physiological buffer are applied to the immobilized antibodies following an incubation period to allow binding of the protein complexes to antibodies. The captured antibody–protein complexes are washed to remove loosely associated proteins and a buffer is applied to release the captured antibodies and the complexes. The released proteins can then be prepared for analysis using SDS-PAGE and, in some instances, electroblotting. Gels can be stained, and bands can be excised and processed for MALDI-TOF spectroscopy (Zhang and Guy, 2005). Electroblots can be stained to identify protein bands that can then be excised and analyzed by Edman degradation to determine N-terminal sequence (Smith et al., 2004). Electroblots can also be probed if antibodies are available that can be used to identify potential partner proteins.

The following is a general protocol that would typically be used for immunoprecipitation:

1. A primary antibody specific for the target protein is incubated with protein A or protein G-sepharose beads either in suspension or in a column packed with the latter complex.

2. A suspension containing a target protein is then incubated with the protein A (or G)-sepharose beads so that target protein can bind to the protein A or G that is later pelleted and washed to remove extraneous proteins or other debris. In the case of immunoaffinity columns, the solution containing the target protein is run through the column, followed by extensive washing.

3. The protein A (or G)-sepharose–target antibody complex is then subjected to conditions that result in the elution of the target protein.

4. The eluted, purified protein can now be subjected to Western blotting, dot blotting, or some other immunodetection method.

10.2.3 Immunohistochemical Staining

Intracellular protein localization can provide valuable information on protein function, potential binding partners, and mode of action. Rudimentary intracellular localization information can be obtained through the analysis of subcellular fractions prepared using centrifugation. The subcellular fractions can then be analyzed using techniques such as enzyme activity assays, ELISA analyses, or immunoblots that permit positive protein identification to establish the relationship between protein and its location in a cell. Immunocytochemistry, the localization of proteins within a cell using antibody reagents, enhances localization information beyond organelles or subcellular fractions by yielding additional detail that could be related to cell structure (Walker et al., 1999, 2001) and cell cycle.

The primary detection reagent for immunohistocytochemistry is an antibody that binds the protein of interest. This bound primary antibody may be tagged or detected using a secondary antibody or protein A or protein G that is tagged with gold particles, enzymes, or fluorophores. Depending on the tag conjugated to the primary or secondary antibody (or protein A or G), specific microscopic methods are used to visualize those tags.

Prior to performing immunohistocytochemistry, tissue samples must be prepared in such a way that subcellular structures remain intact and are able to withstand manipulation, yet are of sufficient porosity to allow the infiltration of the detection antibodies and reagents. Processing is initiated when tissue samples are fixed by exposure to low concentrations (1–2%) of formaldehyde or glutaraldehyde, which cross-link proteins via lysine residues. Following fixation, the tissue samples are washed extensively, infiltrated with an unpolymerized resin, and polymerized. The polymerized resin is sectioned using a microtome and the sections are affixed to a microscope slide for optical microscopy or a metal grid for electron microscopy. The immobilized thin sections are then subjected to a series of incubation steps that include blocking, washing, application of the primary antibody, washing, and application of the tagged antibody (or protein A/G) detection reagent, prior to microscopic visualization.

For observation of secondary antibodies tagged with fluorophore, epifluorescence or confocal microscopy is used. In epifluorescence microscopy, samples are illuminated with a wavelength of light that excites the fluorophore tag and results in the emission of light of a longer wavelength. The emitted light passes through a filter that blocks the shorter excitation wavelength and allows the specific observation of the tagged antibodies. Confocal microscopy also uses light of an appropriate wavelength to excite the fluorophore; however, the excitation light is first passed through a pinhole to yield a point source of illumination. Light emitted by the fluorophore also passes through a pinhole aperture and generates an optically conjugated plane. Unlike epifluorescence microscopy where emitted light from above and below the focal plane is observed, in confocal microscopy only light from the focal plane is observed. As a result, a higher quality image is produced. Typically, confocal images are captured as a series of focal planes captured at different depths through the sample. These focal planes can then be viewed individually or successive focal planes can be combined by a computer to generate a 3D image. Through the careful selection of antibodies and fluorophores, it is possible to simultaneously and independently observe three proteins. Images representing the localization of distinct proteins can be combined to generate false color images that can demonstrate the colocalization of proteins or their association with intracellular structures (Lopez-Molina et al., 2009).

For observation of secondary antibodies conjugated with an enzyme, sample sections are prepared as described earlier and then a chemical substrate is applied to the slide. The conjugated enzyme catalyzes the conversion of the substrate to a colored, insoluble precipitate that can be observed using light microscopy. In a typical application, alkaline phosphatase-conjugated secondary antibody is detected with 5-bromo-4-chloro-3-indolyl phosphate, a substrate that yields a colored product when exposed to alkaline phosphatase (Walker et al., 1999). For applications where electron microscopy is used, gold particle-conjugated secondary antibodies are used (Sonnewald et al., 1990).

When combined, immunoblotting, immunoprecipitation, and immunolocalization provide complementary data that can be employed to identify partner proteins and intracellular protein localization. This information can then be used to aid in the elucidation of the mode of action of introduced proteins *in planta*.

10.3 BREEDING AND BACKCROSSING

The pre-commercial phase of GM crop product development involves conducting extensive laboratory studies and field trials to characterize the product and evaluate its safety and efficacy. Later stages of GM plant product development include event selection of plants expressing the desired phenotype and backcrossing of that event into elite germplasms for commercial development. Insertion of a unique DNA sequence into the plant genome is defined as an event. Numerous events are evaluated and characterized to select the best events for commercial development. Breeders backcross the best events into elite germplasm or germplasm intended for commercial

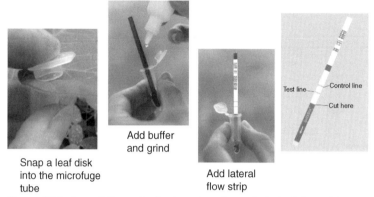

Snap a leaf disk
into the microfuge
tube

Add buffer
and grind

Add lateral
flow strip

Test line   Control line

Cut here

Figure 10.2 Lateral flow strip for detection of protein in leaf tissue (photo courtesy of EnviroLogix, Inc.).

use to assure that the event performs as expected in widely varied maturity groups or when combined with desirable agronomic characteristics such as disease or drought tolerance. The extent and duration of this stage of product development depends upon the number or scale of target markets. Developing varieties with different maturities and varied agronomic characteristics is essential to deliver the traits internationally across a wide range of latitudes and environmental factors, as well as in multiple countries. The GM trait is one of several important considerations for growers who choose to plant the commercial product.

During event selection and backcrossing, many events may be evaluated. Qualitative and quantitative ELISA tests are usually developed earlier in the process or during this period of evaluation and they are validated for the target protein. Qualitative ELISAs are used to determine whether or not an introduced gene is actually expressing the protein of interest during the event screening process. Lateral flow devices (LFDs) may also be used to determine the presence or absence of a protein (Figure 10.2). Relative protein expression levels may be determined using quantitative ELISAs as part of the event selection and ranking process. Functional assays, such as bioassays, may also be used to demonstrate the efficacy of a given protein. Functional assays make it possible to determine whether the expressed protein is functioning as expected. For example, if a target protein is expressed after plant transformation but does not fully fold into its native conformation, it is possible that it would be nonfunctional. A good example of using bioassays in conjunction with qualitative ELISA results is using a bioassay for a specific Cry insecticidal protein to assure that the protein being expressed in the transformed plant is effective against the target insects for which it is known to be toxic, while ensuring that it is not causing harm to nontarget organisms (NTO). These types of studies are a crucial part of the overall assessment of efficacy and safety of GM crops.

Over the past several years, an increasing number of products have been developed that contain multiple traits. These are referred to as "stacks" or "combined event" products. Stacked corn and cotton products containing multiple traits have come to market in recent years, and it is expected that all major crops, including

Seed stacked (multievent)

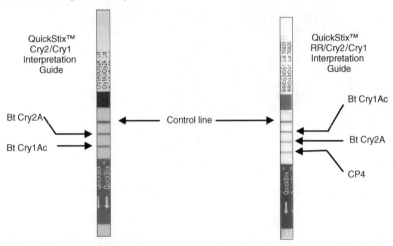

Figure 10.3 Lateral flow strip for detection of multiple proteins in combined trait product (photo courtesy of EnviroLogix, Inc.).

soybean and other oilseeds, will also be marketed primarily as stacked products in the near future. It is expected that anywhere from two to nine traits will coexist in any given GM product over the next few years and there is potential for even more. Immunoassays will continue to play an important role in evaluating protein expression in these complex products (Figure 10.3).

The importance of immunoassays in this scenario is threefold: (1) determining whether the desired protein(s) are being expressed, (2) monitoring the level of expression in both single trait and stacked products during the course of product development, and (3) after commercialization, determining whether or not the subsample of seed being tested from a larger seed pool contains single or multiple traits. This is useful in determining seed purity by enabling the effective detection of admixtures of seed and by supporting appropriate sampling and statistical analytical methods required by both the industry and global regulatory agencies.

10.3.1 Impact of Immunoassays on Variety Selection

Immunoassays are widely used not only in event selection, but also in assessing whether a GM trait has been successfully introgressed into elite germplasm. Currently, commercialized GM traits such as insect-tolerant Bt and glyphosate-tolerant CP4 EPSPS protein have been introgressed into existing crop cultivars such as corn by backcrossing. Backcrossing incorporates the desired trait from the "donor" parent into an already developed cultivar (elite line) in a process that typically takes 3 to 4 years. During backcrossing, other endogenous traits and characteristics from the donor parent may also be transferred to the resulting offspring. It is possible that in certain cases the genes from the donor parent could have inferior agronomic characteristics relative to the elite line parent. Therefore, several rounds of backcrossing are required to remove all or most of the undesirable characteristics while retaining the desirable characteristics in the elite germplasm, including the transgenic trait(s). Immunoassays are frequently used to determine whether or not the transgene is present in order to select the appropriate progeny plants derived from backcrossing.

Conventional breeding, which has some degree of random recombination, provides the added advantage when breeding GM crops by inserting known genes to introduce the desired characteristics. It is also possible to obtain a desired agronomic trait by modifying, removing, or silencing a particular gene or genes. In combination with traditional field observations for agronomic and phenotypic performance and molecular methods for genetic characterization, immunoassays used to evaluate protein expression play an important role in the characterization of GM plants at various points in the breeding and backcrossing stages of product development.

10.3.2 High-Throughput ELISA Analysis from Discovery to Production

ELISA plays an important role in screening large numbers of samples from the many events that are developed and screened as part of the product development process. High-throughput ELISA analysis is applied for quantitation of a protein or proteins of interest in large numbers of samples. High-throughput ELISA systems are typically developed at later stages in the discovery process after there is positive demonstration of the product concept. Some reasons for the delay in deployment of high-throughput ELISA systems are the need for a well-characterized protein standard, the evaluation of sample extraction conditions, and the pairwise evaluation of antibody reagents. While ELISA analysis offers higher throughput and greater sensitivity and facilitates a more robust and reliable quantitation than Western blotting, these distinct advantages come at the expense of a longer and more elaborate assay development process.

The most commonly employed ELISA technique is the sandwich ELISA, where the analyte is sandwiched between an immobilized capture antibody and bound detection antibody complex. A sandwich ELISA is the most appropriate assay for plant tissue extracts that are highly complex mixtures of biomolecules and frequently contain ng/mL quantities of analyte. Sandwich ELISA methodology consists of several sequential steps including plate coating and blocking, sample extract application, plate washing, and the application of detection antibodies and quantitation reagents. Each of the steps is an independent variable in the overall execution of an assay and each step involves specific range finding activities during development. For example, extracts that are prepared from seed tissues may require different extraction conditions and buffers than leaf tissue. In turn, those extracts may require different dilutions prior to loading onto plates that have been uniquely coated and blocked for the analysis of a specific extract.

Plate coating is performed by dispensing a given volume of the capture Ab diluted in an alkaline buffer into the wells of a polystyrene microtiter plate followed by an incubation that permits antibody binding and immobilization to the microtiter plate walls. Depending upon the assay, a blocking step may be required following plate coating. Blocking involves incubation of the bound antibodies with a protein solution to eliminate (block) any unoccupied protein binding sites in the well that could nonspecifically bind proteins contained in the sample extract. Diluted sample extracts are then added to wells and the analyte protein is captured by the immobilized antibodies. The buffers used to prepare sample extracts by and large must be "physiological buffers" in that they cannot contain excessive quantities of detergents or reducing agents that would denature or adversely affect the immobilized antibodies. The sample extracts are removed and the wells of the plate are washed prior to the addition of the detection antibody. Typically, detection antibodies are conjugated to biotin. Following incubation with the detection antibodies, the plates are washed and a reporter tag conjugated to streptavidin is added and allowed to incubate. Depending upon the instrumentation platform used to read the plates, the reporter tag can range from an enzyme such as horseradish peroxidase or alkaline phosphatase to a chemiluminescent or electrochemiluminescent fluorophore. Following additional washes to remove unbound streptavidin conjugate, detection reagent is added that produces a measurable signal proportional to the amount of captured analyte. More details about plate ELISA development are provided in Chapter 4.

Although the use of a biotinylated detection antibody versus a directly tagged detection antibody adds additional incubation and washing steps, it offers two distinct benefits. The first benefit is transferability. One can move from platform to platform with relative ease. If one has multiple ELISA assays each measuring a specific protein, a move to a new platform only requires that a new streptavidin-tag conjugate be obtained. In contrast, if the detection antibodies are themselves tagged with a reporter, a new tagged antibody would be required for each assay that is transferred to a new platform. The second benefit is signal amplification; each biotin-conjugated detection antibody molecule typically has several covalently bound biotin molecules. Each of the biotin molecules has the potential to bind a streptavidin-tag conjugate, resulting in a situation where a single detection antibody may have several bound reporter tags, each of which produces a signal. Directly tagged detection antibodies will have a lower reporter tag to antibody stoichiometry and may yield a lesser signal per captured analyte molecule.

For a sandwich ELISA to be successful, a pair of antibodies that bind to distinct and well-separated epitopes on the surface of the analyte must be available. This pair of antibodies can include monoclonals, a monoclonal and a polyclonal, a pair of polyclonals, or, in certain circumstances, a single polyclonal that is used for both capture and detection. Antibody pairs are selected using a checkerboard assay where a serial dilution of capture antibodies is loaded into pairs of wells running across a microtiter plate. A uniform amount of analyte is added to one well and a buffer blank to the second well of each pair and allowed to incubate. Following a wash step, serial dilutions of the detection antibody are applied going down the plate and the plate is carried through the remaining steps necessary to produce a signal that is measured. The signals from pairs of wells, one containing the analyte and the other a blank, are evaluated and a signal to noise ratio is determined. The combination of coating and detection antibody dilutions that yields the greatest signal to noise ratio is then employed for subsequent stages of ELISA development. As mentioned earlier, since different tissues may require different extraction buffers and conditions, it may be necessary to perform checkerboard assays for specific tissue extracts to confirm that the extract yields acceptable performance parameters.

ELISA analyses provide a means for high-throughput analyte quantitation in tissue extracts. While ELISA methodology is typically not available during earliest stages of biotechnology-derived crop development, rudimentary knowledge of the characteristics of antibody reagents developed for other immunoassays should facilitate assay development when a product concept is validated.

10.4   ROLE OF IMMUNOASSAYS IN REGISTRATION

Global regulatory agencies require extensive data packages as part of the safety and efficacy evaluation process for GM crops as part of the approval process prior to commercialization. The generation of these data for some countries requires multi-seasonal studies and data generation across different geographic areas. One of the major areas of interest is protein expression levels in the various GM plant tissues. These analyses typically include leaves, roots, grain, seed, and other plant tissues to demonstrate the efficacy and the safety of GM crops. Immunoassays play a major role in these analyses. For specific products such as Bt insect-protected crops, additional studies may be necessary to determine the amount of protein in soil (see Chapter 12) and the amount and effect of exposure for soil microbes and insects feeding on plants expressing the target protein. Studies evaluating the effect on predators of those insects may sometimes also be necessary. Immunoassays play an important role in evaluation of some of these parameters and there are many examples of such studies and their use in risk assessment and safety evaluation, including demonstration of the advantages of Ag Biotech plants with respect to food or feed application or impact on the environment.

Validation of immunoassays on all applicable test matrices is required to ensure the accuracy of analytical results. Ag Biotech companies invest a tremendous amount of time and money in the development and validation of detection methods, including immunoassays. Validated methods are a requirement for a number of regulatory bodies internationally, including European Food Safety Authority (EFSA) in the European Union and Korea Food and Drug Administration (KFDA) in Korea.

10.4.1   Global Registration of GM Crops

The goal of regulatory submission is to provide adequate and appropriate data so that regulatory agencies globally can assess the overall safety of GM crops for use as food and feed, as well as assessing their environmental safety. Typically, this includes safety assessment of the whole plant, including analytical evaluations of seed/grain and forage. These assessments are essential to ensuring safety and subsequently are required to obtain final regulatory approval and registration for commercialization. The global economy also creates a need for companies to obtain approvals in key import markets for any given GM crop, in order to prevent interruption of trade when new products are commercialized. Many technology providers have committed to adhere to comprehensive stewardship programs with Excellence Through Steward-ship, which includes an obligation to seek regulatory approvals in all key import markets with functioning regulatory systems. Immunoassays play a key role in the generation of regulatory submission dossiers that contain all data necessary for countries to effectively evaluate and approve GM crops, hopefully resulting in the timely approvals of GM crops in a manner that supports the traders and the global commodity industries.

Sharing of detection methods and reference materials with regulatory agencies globally helps assure that accurate and consistent testing is being conducted as part of the product registration process, as well as during importation. This sharing of methods also makes the test results comparable and is a crucial element to mutual acceptance of data globally. Assuring that the sensitivity and the specificity of various immunoassay test methods are consistent among products and that they meet global regulatory requirements is an important goal in the development of detection methods and in assuring quality processes during product safety assessment.

All regulatory submissions are done in accordance with the regulations of each country or region to which a submission is made. Typically, the data submitted from immunoassays are fairly uniform, and the safety evaluations involve protein characterization, protein expression quantification, and assessment of allergenicity and toxicity.

Immunoassay techniques are an essential tool for effective assessment of the safety of proteins introduced through biotechnology. This assessment of the protein produced as a result of plant transformation requires a multistep approach that includes (1) characterization of each introduced protein's physicochemical and functional properties; (2) quantification of protein expression in plant tissues; (3) evaluation of the structural similarity of the protein to known allergens, toxins, or other biologically active proteins known to have adverse effects to mammals; (4) evaluation of the digestibility of the protein in a simulated gastrointestinal system; (5) documentation of the history of safe consumption of the protein or its structural and functional homologues by humans or animals; (6) investigation of potential mammalian toxicity; (7) in some cases, development of a bioassay; (8) evaluation of the effects the protein has on nontarget organisms (for those proteins with plant-incorporated protectants (PIPs)); (9) compositional evaluation and studies for indirect or pleiotropic effects; (10) and acute and chronic feeding studies with various animal species.

The data generated are then analyzed, typically comparing the transformed crop containing the desired trait to the conventional parental control. The conventional parental control contains the same background genetics as the transformed GM crop and should produce results similar to the transformed crops with the exception of the presence of the trait(s) that have been introduced. Part of this comparison is based on understanding the MOA of the introduced protein(s) and trying to determine what metabolic pathways are affected as a result of the presence of the introduced trait. These changes are assessed by evaluating differences in agronomic and phenotypic traits of the transformed plants compared to the conventional controls, as well as by evaluating factors such as environmental stress tolerance and changes in physiology or changes at the molecular level. These studies, along with toxicology and allergenicity studies, feeding studies, and other NTO bioassays provide the critical data needed to assess overall safety.

Many of the new GM crop products under development include traits with proteins, such as transcription factors that have not previously been utilized in the development of commercial products. Product evaluation in these cases is reliable in these cases, to a large extent, based on the history of the reliability and predictive value of test results from the well-defined and understood test methods that have been used historically to assure safety of GM crops. Immunoassays are still in the technical forefront of providing this reliable and contextual history for testing and validating the safety of GM crops. The safety aspects of these proteins can still be fully evaluated through the same traditional avenues as was used to assess the safety of GM crops historically. These evaluation methods, including the use of qualitative and quantitative immunoassays, are the foundation of effective safety evaluations for GM crops. Although additional technical evaluations will be conducted as appropriate for elucidation of MOAs, immunoassays play a critical role in the direct assessment of the protein itself, and immunoassays can also play a role in the overall evaluation when assessing any potential changes in metabolic processes in GM crops other than detection and quantification of the target protein itself.

Additional conventional reference varieties may also be evaluated for certain types of comparisons, such as composition, feeding studies, and, where appropriate, endogenous allergenicity studies. Data from reference varieties and from historic databases or the published literature are often used to establish whether or not the GM crop being evaluated falls within known reference ranges or tolerance intervals for conventional samples of the same crop. If the references are known to be safe, and the GM crop, when compared to the references, is comparable to the references based on scientifically valid studies that have been analyzed using appropriate statistical methods, then the logical conclusion is that the GM crop is also safe, when these analyses are used in conjunction with fundamental scientific analysis of the data based on scientific observations made in addition to statistics. In light of some increased requirements for evaluations and testing by regulatory agencies on a global scale, immunoassays retain their leadership role among GM crop test methods as a cost-effective and scientifically valid way to evaluate GM crops.

Meanwhile, organizations such as Organisation for Economic Cooperation and Development (OECD) and Codex Alimentarius Commission have laid a firm foundation for the mutual acceptance of data globally, and it will require a concerted effort from industry and trade organizations, as well as other nonprofit organizations such as the International Life Sciences Institute (ILSI) and Crop Life International (CLI), to support countries in following and implementing standards developed by international organizations (see Chapter 15 for more detail). Evolving regulatory systems or rapidly changing requirements in certain countries or regions that are not developed using science-based assessment criteria, can result in slow or inconsistent approval processes.

In particular, the Crop Life International Detection Methods Team and Environmental Assessment Team are just two of the several CLI teams that play major global roles in coordinating and harmonizing international standards, as well as addressing global concerns regarding GM crops. The CLI Detection Methods Team works globally to address and evaluate detection method needs, including immunoassays and DNA-based detection methods. These teams facilitate global discussions that result in the resolution of global issues and the development of government and industry-wide approaches to GM crop testing on a global scale. In addition, there are regional and in-country CLI teams that make sure that both international issues and issues within their own countries are heard and dealt with in both directions. These teams play an invaluable role in quickly initiating global discussions that affect the entire Ag Biotech industry, including any issues related to immunoassays.

Review processes are sometimes unpredictable and may take many months or many years. Recently, the EU has requested evaluation of some GM products using 2D gel immunoelectrophoresis, in addition to the ELISA data that have been generated historically for assessing reactivity of sera from patients known to have soy allergies. The basis for this request was to determine whether any new allergens had been produced by the transformed plants as a result of the transformation process. ELISA-based methods alone were no longer deemed adequate by the EFSA sub-committee for this purpose. While these types of assays are important in assessing allergenicity of the proteins being evaluated for safety, these types of changing requirements may occur very late in a multiyear review process and this is something that can occur at any time globally due to evolving regulatory systems.

Many countries also require technology providers to supply validated detection methods and reference materials as part of the regulatory submission package or prior to product approval. For some applications, it is necessary to have event-specific methods, so DNA detection methods are required. However, ELISA and lateral flow devices can be used in addition to DNA detection methods to meet requirements in many countries.

10.5 IMMUNOASSAYS IN SEED PRODUCTION AND QUALITY ASSURANCE

Seed production (manufacturing) and quality control functions of technology provider companies also use quantitative and qualitative ELISA methods to determine purity (% of target event) and evaluate low-level presence or adventitious presence (AP = levels of nontarget events) during scale-up and commercial seed production processes. A common procedure used to determine seed purity is to test single seeds using a qualitative ELISA. To statistically confirm high levels of purity with a high level of confidence, a large number of single seeds must be tested. Pooling strategies are commonly used when testing for unintended events, so the total number of tests samples per lot may be smaller than that with purity testing. This approach for unintended events is generally accepted because of the high level of sensitivity and specificity that can be achieved using immunoassay test methods. For example, if a pool of 500 seeds contains 1 positive seed for a given trait, it could be detected. A statistical package such as SeedCalc (www.Seedtest.org) could be used to determine whether that level of purity is acceptable. However, since the number of commercial and pre-commercial events continues to increase significantly, the number of tests conducted per pool can be large because any given lot of seed must be tested for multiple events.

Because of the high numbers of samples and tests required to assure product purity, seed production and quality assurance applications are amenable to automation. Automation not only enables high-throughput analytical and production methods, but also reduces the chance for human error during sample preparation and testing. Liquid handlers, automatic plate washers, readers, and stackers are commonly used in the immunoassay process. Data gathering, interpretation, and archiving systems are also important considerations for high-throughput ELISA analyses. In some cases, laboratories may choose to package all the data handling into a LIMS (Laboratory Information Management System). LIMS for handling immunoassay data can be very sophisticated or relatively simple depending on the requirements of the particular laboratory.

LFDs are also used for some applications in seed purity and testing for levels of nontarget events during seed production and quality assurance. An example of lateral flow applications is spot-checking to confirm the identity of a seed lot, or to quickly confirm an ELISA result. To confirm the result of an ELISA test, the LFD can be inserted directly into the sample that has already been prepared and used for the ELISA (Figure 10.4).

Quantitative ELISAs are also used in some production situations to confirm levels of protein expression. Quantitative applications would require much smaller numbers of tests because the sampling and testing can be based on statistical methods that provide some measure of confidence in whether or not a given lot of seed contains the trait for which it is being tested. Quantitative ELISAs for GM crops are used on a large scale in the United States and globally to manage the sale and distribution of grain. For example, these tests are used to determine whether concentrations of biotech grain in a sample exceed or fall below defined threshold limits. Threshold determination is expressed as statistical confidence levels.

QC labs

Seed quality testing identity preservation

Figure 10.4 Lateral flow strips used in high-throughput application (photo courtesy of EnviroLogix, Inc.).

Immunoassays are used in combination with identity preservation systems (IDP) to effectively characterize raw and processed GM grain as part of the food production system, and this enables food producers to meet labeling requirements where they exist. The IDP systems are designed to enable producers and processors to channel grain for specific uses while reliably preserving the identity of the grain, thereby preventing grain containing a trait intended for a specific use from entering the general commodity stream. An example might be soybean containing higher levels of high oleic fatty acid than its conventional counterpart. Although this grain is as safe and efficacious as conventional soybean, and would cause no harm if it enters the general commodity stream, it has a premium market value in certain parts of the food industry and there is some economic benefit to commercializing it under an IDP system. Non-GM seed and grain is the most prominent IDP system and it relies heavily on immunoassay testing.

10.6 PRODUCT STEWARDSHIP AND STAKEHOLDER SUPPORT

Product stewardship has continued to command an increasingly more important role globally since the requirement for approval of GM products in key import countries with functional regulatory systems prior to commercial product launch has become a major issue for traders and processors. In addition, as the Ag Biotech industry matures, the discontinuation and product withdrawal process assumes a much more significant role than it has in the past. Product stewardship plays an essential role during the global product approval and commercialization process, and LFDs may be developed and used by technology providers earlier in the product development process for breeding and in support of production. However, the primary use of LFD is in support of product stewardship and stakeholders when the trait is commercialized and enters commerce.

The need to have a test that uniquely identifies a specific trait or event in a field situation is driven by a number of factors. Grain traders and food processors are required to meet regulatory obligations by avoiding shipping materials that contain a trait or event not registered in a given country. In addition, in certain countries, although the traits have been judged to be safe and legal through extensive regulatory and safety assessments, food processors need to test for the presence of GM traits in order to meet labeling requirements. DNA detection methods are also commonly used for these applications, but the cost and time factors associated with laboratory-based PCR methods make field use protein-based testing a very attractive alternative for this type of testing, assuming that the appropriate data can be obtained with a protein-based test.

Technology providers generally also make ELISA kits commercially available to the public at the time of commercialization by working with third-party kit developers and manufacturers. Some commercial kit suppliers include EnviroLogix, Strategic Diagnostics, AgDia, and Neogen (www.Envirologix.com, www.Sdix.com, www.Agdia.com, and www.Neogen.com). These companies work with the Ag Biotech companies to commercialize the new immunoassay products that are being developed.

For example, before EFSA issues a positive scientific opinion for a GM crop, the manufacturer of that crop is required to provide documentation that both test methods and certified reference materials (CRM) are available to the public at that time and prior to final EU approval. Typically, these methods and CRM can be purchased from American Oil Chemists' Society (AOCS) or the Institute for Reference Materials and Measurement (IRMM) (see Chapter 7 for more detail).

To harmonize trade, fulfill regulatory requirements, and meet consumer preferences, it is extremely important that the tests being employed give the same results. If the contents of a shipment of grain or other material test negative leaving the port of the exporting country but that same material tests positive for the same trait in the importing country, this can cause trade disruptions and disagreements and may be considered "illegal" in some cases. A number of international organizations are working to harmonize the regulations and policies regarding standardization of test methods and reference materials for grain trade applications of immunoassay methods, including OECD and Crop Life International.

There are also many organizations working to standardize the technical parameters for detection methods and reference materials associated with trade. Two of the major technical challenges are distinguishing breeding stacks from mixtures of two or more single events and the variability of the final interpretation of the test results.

Because a breeding stack is simply a cross between plant materials containing one of the two single events, the proteins expressed and the DNA insert in the stack are identical to those present in the single events. For these reasons, if the test sample is pooled seed, it is not possible to determine whether the pool contains the stack, a mixture of the two events, or both. It is possible to determine the identity of the event(s) by testing single seeds. If both events are detected in a single seed, the seed contains the stack. Single seed testing can be expensive and time-consuming, especially if the goal is to detect low levels of the event or event combinations.

Copy number, variability in expression levels, units of measure (%GM DNA versus %GM seed) have made quantitative detection methods extremely difficult to use and interpret for seed and grain testing applications. Copy number and units of measure have more impact on DNA-based methods, but variability of protein expression levels makes quantitative interpretation of immunoassays difficult. Increasing numbers of commercial events, especially stacks and events with multiple copies of some genes, will make interpretation of quantitative detection methods even more difficult in the future.

For the reasons discussed previously, pooling strategies using qualitative test methods in conjunction with statistical sampling and data analysis (such as SeedCalc) are much more practical and give much more reliable data for testing seed, grain, and other materials for the presence of biotech traits. This approach is much more reliable because the estimate is based on the number of positive pools rather than a quantitative measure. Since the detection method is validated to detect one positive unit in a given sized pool with a high degree of confidence, the analytical accuracy is quite good and the issues with interpreting quantitative results are eliminated. Although costs of seed and testing may be considerable, it is possible to test with high sensitivity and/or high accuracy if enough pools are tested. The SeedCalc program can be downloaded free of charge from www.Seedtest.org.

Immunoassays have also been used for other product support and stewardship applications such as helping growers or technology provider sales personnel. Lateral flow strips are sometimes used to determine which type of crop (i.e., herbicide resistant or conventional) was planted in a field before the herbicide is applied. This application of immunoassay technology can be completed with reliable results, quickly, and cost effectively, preventing major losses that would occur if a conventional crop were accidentally sprayed with herbicide.

Effective test systems are crucial to "point of delivery value sharing" systems where a commodity grain can be tested at the elevator to evaluate whether farmers harvested grain containing a patented technology and assure that appropriate licensing payments are made for use of this technology. In return, the farmers get the benefits associated with the biotechnology-derived trait. Testing as part of a value sharing or indemnity system helps identify those farmers who save seed or otherwise obtain seed illegally without appropriate payment of technology fees or royalty.

The second generation of GM soybean, Roundup Ready 2 Yield™ (RR2Y), was commercially released in 2009 and it will replace Roundup Ready soybean 40-3-2 (RR 40-3-2) over a number of years. It will be important to be able to distinguish between the two products because the two products are not approved for cultivation in all of the same countries. In this case, immunoassay-based methods will not distinguish between the two products because they contain the same protein CP4 EPSPS. Consequently, a DNA-based detection method is required to distinguish between the two. This is one of the few times that has occurred for GM crops, where a new product did not contain at least one protein that was uniquely distinguishable by immunoassay, and it reflects one of the many challenges facing the agricultural biotechnology industry.

As with any other analytical method, immunoassays for GM crop products or traits require reference materials. Initial reference materials for assay development, validation, and transfer of the methods to regulatory agencies are produced and supplied by the technology providers. CRM are developed and certified according to ISO standards by a third party as contracted by the technology providers. CRM are generally made available to the public at the time the commercial product is introduced.

In summary, it is clear that immunoassays remain an essential tool in the development, commercialization, and effective stewardship of GM crops. These methods are used in conjunction with molecular and other experimental test methods to ascertain the safety and efficacy of GM crops and will continue to do so for the foreseeable future.

REFERENCES

Ares, X., Calamante, G., Cabral, S., Lodge, J., Hemenway, P., Beachy, R. N., and Mentaberry, A. Transgenic plants expressing potato virus X ORF2 protein (p24) are resistant to tobacco mosaic virus and Ob tobamoviruses. *J. Virol.*, **1998**, *72*, 731–738.

Doran, P. M. Foreign protein degradation and instability in plants and plant tissue cultures. *Trends Biotechnol.*, **2006**, *24*, 426–432.

Duncan, K. A., Hardin, S. C., and Huber, S. C. The three maize sucrose synthase isoforms differ in distribution, localization, and phosphorylation. *Plant Cell Physiol.*, **2006**, *47*, 959–971.

Gershoni, J. M. and Palade, G. E. Protein blotting: principles and applications. *Anal. Biochem.*, **1983**, *131*, 1–15.

Goding, J. W. Use of staphylococcal protein A as an immunological reagent. *J. Immunol. Methods*, **1978**, *20*, 241–253.

Lopez-Molina, L., Mongrand, S., Kinoshita, N., and Chua, N.-H. AFP is a novel negative regulator of ABA signaling that promotes ABI5 protein degradation. *Genes Dev.*, **2009**, *17*, 410–418.

Padgette, S. R., Re, D. B., Barry, G. F., Eichholtz, D. A., Delannay, X., Fuchs, R. L., Kishore, G. M., and Fraley, R. T. New weed control opportunities: development of soybeans with a Roundup Ready gene. In Duke, S. O. (Ed.), *Herbicide-Resistant Crops: Agricultural, Economic, Regulatory and Technical Aspects*, CRC Lewis Publishers, 1996, pp. 53–84.

Smith, C. S., Morriceb, N. A., and Moorhead, G. B. G. Lack of evidence for phosphorylation of *Arabidopsis thaliana* PII: implications for plastid carbon and nitrogen signaling. *Biochim. Biophys. Acta*, **2004**, *1699*, 145–154.

Sonnewald, U., von Schaewen, A., and Willmitzer, L. Expression of mutant patatin protein in transgenic tobacco plants: role of glycans in intracellular location. *Plant Cell*, **1990**, *2*, 345–355.

Towbin, H. and Gordon, J. Immunoblotting and dot blotting—current status and outlook. *J. Immunol. Methods*, **1984**, *72*, 313–340.

Towbin, H., Staehelin, T., and Gordon, J. Electrophoretic transfer of proteins from polyacrylamide gels to nitrocellulose sheets: procedure and some applications. *Proc. Natl. Acad. Sci.*, **1979**, *76*, 4350–4354.

Walker, R. P., Chen, Z.-H., Tecsi, L., Famiani, F., Lea, P. J., and Leegood, R. C. Phosphoenolpyruvate carboxykinase plays a role in interactions of carbon and nitrogen metabolism during grape seed development. *Planta*, **1999**, *210*, 9–18.

Walker, R. P., Chen, Z. H., Johnson, K. E., Famiani, F., Tecsi, L., and Leegood, R. C. Using immuno-histochemistry to study plant metabolism: the examples of its use in the localization of amino acids in plant tissues, and of phosphoenolpyruvate carboxykinase and its possible role in pH regulation. *J. Exp. Bot.*, **2001**, *52*, 565–576.

Zhang, C. and Guy, C. L. Co-immunoprecipitation of Hsp101 with cytosolic Hsc70. *Plant Physiol. Biochem.*, **2005**, *43*, 13–18.

CHAPTER 11

IMMUNOASSAY APPLICATIONS IN GRAIN PRODUCTS AND FOOD PROCESSING

Gina M. Clapper
Lulu Kurman

11.1   INTRODUCTION

There are several applications of immunoassay (IA) in agriculture. These include water quality, veterinary diagnostics, and pharmaceutical development to name a few outside the food industry. With regards to food, there are a plethora of applications for detecting mycotoxins and other microbiologicals, allergens, hormones, toxins, drug residues, pesticides, and other residues. However, this chapter will focus on the application of immunoassays in grain products and food processing.

Immunoassay for genetically engineered (GE or GM) proteins is used during several points in product development and by feed and food suppliers for compliance

Immunoassays in Agricultural Biotechnology, edited by Guomin Shan
Copyright © 2011 John Wiley & Sons, Inc.

and contractual purposes. Stakeholders from the food and feed supply chains, such as commodity, food, and feed companies, as well as third-party diagnostic testing companies, also rely on immunoassays for a number of purposes (Grothaus et al., 2006). The primary application of IA in grain products and food processing is to verify the presence or absence of GM material in a product. The two most commonly used formats are plate-based enzyme-linked immunosorbent assays (ELISA) and lateral flow devices (LFD), which have been further discussed in Chapters 4 and 5, respectively.

Global acreage dedicated to GM crops has increased steadily since they were first put onto the market in 1996. These crops are the result of the incorporation of specific agronomic, crop protection, or quality traits into plants grown for agricultural production. These modified or transgenic plants differ from conventional plants in that they contain modified native genes or additional genes that produce a corresponding number of new proteins, the presence of which give the modified plants the desired beneficial traits. With the introduction of transgenic plants, analytical technology is rapidly evolving to measure the introduced proteins.

The increase in cultivation of GM crops has led to a corresponding increase in the introduction of GM-derived foods into the grain and food processing supply chain (Table 11.1). As of 2009 more than half of the world's population and more than half of the world's cropland are in the 25 countries where GM crops are approved for cultivation (James, 2009). The accumulated acreage of GM crops exceeded 2 billion acres (800 million hectares) in 2009 (James, 2009). Maize, soybeans, canola, and cotton are the major commodity crops that are available in GM format on a massive scale. From 1996 to 2009, the global area of GM crops increased 80-fold, making GM crops the fastest adopted crop technology in recent years (ISAAA Brief 41-2009). The impact of GM crops, globally, for 1996–2007 was 51.9 billion U.S. dollars (49.6% for substantial yield gains and 50.4% due to a reduction in production costs) for developing and industrial nations in terms of net economic benefits to GM crop farmers (James, 2009).

TABLE 11.1 Input and Output Traits that are Commercially Available or Under Development

Input traits	Output traits
Insect resistance	Quality-enhanced fibers
Herbicide tolerance	Trees that make it possible to produce paper with less environmental damage
Resistance to diseases caused by viruses, bacteria, fungi, and worms	Nicotine-free tobacco
Protection from environmental stressors such as heat, cold, and drought	Ornamental flowers with new colors, fragrances, and increased longevity
Tolerance of high salt concentrations	Nutritionally enhanced foods
Improved taste	More starch or protein
Increased shelf life	More vitamins
Better ripening characteristics	More antioxidants

Source: Vines 2001.

In the ensuing years since the introduction of GM seeds, governments have promulgated regulations concerning the cultivation, movement, and sale of food/feed derived from GM crops. The regulations governing GM crops vary between trading partners and regions, as do consumer attitudes toward the foods derived from them, leading many governments to require labeling of GM-derived foods and ingredients when they are present in a food product. Countries with GM labeling regulations will generally exempt products from labeling if the presence of GM is incidental, presence is below an established threshold, and all reasonable measures have been taken to prevent the introduction of GM into the product.

Most countries that produce or engage in trade of GM seed, crops, or foods require approval of an event before it can be sold. Regulatory approval of new events can be asynchronous, which can affect grain handling procedures at processing operations that serve a market where the event has not yet been approved, but is legally cultivated in the grain processor's market. In the United States, three independent authorities are involved in the regulatory process: the Animal and Plant Health Inspections Service (APHIS) under the United States Department of Agriculture (USDA), the Food and Drug Administration (FDA), and the Environmental Protection Agency (EPA). The current position in the United States is that the composition of food consisting of, or derived from, GM crops does not differ significantly from its conventional counterpart; therefore, labeling is not mandatory in the United States (Jasbeer et al., 2008). By comparison, Regulation (EC) No. 1830/2003 within the EU sets out to ensure that the relevant information regarding any GM products placed in the market is available at each segment of the supply chain and that those foods and feeds are accurately labeled. Japan has a 5% tolerance threshold labeling policy for GM products (USDA-FAS, 2008), though labeling is mandatory at and above 5%. Processed foods (i.e., soy sauce, corn starch, salad oil) are exempted from this requirement if the protein and recombinant DNA are removed and/or the ingredient is not one of the first three processed foods in terms of weight or its proportion is not 5% or more by weight. Australia and New Zealand (USDA-FAS, 2008) have mandatory labeling of food products containing GMs at or above 1%. South Korea (USDA-FAS, 2008) also has mandatory labeling of food products if the top five ingredients contain GM material at or above 3%. China's labeling regulations are governed by the Ministry of Agriculture Decree 10 (CH2002), which requires agricultural GM products listed in the regulations (see Table 11.2) to be labeled and prohibits the import and sale of any unlabeled or mislabeled products.

As a result of regulatory and customer requirements, IA testing is used at various stages of the supply chain, from receipt of seed at farm level to finished retail product, in order to detect and quantify the presence of GM products. Several critical control points for GM testing by IA are depicted in Figures 11.1 and 11.2, and are examined in detail later.

11.1.1   Advantages of Immunoassays

Immunoassays provide a flexible format for diverse applications, which are depicted thoroughly in Chapters 4 and 5. They are qualitative, semiquantitative, or quantitative. Commercial methods are available for agricultural commodity crops as well

TABLE 11.2 Agricultural Products listed in the Chinese Labeling Regulations[a]

Soybean seeds	Corn seeds	Rape seed for planting	Cotton seed for planting	Tomato seed
Soybean	Corn	Rape seed		Fresh tomato
Soybean powder	Corn oil	Rape oil		Tomato jam
Soybean oil	Corn flour (including the corn flour with harmonized schedule codes 11022000 11011300 11042300)	Rape meal		
Soybean meal				

[a]*Source*: United States Department of Agriculture (USDA). GAIN Report Number: CH7053 (2007). Foreign Agricultural Service Attaché Reports.

as processed food. Immunoassays are becoming attractive tools for rapid field monitoring for the integrity of agricultural commodities in identity preservation (IP) systems, where nonspecialized personnel can employ them in a cost-effective manner.

Currently, the commercialized varieties of four major commodity crops, canola, cotton, maize, and soybeans, all consist of input traits that directly benefit the grower and indirectly benefit the environment. The GM protein can be expressed in the grain/seed, leaf tissue, and other organs, thus presenting in food ingredients, and processed foods. Among the most important advantages of immunoassays are their speed, sensitivity, selectivity, and cost-effectiveness. Immunoassays can be designed as rapid, field-portable, semiquantitative methods, or as standard-quantitative laboratory procedures. They are well suited for the analysis of large numbers of samples and often do not involve lengthy sample preparations. Immunoassays can be used as screening methods to identify samples needing further analysis by sophisticated analytical methods and are especially applicable in situations where analysis by conventional methods is either not possible or is prohibitively expensive.

The LFD is widely used at the field, the grain elevator, transportation vessel(s), the primary processor, the secondary manufacturer, and the final customer or importer. Immunoassay, when coupled with Identity Preservation and Quality Management Systems, provides an effective means of characterizing the trait (s) of interest.

11.1.2 Disadvantages of Immunoassays

If the structure of GM protein has been altered during sample processing, an immunoassay may not be accurate in the analysis. Sample processing has varied effects on protein conformation (i.e., extreme heat denatures protein, product fractionation removes the protein), which may impact the antibody–antigen bind. Some products may not express a detectable level of protein in the harvested crop and/or the antibodies may be described as cross-reactive because they cannot discern the difference between different GM events expressing the same protein (Grothaus et al., 2006). When the situation involves government regulations for a labeling threshold, IA is not the ideal tool to measure the exact level of GM due to protein expression variability. In these situations, a test that measures DNA, such as a PCR method, may be more applicable (Lipp et al., 2005).

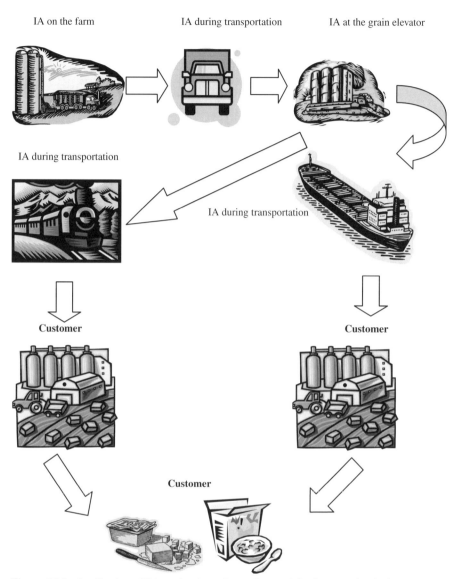

Figure 11.1 Application of IA testing in grain product and food processing industry.

11.2 APPLICATION OF IA TESTING IN GRAIN PRODUCT AND FOOD PROCESSING INDUSTRY

11.2.1 Supply Chain and Quality Management Systems for GM Identity Preservation

Although most of the testing in commodity grains has been to restrict GM grains either at zero tolerance (unapproved events) or at specific thresholds (approved events), in

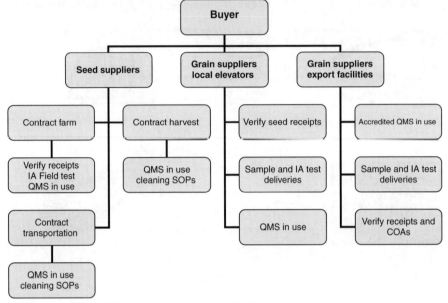

Figure 11.2 Flow of food production and supply chain.

some applications GM commodities are valued over non-GM and tests are used to make certain that buyers receive the added value that they purchased.

11.2.2 Testing for the Presence of High-Value GM Commodities: Cotton

Currently, quality control of cottonseed production is the most common application of immunoassay. This commercial application of testing for GM commodities was introduced in 1997. Cotton production relies heavily on herbicides to control weeds, often requiring applications of two or more herbicides at planting and postemergence herbicides later in the season (Culpepper and York, 1998). The USDA reported in 1997 that the 12 major cotton producing states applied close to 28 million pounds of herbicides to 97% of the 13 million acres devoted to upland cotton production (USDA NASS, 1998).

Herbicide-tolerant and insect-resistant cotton varieties offer farmers many benefits, including decreased pest management costs, increased yields, and greater crop production flexibility. GM seed was more expensive than conventional seed due to technology fees; therefore, the growers needed verification of the product claim. The GM seed producers implemented a strict sampling regimen of individual seeds, based on statistics with a high degree of confidence, coupled with LFD test methodology to validate the label claims.

The expected benefits and performance of these crops varies greatly by region, mostly depending on pest infestation levels, the development of popular regional varieties containing these genes to ensure yield advantages, and seed and technology costs. Some studies (Klotz-Ingram et al., 1999; Culpepper and York, 1998) show that benefits appear to outweigh expected costs for many farmers, translating into rapid adoption of these GM cotton varieties.

The study by Culpepper and York 1998 found that the greatest advantage of planting herbicide-tolerant varieties was the reduced herbicide use. More recently, in 2006, Bollgard® cotton plantings were associated with significantly higher lint-yields and lower pesticide use in all cotton producing states. Bollworm and the bollworm pest complex were again ranked as the number one pest problem in cotton. The use of Bollgard® cotton varieties was credited with the reduction in this pest. Of the total cotton crop, the bollworm/bollworm pest complex reduced yields by about 0.9%, the lowest in recent years (Williams, 2007). The impact of Bollgard® is not only in increased yields but also in reduced sprayings and lower amounts of active ingredients on cotton acreage.

Today, there is a wide array of GM cotton varieties expressing multiple, or stacked, unique GM traits. Immunoassay is the most efficient and economical method for detecting the presence or absence of these traits. Some manufacturers of LFD have developed tests that are capable of detecting all the unique traits on a single strip (Chapter 5). This same approach is likely to be used in the future as value-added GM grains begin to be commercialized and IP programs are used to manage their increased value through the supply chain.

11.2.3 Supply Chain and Quality Management Systems for Non-GM Identity Preservation

Genetically engineered foods have become more and more prevalent in our daily purchases over the past decade. The food system is global. Soybeans may be grown in Central Illinois, United States, from seed that was produced in Argentina. The soybeans may then be shipped to Japan, where they could be combined with wheat from Canada, to make soy sauce, which could then be sold to consumers anywhere in the world. Surprising as it may be, a product that can be purchased from the grocery store in Ireland could have ingredients from two or more continents. This is not a food safety issue; however, it is a labeling issue in some regions of the world with regards to GM. Many food manufacturers, producers, packagers, processors, and farmers have found a niche market by segregating products that contain GM from those that do not.

Traceability is not an objective in itself, but rather a tool, which may assist companies to demonstrate that products meet safety, legal, and/or quality requirements, or consumer preferences. There are a variety of traceability system documents available from international and regional bodies that companies can implement to ensure the identity of their product(s) (Table 11.3). Also, there are several organizations that facilitate supply chain transparency from the farm to the grocer's shelf, by assisting with the implementation of a particular system, or by auditing the current system. The traceability systems are more urgent with regards to food safety issues (microbiological contamination, mycotoxins, previous cargo contamination, etc.) but can also be used to facilitate differentiation between Identity Preserved, non-GM grain/oilseeds, and commodity grain/oilseeds or other niches, such as certifications based on religious beliefs including Kosher, "fit" or "proper" as relates to Jewish dietary law, and Halal, "permissible" or "lawful," and refers to all matters of life for

TABLE 11.3 Summary of Some International Labeling Regulations for GM Foods

Country	Voluntary labeling or mandatory labeling	Threshold for unintended GM material (%)
Argentina	Voluntary	NA[a]
Australia and New Zealand	Mandatory	1%
Canada	Voluntary	NA[a]
China	Mandatory	None
European Union	Mandatory	0.90%
Japan	Mandatory	5%[b]
Russia	Mandatory	0.90%
South Korea	Mandatory	3%[b]
United States	Voluntary	NA[a]

Source: Various USDA Foreign Agricultural Service Attaché Reports.

[a]Not Applicable.

[b]Top three ingredients in Japan and top five ingredients in South Korea.

Muslims (Leake, 2008), or organic. These tools can be particularly helpful for companies that choose to export their products to countries that have labeling requirements for GM. Product testing, in addition to document control and data management, can have benefits at all levels of production because of GM's appearance at every level of the supply chain.

11.2.4 Programs for Identity Preservation

The food industry is rapidly transforming. At times, ingredients are sourced from several different continents for a processed food product. Food and agricultural processing companies are looking to differentiate themselves from their competitors by turning toward specialty and value-added products. These products usually depend on a complex supply chain. Identity preservation programs assist the food industry by outlining a plan to verify the integrity of their product from farm to fork. Table 11.4 is a list of the current international programs that have been actively involved in this field. Below is a brief introduction to the International Organization for Standardization (ISO) and the Codex Alimentarius Commission (CAC).

11.2.4.1 International Organization for Standardization ISO Technical Committee 34, Food Products, has a "family" of standards dedicated to food management systems. ISO 22000 and ISO/TS 22004 are the output of working group (WG) 8–*Food safety management systems*. Experts from 23 countries participated, together with international organizations with liaison status [Codex Alimentarius Commission, Confederation of the Food and Drink Industries of the European Union (CIAA), and the World Food Safety Organization (WFSO)]. Working Group 8 and its liaison organizations have partnered with experts from the ISO committee on conformity assessment, ISO/CASCO, the International Accreditation Forum (IAF), and the IQNet international certification network for the development of ISO/TS 22003. ISO 22005 was

TABLE 11.4 International Programs

Organizations and the Standards offered	
International Organization for Standardization	ISO 22000:2005, Food safety management systems—requirements for any organization in the food chain
	ISO/TS 22004, Food safety management systems—guidance on the application of ISO 22000:2005
	ISO/TS 22003, Food safety management systems—requirements for bodies providing audit and certification of food safety management systems
	ISO 22005, Traceability in the feed and food chain—general principles and guidance for system design and development
	ISO/DIS 22006, Quality management systems—guidelines for the application of ISO 9001:2000 in crop production
Codex Alimentarius	CAC/GL 60-2006, Principles for traceability/product tracing as a tool within a food inspection and certification system
GLOBALGAP[a]	GAP, Good agricultural practices is a single integrated standard with modular applications for different product groups, ranging from plant and livestock production to plant propagation materials and compound feed manufacturing
British Retail Consortium	Global Food Standard[b]
Safe quality food (SQF)	SQF 1000 Code, a HACCP-based supplier assurance code for the primary producer
	SQF 2000 Code, a HACCP-supplier assurance code for the food industries
International Committee of Food Retail Chains (CIES)	*The Global Food Safety Initiative Guidance Document*, 5th edition[c]

[a]http://www.globalgap.org/cms/front_content.php?idcat=48; http://www.globalgap.org/cms/front_content.php?idcat=49; http://www.globalgap.org/cms/front_content.php?idcat=50.

[b]http://www.brcglobalstandards.com/standards/food/.

[c]http://www.ciesnet.com/2-wwedo/2.2-programmes/2.2.foodsafety.gfsi.asp.

developed by WG 9, *traceability in the feed and food chain*. ISO/FDIS 22006, *Quality management systems—guidelines for the application of ISO 9001:2000 in crop production*, is currently under development by WG 12. Genetically engineered products are not a food safety issue, though labeling these products is a legal issue in some regions of the world. Therefore, implementing the strategies outlined below can set the stage for an identity preserved, non-GM grain/oilseed supply chain.

ISO 22005:2007 is the international standard for traceability in the food and feed supply chain, and is applicable to IP of non-GM products. It provides tools to implement an effective traceability system as part of a food safety/quality management system. ISO 22005:2007, Traceability in the feed and food chain—general principles and basic requirements for system design and implementation, "gives the principles and specifies the basic requirements for the design and implementation of a feed and food traceability system. It can be applied by an organization operating at any step in the feed and food chain. It is intended to be flexible enough to allow feed organizations and food organizations to achieve identified objectives. The traceability system is a technical tool to assist an organization to conform within its defined objectives and is applicable when necessary to determine history, or location of a product, or its relative components" (scope, ISO 22005).

When implemented properly, this standard will allow organizations operating at any step of the food chain to trace the flow of products by identifying necessary documentation and tracking each stage of production, ensuring adequate coordination between the different segments of the supply chain involved, and requiring that each party be informed of direct suppliers and clients (Groenveld, 2009).

The standard uses the same definition of traceability as the CAC and provides a complement for organizations implementing the ISO 22000:2005 standard, giving the basic requirements for a food safety management system to ensure safe food supply chains. ISO 22000 incorporates the principles of the CAC's "Hazard Analysis and Critical Control Point" (HACCP) system for food hygiene.

11.2.4.2 Codex Alimentarius Commission The CAC was created in 1963 by FAO and WHO to develop food standards, guidelines, and related texts such as codes of practice under the Joint FAO/WHO Food Standards Programme. The Codex Alimentarius is recognized by the World Trade Organization as an international reference point for the resolution of disputes concerning food safety and consumer protection. "The CAC/GL 60-2006 is a set of principles to assist competent authorities in utilizing traceability/product tracing as a tool within their food inspection and certification systems. These principles cover the context, rationale, design, and application of traceability/product tracing as a tool to protect consumers against food-borne hazards, deceptive marketing practices, and facilitates trade on the basis of accurate product descriptions" (CAC/GL 60-2006).

Traceability/product tracing is a tool that when applied in a food inspection and certification system can contribute to the protection of consumers against deceptive marketing practices and facilitation of trade on the basis of accurate product description. For example, by reinforcing confidence in the authenticity of the product and the accuracy of information provided on the products (e.g., country of origin, organic farming or non-GM farming, religious concerns such as kosher or halal). The design of the traceability/product tracing system should apply to either all or specified segments of the supply chain from production to distribution. Each vendor in the supply chain should know and communicate with the vendor one step behind and one step ahead in the supply chain. Most importantly, the system should be transparent and available, upon request, to competent authorities.

11.2.5 Implementation of IP Program and Immunoassay Use at Critical Hold Points in Processing

What are the first steps involved in implementing an IP program? First, ideally, one would identify a need in the marketplace and then set a plan to meet that need. For example, there is a strong desire in Japan for non-GM, food-grade soybeans. What is the best way to ensure that soybeans grown in the Midwest section of the United States meet the need of the Japanese buyers? For this scenario, the company providing the non-GM, food-grade soybeans to Japan will be called Illinois Soy, a company that specializes in non-GM soy for food and feed applications. Agreements need to be forged with the buyers in Japan, as well as the growers in the United States with Illinois Soy. Terms would be set for the condition of the product received in Japan. Farmers would be required to sign contracts regarding production practices and possibly provide a Quality Management System manual. Appropriate seed would need to be sourced so the farmers could meet their objectives. A testing scheme would need to be organized and agreed to by all parties involved. Transportation, storage, and distribution would also need to be determined.

Many farmers today have diversified their agricultural portfolios by contracting to grow food-grade crops, as well as non-GM and organic crops. To do this, the farmer agrees to only purchase and plant certain varieties of seed prescribed by the company. This is verified by reviewing seed sales receipts as well as by immunoassay testing the seeds delivered to the farm to verify the absence of GM. In many cases, the farmer agrees not to plant any crop with GM in exchange for higher premiums on the niche, non-GM crop. Protocols are established, prescribed, and followed for cleaning the planter, as well as researching what crops will be planted in neighboring fields. This is less of an issue with soybeans because of self-pollination, it is more so an issue when growing non-GM corn because of wind- or insect-mediated cross-pollination. Buffer rows and alternate planting schedules are two of the most common strategies for avoiding cross-pollination.

A representative from Illinois Soy may visit the farm to oversee the planting and administer the immunoassay tests. Site visits are necessary to ensure IP programs are followed. Once the non-GM crop has been planted, it is important to test the leaf tissue with immunoassay in the field to verify the absence of GM. Cleaning measures for harvesting equipment and transportation (trucks, wagons, etc.) will also be prescribed and verified by a representative from Illinois Soy. If the harvested crop is to be stored on the farm, the storage space will need to be cleaned and monitored to ensure that crops are not mixed. If the farmer has already agreed to not grow any GM crops, that is one less possibility for contamination. However, if some of the farmer's crop is feed-grade soy, it would be a violation of the contract if the crops comingled.

To avoid this situation, Illinois Soy could insist that all crops be delivered from the field directly to Illinois Soy. Once there, the load would be sampled (according to USDA-GIPSA protocols) for the standard quality verifications (hilum color; bean color uniformity; cracked, damaged, or broken beans; foreign material; moisture; mycotoxins; etc.) and lastly, the load will be tested again for GM by immunoassay. If the load were to fail either test, it would be rejected. A load that passes both tests would then be allowed to deliver to the facility. Each load that arrives would be subjected to this scrutiny to further the IP program.

When the customer in Japan is ready to receive a shipment, Illinois Soy will utilize immunoassay tests as each load of soybeans is removed from the warehouse. The sampling and testing plan will be based on the size of the shipment and the mode of transportation. If the soybeans are bulked in baffle bags, each bag would be sampled and tested according to ISTA's International Rules of Seed Testing: Seed Science and Technology Rules. This protocol is not specifically for GM testing, but for overall quality of the lot in question.

The soybeans are immunoassay tested before the bulk, baffle totes are sealed. The bulked material would then be transported by truck or rail to an export terminal. Trucks and train cars would need to follow the cleanliness standards as well. Bags would be inspected for visible tears or weakness in the fabric before being loaded on the ship. If damage is noted, the soybeans would be rebagged and the immunoassay tests would be repeated. Once the soybeans are placed on the ship, a complete history of the shipment will be sent to the customer in Japan. This history would include verifications of the seed receipts as well as all the immunoassay test results from the seed through placement on the ship. Once at the destination, each bulk bag would be immunoassay tested again to verify the shipment before further processing. Illinois Soy would retain the samples from all tests until the customer has verified the shipment meets expectations.

In October 2004, World-Grain.com published an article written by Emily Buckley entitled "GMO Testing: A key element of integrated IP programs." She had the opportunity to interview three company executives regarding IP programs and their approaches to serving the non-GM marketplace. James Stitzlein, Consolidated Grain and Barge Company (CGB), manager of market development, licensing and storage, was quoted, "We have always been willing to handle grains produced from GM seed and were somewhat surprised when some users asked us to separate non-GM crops for them. Despite our own favorable opinions about the technology, we realized that we needed to listen to our customers' concerns (Buckley, 2004)." Traceability measures, grower programs, and new standard operating procedures were imple- ments. Growers were contracted based on assessment of capability and willingness to accept the concept and then trained to handle and segregate non-GM crops according to set procedures. A combination of immunoassay and DNA testing is applied as part of the CGB IP program. The level of coordination convinced CGB to pursue ISO 9001:2000 certification in 2001. Eighteen of the major river terminals that handle IP grain have been certified. At press, CGB was handling 30 different IP grain programs.

National Starch & Chemical is a worldwide leader in specialty starch tech- nology and manufacturing, serving the global food industry. According to Joe Emling, senior project supervisor of agribusiness, "while the company fully supports biotech's responsible use within industry, it currently sees non-GM maize use as the best solution for its customers' needs at the time being (Buckley, 2004)." In order to achieve a non-GM raw material supply chain, National Starch based its own grain quality and traceability system, TrueTrace, on the British Retail Consortium technical standard for supplying non-GM food ingredients and products. The TrueTrace system includes documentation, testing, and auditing as its main components. Immunoassay strip tests are administered routinely on loads of maize delivered to the country elevators and upon delivery to a National Starch processing facility. Composite samples are sent for DNA testing to verify TrueTrace is working effectively.

An independent grain certification company in Brazil, IQS, guarantees the identity and quality of a product from production to export. IQS works with customers in Europe, Asia, and the Middle East. Lateral flow strips are used exclusively in the "Non-GM Hard IP Programs" offered by IQS for soybeans. Beans are tested at the local elevator and the transportation vehicles travel to the export terminal with a sealed load and complete identification documentation. "During the 5 years that we have been using our Hard IP program, it had to be upgraded due to the requests of our customers," Faifer, executive director, said, "The customers wanted a more sensitive test than was currently available from the testkit manufacturers." At the time the article was written, most commercially available testkits were operating at the one GM seed in 1000 seeds sensitivity level and the IQS customers wanted an increased sensitivity of one GM seed in 4000 seeds (Buckley, 2004). IQS worked with a specific immunoassay testkit provider to increase the sensitivity of the test for their purposes.

It should be reinforced that developing an IP or traceability system is a separate process from certification of the system. Developing the system involves restructuring the production, storage, shipping, processing, manufacturing, distribution, and so on, and take segregation, documentation, sampling, and testing into account for identity preservation and traceability. Certifying your IP or traceability system by a third-party audit, to verify if it meets your goals and objectives, is the next step.

11.3   VALIDATION AND VERIFICATION OF NON-GM IP PROGRAMS

As with any quality assurance (QA) system, non-GM IP production systems and supporting analytical testing programs must be validated and verified on a regular basis. Frequent internal and third-party audits of processing facilities, including an extensive review of program documents and observation of practices, are often used to ensure the non-GM IP program has been properly designed and executed. Third party, independent certification systems are a powerful tool that the food industry can use to provide reassurances. In addition, immunoassay biotechnology testing at critical points in the supply chain is a necessary component to verify the effectiveness of a non-GM IP program. In-house and contract laboratories providing immunoassay testing services, as well as the immunoassay testing devices, should be subject to audits and proficiency testing trials to verify the precision and accuracy of analytical data.

Internal audits of processing facilities should be conducted at planned intervals to ensure the non-GM IP program meets its stated goals, is being executed properly, and the program is updated, as needed (ISO 22000 and ISO 22005). By providing a documented assessment of the performance of the non-GM IP program, internal audits can be an effective tool to help minimize the risk of producing a nonconforming product (detection of the presence of an unapproved GM event or GM above the program specified threshold) or provide additional data during the course of the investigation of nonconformity. In addition, the results from internal audits can be used as a source of guidance when assessing the effectiveness of the non-GM IP program for continuous improvement initiatives.

Just as a Quality Management System is integral for business practices, a comprehensive scheme of quality assurance throughout the supply chain's analytical facilities is also necessary. The QA should include proficiency testing, method validation if applicable, incorporation of certified reference materials (CRMs) where available, and routine internal quality control. Analytical facilities with no external reference could operate for long periods with biases or random variations. Proficiency testing is a means of detecting error and remedying the situation. Its main virtue is that participants can obtain an external and independent assessment of the accuracy of their results. One of the major challenges with detecting the presence of GM grains and oilseeds is that different laboratories do not always obtain comparable results when testing identical samples. Proficiency testing provides laboratories with a means of assessing their capabilities and improving their overall performance, repeatability, and reproducibility, as well as help them meet one of many requirements for ISO 17025 accreditation (Leake, 2006).

11.4   CONCLUSION

A large proportion of major crops currently grown in the United States have been enhanced with one or more traits through the use of modern biotechnology. Immunoassays offer simple, specific, and sensitive protein detection to address and/or meet a wide range of needs of technology providers, food and commodity distribution channels, and regulatory bodies.

To date, commercialized GM crops have addressed only the simplest of problems (herbicide tolerance and insect resistance) for crop growth. Cold tolerance, drought resistance, and increased yield are on the horizon. Flavor and shelf life, traits with human health benefits, are other issues that one would have to believe will be addressed by GM crops that might be more rapidly adopted if available.

11.5   FUTURE PERSPECTIVES

Immunoassay biotechnology testing has been used almost exclusively as a verification tool in non-GM IP grain and food processing supply chains to ensure GM levels are below established thresholds or to verify the absence of unapproved GM events. This practice may change, however, with the marketing of events that possess traits desired by grain processors and retail consumers. In the future, immunoassay testing may be used on a large scale by grain and food processing facilities to ensure the presence of specific GM events.

For example, immunoassay testing may be required at critical points along the supply chain to ensure the presence of a trait for a food or food ingredient that has been genetically modified to enhance its nutritional profile to help the food manufacturer verify the product meets nutritional specifications. It would be essential to be able to verify the presence of the trait to ensure any advertising and labeling claims related to the enhanced nutritional qualities can be substantiated.

For grain processors, immunoassay testing may be required to ensure the presence of a trait, such as increased starch levels in corn for ethanol production, so the premium is properly paid to the grain seller and the operators at the facility will be able to make adjustments as needed based on the quality characteristics of this critical input material. In both examples, immunoassay biotechnology testing may be an invaluable tool in carrying out quality assurance and quality control activities by providing a rapid and accurate detection method for grain processing and food production facility operators.

ACKNOWLEDGMENTS

The authors would like to share their sincere appreciation to Guomin Shan for offering the opportunity to participate in this project and for his patience throughout the process. It is an honor to be included as authors in this group of highly respected scientists. Gratitude is also expressed to AOCS and Eurofins Scientific for allowing us to spend time writing this chapter.

REFERENCES

Buckley, E. GMO testing: a key element of integrated IP programs, available at http://www.non-gm-farmers.com/news_print.asp?ID=1700, October 1, 2004.

CAC/GL 60-2006. Principles for traceability/product tracing as a tool within a food inspection and certification system, available at http://www.codexalimentarius.net/web/more_info.jsp?id_sta=10603.

Culpepper, A. S. and York A. C. Weed management in glyphosate-tolerant cotton. *J. Cotton Sci.*, **1998**, *4*, 174–185.

European Commission Regulation (EC) No. 1830/2003. Official Journal of the European Union L268.

Groenveld, C. A new International Standard for traceability, Businessassurance.com, available at http://businessassurance.com/a-new-international-standard-for-traceability, March 4, 2009.

Grothaus, G. D., Bandla, M., Currier, T., Giroux, R., Jenkins, G. R., Lipp, M., Shan, G., Stave, J. W., and Pantella, V. Immunoassay as an analytical tool in agricultural biotechnology. *J. AOAC Int.*, **2006**, *89*, 913–928.

ISO 22000:2005(E). Food safety management systems—requirements for any organization in the food chain.

ISO/TS 22004:2005(E). Food safety management systems—guidance on the application of ISO 22000:2005.

ISO/TS 22003:2007(E) Food safety management systems—requirements for bodies providing audit and certification of food safety management systems.

ISO 22005:2007(E). Traceability in the feed and food chain—general principles and basic requirements for system design and implementation 2007.

ISO/FDIS:22006 (E). Quality management systems—guidelines for the application of ISO 9001:2000 in crop production.

James, C. Highlights of ISAAA Briefs No. 41-2009. Global status of commercialized biotech/GM crops, available at http://www.isaaa.org/Resources/Publications/briefs/35/executivesummary/default.html, 2009.

Jasbeer, K., Ghazali, F. M., Cheah, Y. K., and Son, R. Review paper: application of DNA and immunoassay analytical methods for GMO testing in agricultural crops and plant-derived products. *ASEAN Food J.*, **2008**, *15*, 1–25.

Klotz-Ingram, C, Jans, S., Fernandez-Cornejo, J., and and McBride, W. Farm-level production effects related to the adoption of genetically modified cotton for pest management. *AgBioForum*, **1999**, *2*, 73–84.

Leake, L. Improving proficiency of GMO testing. *Food Technol.*, **2006** (April), 74–77.

Leake, L. Answering to a higher authority. *Food Quality*, **2008**. (October/November) 14–23.

Lipp, M., Shillito, R., Giroux, R., Spiegelhalter, F., Charlton, S., Pinero, D., and Song, P. Polymerase chain reaction technology as an analytical tool in agricultural biotechnology. *J. AOAC Int.*, **2005**, *88*, 136–155.

National Agriculture Statistics Service. Quick Stats, Crops, Cotton, available at http://www.nass.usda.gov/.

SQF 1000 Code. A HACCP based supplier assurance code for the primary producer. Safe Quality Food Institute, available at http://www.sqfi.com/sqf_documents.htm.

SQF 2000 Code. A HACCP supplier assurance code for the food industries. Safe Quality Food Institute, available at http://www.sqfi.com/sqf_documents.htm.

USDA-FAS. United States Department of Agriculture, Foreign Agricultural Service Attaché Reports for Argentina, Australia, Canada, China, European Union, Japan, Russia, and South Korea. Retrieved on June 2009 from http://www.fas.usda.gov/scriptsw/attacherep/default.asp, 2008.

USDA NASS. National Agricultural Statistical Service (NASS). Agriculture chemical usage: 1997 field crops summary, USDA/NASS, Washington, DC, **1998**.

Vines, R. Plant biotechnology. Virginia Cooperative Extension Publication 443-002. Biotechnology Information, available at http://pubs.ext.vt.edu/443-002/, 2001.

Williams, M. Cotton insect losses—2006, available at http://www.ncfap.org/documents/2007biotech_report/Quantification_of_the_Impacts_on_US_Agriculture_of_Biotechnology.pdf, 2007.

CHAPTER 12

IMMUNOASSAY APPLICATIONS IN SOIL MONITORING

Guomin Shan

12.1   INTRODUCTION

Since their introduction in mid-1990s, the cultivation of genetically engineered (GE or GM) crops has become widespread. In 2009, over 134 million hectares of GM crops were planted by over 14 million farmers globally (James, 2009). Despite their increasing domination in crop productions systems, some public concern and scientific debate remain regarding the potential ecological impacts of these crops, particularly with respect to fate in soil. Genetically engineered crops expressing insecticidal proteins are of particular interest and the US EPA has requested product developers to provide more environmental risk assessment data for registered GM traits such as field accumulation studies that require monitoring the residue of the transgenic protein in soil after three consecutive plantings of the GM crop.

In the past decade, research groups across the world have conducted numerous studies on the fate of *Bacillus thuringiensis* (Bt) insecticidal proteins in the environment, especially in soil (West et al., 1984; Venkateswerlu and Stotzky, 1992; Tapp et al., 1994; Donegan et al., 1995; Palm et al., 1996; Sims and Holden, 1996; Sims and Ream, 1997; Tapp and Stotzky, 1998; Saxena and Stotzky, 2002; Zwahlen et al., 2003; Accinelli et al., 2008). There are two main paths by which plant-expressed Bt proteins enter the soil ecosystem: by decomposition of crop residue (Palm et al., 1994;

Immunoassays in Agricultural Biotechnology, edited by Guomin Shan
Copyright © 2011 John Wiley & Sons, Inc.

Blackwood and Buyer, 2004; Marchetti et al., 2007) and possible exudation from roots during the growing season (Saxena et al., 2002; Saxena and Stotzky, 2002).

Laboratory studies from Stotzky's research group have shown that Bt proteins are readily and strongly bound to clay in some soils, thereby significantly slowing the microbial degradation process (Tapp et al., 1994; Tapp and Stotzky, 1998). The bound proteins reportedly retained their insecticidal activity and persisted in soil for up to 6 months (Crecchio and Stotzky, 1998; Stotzky, 2004). In a degradation study of the Cry1Ab protein in Bt corn leaves under field condition, a trace level of Cry1Ab residue (0.3% of the initial level) was detected in soil *in situ* up to 200 days (Zwahlen et al., 2003). These findings have led to a hypothesis that with repeated GM crop plantings Bt proteins may accumulate and reach biologically active levels. However, other studies have reported rapid decay rates of Bt proteins in soil under either laboratory or field conditions (Sims and Holden, 1996; Sims and Ream, 1997; Head et al., 2002; Shan et al., 2008; Herman et al., 2001, 2002b), thus indicating that the risk of bioaccumulation in soil after continuous cultivation of Bt crops is low. Herman et al. (2001, 2002a) reported that the half-life of microbe-derived Cry1F in a light clay soil was 0.6 days. Half-lives of Cry1Ac protein when added to a silt loam soil as corn residues or as purified protein were estimated at 8.3 and 1.6 days, respectively (Sims and Holden, 1996). Palm et al. (1994, 1996) concluded that half-lives of purified core Cry1Ac protein in soil ranged from 18 to 40 days dependent upon soils, while Cry1Ab had an approximate half-life of 11 days in a fine sandy loam soil. Laboratory testing of Cry3Bb1 expressing Bt maize had no detectable protein after 21 and 40 days in monotmorillonite-amended and kaolinite-amended soils, respectively (Icoz and Stotzky, 2008a). Similar results were reported by other authors as well (Herman et al., 2002b; Wang et al., 2006; Accinelli et al., 2008). Field-accumulation studies during multiple years of GM crop planting were consistent with laboratory results and also showed that Bt proteins degrade rapidly and do not persist in the soil. For example, after 3–6 years of consecutive Bt cotton cropping in multiple locations, no detectable levels of Cry1Ac protein was found in any soil samples collected from the fields (Head et al., 2002). Dubelman et al. (2005) investigated the persistence and accumulation of Cry1Ab protein from GM maize plant and no Cry1Ab protein was detected in soils from fields planted for three or more consecutive seasons with Bt maize. Gruber et al. (2008) monitored Cry1Ab protein in soil samples collected from multiple field sites in Germany where Bt maize (Mon810) was grown for up to 8 years and no Cry1Ab was detected. Ahmad et al. (2005) found no detectable Cry3Bb1 protein in soil collected from field planted with Bt maize for three consecutive seasons. Similar results were obtained from accumulation studies with Cry1F corn (Shan et al., 2008), Cry1Ac, Cry1F cotton (Shan, unpublished data), and Cry34Ab1/Cry35Ab1 corn (Shan, unpublished data), where no Cry protein was detected in either bulk soil or rhizosphere soil samples from fields planted with Bt maize or Bt cotton for three consecutive seasons in multiple locations in the United States.

Various factors might have contributed to these contrasting results including the difference in soil type, experimental conditions, and type of protein tested (Icoz and Stotzky, 2008b). However, one major source of the discrepancy is the lack of efficient and reliable quantitative detection methods due to low extractability of Bt proteins from soil (Clark et al., 2005; Shan et al., 2003; Accinelli et al., 2008).

12.2   MONITORING METHODS

Two analytical methods have been used for quantification of Bt protein in soil: insect bioassay and immunoassay.

Insect bioassay was historically the primary method for measuring active Bt protein in soil (Sims and Holden, 1996; Sims and Ream, 1997; Saxena et al., 2002; Head et al., 2002; Herman et al., 2001, 2002a, 2002b). Normally, soil samples were suspended in agar solution and then applied to the top of insect diet. The target neonate insects were placed on and fed on treated diet for a period of time (e.g., a week) and the end point results, such as mortality or/and insect weights, were recorded after the incubation. Degradation rates were determined by modeling the reduction in the LC_{50} (concentration estimated to kill 50% of the insects) over time (Herman et al., 2001, 2002a, 2002b). In general, this technique is very sensitive and measures direct biological activity of protein in the soil matrix and no extraction step is needed. The limit of detection (LOD) of Cry1Ab in the European corn borer bioassay was 0.03 μg/g soil (Dubelman et al., 2005). The LOD of Cry1Ac using tobacco budworm can be as low as 8 ng/g (Head et al., 2002). Similarly, insect bioassay detects Cry1F in soil at 8 ng/g level with tobacco budworm. However, in some cases, bioassay is of low sensitivity. For example, the detection limit of Cry34Ab1 in the Southern corn rootworm is >1 mg/g soil (Herman et al., 2002b). Due to the resource-intensive nature of the assay and in some cases due to poor sensitivity, it is not practical to routinely monitor the residue level of protein in soil by insect bioassay.

In addition, due to the nature of insect bioassay, bioassay often provides semiquantitative results and may be affected by toxic components in the soil matrix. Another challenging issue for bioassay is the selection and maintenance of target insect species. To achieve a better sensitivity, the most susceptible insect species must be used for each toxin. Different proteins may require different insect species as the target insect. For example, *Heliothis virescens* (tobacco budworm) or *Ostrinia nubilalis* were used for Cry1 and Cry2 proteins (Herman et al., 2001; Sims and Holden, 1996; Zwahlen et al., 2003; Dubelman et al., 2005), and *Diabrotica undecimpunctata* Howardi (southern corn rootworm) was used for Cry34/Cry35Ab1 proteins (Herman et al., 2002b). It is a resource-intensive task to maintain multiple susceptible insect species in the laboratory and even more difficult to be consistent across the research and scientific community and industry (Clark et al., 2005). Therefore, bioassays are time-consuming, expensive, and difficult to standardize.

An alternative analytical method involves extraction of the transgenic protein from soil followed by chemical detection technology such as Western blot (Sims et al., 1996) or Enzyme-Linked Immnosorbent Assay (ELISA) (Palm et al., 1994, 1996; Shan et al., 2003). As described in previous chapters, immunoassay is the primary method to qualitatively and quantitatively detect GM proteins in grain and plant tissue matrices (Grothaus et al., 2006). It provides a sensitive, specific, fast, easy-to-operate, and cost-effective analytical tool for the protein detection needs during discovery, product development, and product stewardship. In addition, immunoassay is suitable for high-throughput application and automation. Commercial ELISA kits are usually available for GM traits and are convenient for gaining access to the technology. Therefore, immunoassay would be an excellent choice to monitor GM protein in the environment and to serve as an analytical tool for risk assessment (Lipton et al., 2000). In the past few years, scientists have been actively using this technology for monitoring transgenic proteins in soil, including Cry1Ab (Hopkins and Gregorich, 2003; Wang et al., 2007; Gruber et al., 2008), Cry1Ac (Hopkins and Gregorich, 2003), Cry1F (Shan et al., 2003, 2008), and Cry3Bb1 (Icoz and Stotzky, 2007). However, the application of ELISA in soil detection has been hindered by the lack of satisfactory chemical extraction techniques. The extraction efficiency of transgenic proteins from soil was poor, and sometimes less than 10% of the protein was extracted with the selected assay buffer system (Head et al., 2002; Palm et al., 1994; Hopkins and Gregorich, 2003).

It was important to develop an efficient satisfactory extraction system, which could be used in tandem with ELISA for protein soil monitoring. In this chapter, we will review the mechanism of protein binding and biodegradation in soil and then use examples to discuss the potential practical solutions for soil extraction and quantitative detection with immunoassays.

12.3 MECHANISM OF PROTEIN BINDING AND DEGRADATION IN SOIL SYSTEMS

12.3.1 Protein Binding in Soil

Soil particle size is frequently used to describe soils on the basis of their sand, silt, and clay content (Brady and Weil, 2002). According to the US classification system, particles with a size smaller than 2 mm but larger than 0.05 mm are sand and particles smaller than 0.002 mm are classified as clay. Others with a size between sand and clay are called silt (Table 12.1). Clay soil particles have very large specific surface area of $80–100\,m^2/g$ of soil and therefore have a huge capacity to adsorb water and other substances such as proteins. Soil consists of the various-sized aggregates formed through the combination or arrangement of these primary soil particles. According to U.S. Department of Agriculture (USDA) classification system, there are three broad groups of textual classes: sandy soils, clayey soils, and loamy soils (Brady and Weil, 2002). Within these groups, 12 specific textural classes were recognized as illustrated in the soil textural triangle (Figure 12.1). Each class name represents the distribution of particles and indicates the general nature of soil physical property. In

TABLE 12.1 Classification of Soil Particles According to Their Sizes

ISSS	Clay	Silt	Sand		Gravel
			Fine	Coarse	
Particle size (mm)	0.002	0.02	0.2	2.0	
	0.002		0.05 0.1	0.25 0.5 1.0 2.0	
USDA	Clay	Silt	Very fine	Fine Med. Coarse Very coarse	Gravel
				Sand	

ISSS: International Society of Soil Science; USDA: U.S. Department of Agriculture.

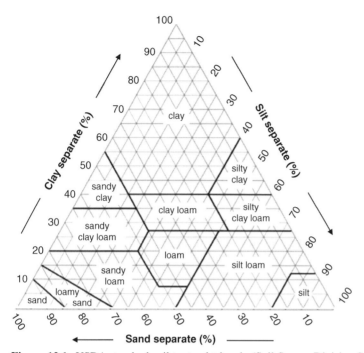

Figure 12.1 USDA standard soil textural triangle (Soil Survey Division Staff, 1993).

addition, various sizes of pores exist between and within soil aggregates and are occupied by water and air that provide the habitat for soil organisms and organic substances. Regarding the physical and chemical binding properties of soil with other substances, clay particles play an important role. Other than its large specific surface area, most types of clay particles have electronically charged surface, which bind charged particles or substances. Soils will vary widely in their mineralogical composition, and this is especially important with regard to the clay-sized particles, since the nature of the clay-minerals present will impart vastly differing chemical reactivity to soils. The major chemical reactions in soil including the binding of molecules take place at these interfaces and the same apply to GM proteins. When a protein enters the soil, it will quickly bind to the soil particles through multiple interactions including ionic bonding, hydrogen bonding, and hydrophobic interaction. The mechanism of adsorption and binding of Bt proteins on clay minerals has been investigated by Stotzky's research group (Tapp et al., 1994; Venkateswerlu and Stotzky, 1992; Fiorito et al., 2008). Several key parameters may affect the adsorption and binding process and need to be considered during the protein extraction method development. First, the soil type: the greater the surface area of the soil particles, the more protein is bound to the soil particles. Soils with higher clay percentage usually have higher capacity for protein adsorption and binding. Second, the isoelectric point (pI) of protein and pH environment: since most clay particles have negatively charged surfaces, at a pH near the protein pI, the repulsion between a net neutral-charged protein and soil particles is minimal and thus results in increased adsorption. In some cases, if the protein has a net positively charged surface, it would further accelerate the adsorption process and formation of ionic bonds, and thus enhances the protein-particle binding. Third, soil organic matter, which is also highly chemically reactive and will occur as coatings on the soil particle and which affects their aggregation. Therefore, a higher level of organic matter in soil will lower its protein binding ability. Last, the protein size and its tertiary structure: other than the net charges on protein surface, the shape and size of protein may significantly affect the interaction between protein and soil particles. Compared with large proteins, smaller proteins (e.g., 10–20 kDa) usually have a relatively large linear region that provides greater opportunities for binding to charged soil surfaces. Binding may be tighter than for large proteins resulting in reduced extractability. The Bt protein Cry34Ab1 is a small protein with a molecular weight of 14 kDa (Gao et al., 2004). Studies showed that it is extremely difficult to extract this protein from soil, and spike-recovery of Cry34Ab1 in a loam soil with a regular buffer such as PBST or tris-borate buffer is virtually zero (Shan, unpublished data). Similarly, maize expressed *Escherichia coli* heat-labile enterotoxin subunit B (LTB) has a small size at 11.6 kDa. When spiked into soil, five out of six published extraction buffers gave a 0% recovery. After an extensive study, the optimized extraction buffer still only gave a 17.3% recovery in clay loam soil for LTB (Kosaki et al., 2008).

12.3.2 Protein Degradation in Soil

Protein binding to soil particle surfaces may be physically or functionally irreversible. In other cases this process may be reversible depending on protein properties and the soil system. An adsorption and dissociation process may exist at the soil particle surface. Tapp and Stotzky (1998) reported that the Bt proteins retain their insecticidal toxicity after being adsorbed to soil particles, a property which is common with endogenous soil enzymes that are stabilized and retain activity when adsorbed to soil particles.

Once the GM protein enters the soil system, other than adsorption and binding to soil particles, protein will be dissipated by biodegradation and physical/chemical denaturation (Figure 12.2). Microorganisms in topsoil are the dominant source of GM protein biodegradation, which may occur through plant tissue decomposition and direct protein mineralization (Zwahlen et al., 2003). The breakdown of plant tissue is an important component of the degradation process of plant-produced Bt proteins. The released protein may be adsorbed and bound to soil and thus have reduced bioavailability to microorganisms. On the other hand, soil-bound protein may be freed and become readily available for biodegradation. The physical and chemical degradation may include thermal denaturation and chemical interaction, which involves the disruption and possible destruction of both secondary and tertiary structures that cause protein precipitation or aggregation.

Protein structure may be altered by the disruption of hydrogen bonds and nonpolar hydrophobic interactions at elevated temperature. Chemical causes of protein denaturation in soil may include acidic or basic environments, or the presence of heavy metals. Soils may contain heavy metal salts such as Hg^{2+}, Pb^{2+}, Ag^+, Tl^+, Cd^{2+}, and other metals. These metals can easily react with a protein by disrupting salt bridges and forming an insoluble metal protein salt. In addition,

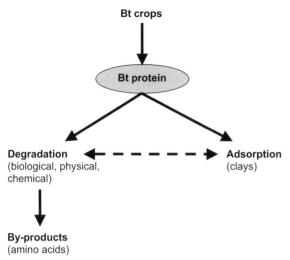

Figure 12.2   Fate of Bt protein in the environment.

heavy metals usually have a high affinity and attraction to sulfur. Therefore, they may disrupt disulfide bonds leading to protein denaturation. Finally, high or low soil pH may cause protein denaturation. The acids and bases disrupt salt bridges held together by ionic charges.

12.4   MONITORING PROTEIN IN SOIL WITH IMMUNOASSAYS

In general, once the protein enters the soil system, other than degradation, it may be present in four forms: (1) free and functional form, (2) soil bound and functional form, (3) soil bound and inactive form, and (4) free and denatured or degraded form. The meaningful target for monitoring is the first two forms, which are biologically active and relevant for any risk assessment. This implies that a good detection method should be able to extract both free and soil-bound functional form proteins. For a chemical detection method, a proper extraction method is critical. Numerous laboratories have reported their effort to develop an efficient extraction buffer system and all found it challenging and almost impossible to have a one buffer fit for all proteins. Two laboratories took different approaches and developed buffer systems that are able to effectively extract Cry1 Bt proteins from soils even with some clay rich soil types (Palm et al., 1994; Shan et al., 2003). Palm et al. (1994) optimized the extraction buffer by varying different buffer components including salt types and concentrations, buffer pH, and surfactant levels. The resulting extraction buffer (PALM buffer) consists of 50 mM sodium borate with pH of 10.5, 0.75 M KCl, 0.075% Tween, and 10 mM ascorbic acid. The antioxidant ascorbic acid was added to lower the high ELISA background reading caused by plant materials in the sample. With this optimized extraction buffer, spike recoveries of up to 60% were achieved for Cry1Ac protein. Shan et al. (2003) adopted a novel biorational path and developed a buffer system that effectively extracted Cry1 proteins from soils (extraction recoveries of >60% were routinely achieved in soils of varied composition). As mentioned earlier in this chapter, under conditions where Bt protein extractability in soil was very low, however, insect bioassay was able to detect the activity of Bt proteins spiked into soil (Herman et al., 2001, 2002a, 2002b; Stotzky, 2000; Tapp and Stotzky, 1998). This indicates that the insect has the ability to recover adsorbed protein from soil and the adsorbed protein has retained biological activity. This subsequently led Shan and coworkers to a hypothesis that insect gut fluid could effectively extract protein from soil. Unlocking the components of insect gut fluid led to the development of a liquid system for protein soil extraction. Considering the need to supply large quantities of gut fluid (up to 1 mL per worm) and the importance of similar feeding habits compared with agricultural insect pests, bottom feeding marine invertebrates were collected and used for the study. Analyzing the physical and biochemical properties of selected worm gut fluid that had the best extraction recovery from soil, led to the development of an artificial gut fluid (Shan et al., 2003). The final fluid (SHAN buffer) used an artificial seawater (see receipt in Parsons et al., 1984) with a neutral pH as a base solution, containing high concentrations of salts (3–5%) including NaCl, KCl, Na_2SO_4, $NaHCO_3$, $CaCl_2$, and $MgCl_2$, and added surfactant sodium taurocholate (10 mM, final concentration) as well as protein bovine serum albumin (BSA) (0.5%, final concentration). With this model artificial fluid, the spiked Cry1 Bt protein can be efficiently extracted from soils (>60%) including clay rich soils (Shan et al., 2003).

Both Palm et al. and Shan et al. studies pointed out three key common components of a successful extraction buffer: high salt concentration, high level of surfactant and neutral, or higher pH. Palm et al. (1994) found that other than high concentration of salts, monovalent cation K^+ (on KCl form) was the most effective ion. Similarly, the SHAN buffer is dominated by monovalent cations Na^+ and K^+. As described earlier, clay particles have a negatively charged surface and it attracts cations in the solution. Multivalent cations such as Mg^{2+}, Ca^{2+}, and Al^{3+} are very effective to neutralize the negative surface of clay particles; however, they can serve as a bridge to draw two particles together and form small stable soil aggregates. Some bound protein on soil particle surface may be folded in the aggregate and become unavailable for extraction. On the other hand, these multivalent cations usually have smaller ionic radii (0.066 nm for Mg^{2+} versus 0.133 nm for K^+) (Table 12.2) (Brady and Weil, 2002), which suggest that it may be less competitive to replace adsorbed protein on the surface than monovalent cations. Moreover, monovalent cations are weakly charged and large in size (ionic radii) and thus are ineffective to promote clay stacking and aggregate formation. It may be possible to form a more seamless cation layer on the surface of clay particle and leave less room for protein binding. In other words, monovalent cations are more competitive to replace the adsorbed protein from the clay surface.

The appropriate type and concentration of surfactant in extraction buffer can significantly improve protein extractability from soil. Surfactant may form micelles in the solution at or greater than the critical micelle concentration (CMC), which is particularly helpful to solubilize proteins. This will influence the hydrophobic interaction between soil and protein. Tween 20 and Tween 80 are common surfactants in laboratories, which have low CMC at 0.0074% and 0.0016% (w/v), respectively. PALM buffer used 0.075% Tween 20 as surfactant and SHAN buffer used 10 mM of sodium taurocholate (CMC: 2.3 mM) (Voparil and Mayer, 2004).

TABLE 12.2   Ionic Radii of Elements Found in Clays

Ion	Radius (nm)
Si^{4+}	0.042
Al^{3+}	0.051
Fe^{3+}	0.064
Mg^{2+}	0.066
Fe^{2+}	0.070
Na^+	0.097
Ca^{2+}	0.099
$K+$	0.133
OH^-	0.155

Source: Brady and Weil, 2002.

Third, buffer pH plays a significant role in protein extraction. If a buffer is used with pH near or below the protein pI, the protein will be more tightly bound to soil and more difficult to extract. At the pI, the protein surface carries no net electrical charge and results in a hydrophobic adsorption with soil particles. Cry1 proteins have a pI at 5–6; a neutral or basic pH buffer environment will help the recovery of protein from soil binding. Palm et al. (1994) showed that protein recovery is about 20–30% higher at neutral or basic pH than under acidic conditions. Shan et al. (2003) found neutral pH buffer functions well. In some cases, a positive charge of the protein surface is helpful to repulse the protein from soil particles that are covered by monovalent cations. Similarly, working with *E. coli* heat-labile enterotoxin LTB, Kosaki et al. (2008) found pH 9.5 PALM buffer is the most effective for soil extraction.

Finally, protein is often added to an extraction system to solubilize other hydrophobic molecules including proteins in the solution (Voparil and Mayer, 2004). In addition, adding a sacrificial protein stabilizes the target protein in matrix extract by shielding it from enzymatic degradation especially when its concentration is low. In soil extraction, additional proteins in solution may support extraction by replacing the target protein on the soil surface held by hydrophobic binding. Commercially available bovine serum albumin, chicken egg albumin (OVA), and casein are common selections. Shan et al. (2003) used 5 mg/mL (0.5%) of BSA in extraction buffer and were effective in protein extraction. In summary, a few simple steps are recommended to consider, if one plans to develop an extraction buffer and detection method for a protein from soil. These include selection of buffer pH, cation salt, surfactant, protein, and other additives.

12.5   A CASE STUDY

Cry34Ab1 Bt protein is one of the proteins expressed in Herculex[TM] RW[*] maize, which confers resistance to corn rootworms. The bioassay for this protein with

[*]Trademark of DOW AgroSciences LCC

rootworm larvae is extremely insensitive and it is about 1000 times less sensitive than other Cry protein bioassays using tobacco budworm as the testing insect. To monitor this protein in plant matrix or soil, a chemical extraction in tandem with ELISA is the method of choice. However, the major issue with applying ELISA for soil detection is the low extractability of the protein. Almost no Cry34Ab1 protein was recovered when spiked into soil and extracted with published extraction buffers (0–5%) (unpublished data). In an effort to develop an effective extraction method, the protein properties that might be helpful for extraction system design were investigated. The unique features of this protein are its small size containing 123 amino acids with a molecular weight of ~14 kDa (Gao et al., 2004) and a theoretical pI of 6.17. It has more linear peptide regions exposed, which will enable the protein to bind readily to soil particles. In addition, a high percentage of charged amino acids in this protein (24%) may impact its binding with particles and matrices. This protein is more soluble in acidic conditions such as citrate buffer (pH 3.3). Based on its unique properties and the key components needed to produce a successful extraction buffer described above, conditions such as different ion components, pH, detergent, and protein were studied for potential buffer combination. Although acidic pH may be slightly better for Cry34Ab1 protein solubility than basic conditions, to facilitate flexible ion selection, a basic pH borate buffer was chosen for the system. Due to its large radii, monovalent cation K^+ was included. Again, surfactant Tween 20 was selected to form necessary micelles to assist protein solubilization. Lastly, based on a number of marine worm gut fluids that were screened for Cry34Ab1 extraction (Shan, unpublished data), a high concentration of a sacrificial protein was found to be necessary for a better extraction. As described in previous section, proteins can solubilize target protein as well as displacing them from soil binding sites. Therefore, both protein size and structure need to be considered for extraction buffer selection. After extensive screening, trypsin was selected for Cry34Ab1 extraction, which has a similar size to Cry34Ab1 with an MW of ~22 kDa. However, trypsin is a proteomic enzyme and may introduce an enzymatic reaction during extraction. To avoid unnecessary cleavage, protease inhibitor cocktail was included in the final buffer solution. The final extraction buffer (Cry34Ab1 buffer) contained pH 9.5 borate buffer, 0.075% Tween 20, 0.2 M KCl, 2 mg/mL trypsin, and 1 mg/mL of cocktail protease inhibitor. This buffer was then applied to different types of soils and was able to effectively extract Cry34Ab1 from a variety of soil matrices (Table 12.3). Greater than 70% recovery was obtained for most

TABLE 12.3 Comparison of Cry34Ab1 Extraction from Soil

	Mean recovery[a] (%)	
	SHAN buffer and PALM buffer	Cry34Ab1 buffer
Autoclaved soil	0–5	
Silt loam soil	0–5	74.8 ± 6.8
Loam soil	0	34.5 ± 4.8
Sandy loam soil	0–6	96.2 ± 11.2

[a]Mean recovery: three different fortification levels.

of the soils. Although the high clay content soil gave a relatively low recovery at 34.5%, it is a breakthrough improvement compared with 0% recovery with other buffer systems. Moreover, the interanalyst and interday precision was very good, which suggests that this system with ELISA is reliable and suitable for soil monitoring.

12.6 APPLICATION OF IMMUNOASSAY IN SOIL ACCUMULATION STUDIES

ELISA assisted soil detection methods have been used in field accumulation studies including studies with GM crops associated with Cry1Ac, Cry1Ab, Cry1F, Cry3Bb1, and Cry34/Cry35Ab1 proteins (Head et al., 2002; Ahmad et al., 2005; Shan et al., 2008; Gruber et al., 2008). In a study to investigate the accumulation of Cry1Ac protein (Head et al., 2002), Bollgard cotton was planted each year (one or two times per season) at six different locations in the United States for 3–6 years. In each season, after harvest of the cotton lint and seeds, the plants were shredded and tilled into the soil. The soil samples were collected 3 months after the last season when a majority of plant residues from previous seasons had decayed. Both ELISA and bioassay were used as detection methods. The PALM buffer was used to extract Cry1Ac protein from soil, which was combined with an ELISA with a limit of detection of 3.68 ng Cry1Ac/g of soil. Tobacco budworm was used as the test species of insect bioassay with a detection limit of 8 ng/g soil. Both methods revealed that no detectable Cry1Ac protein was present in any of the soil samples collected (Head et al., 2002). A few years later, Ahmad et al. (2005) conducted a similar study by planting YieldGard rootworm hybrid maize, which expresses Cry3Bb1 protein, at three different sites for three consecutive years. Again after each season, the stalks were incorporated into the soil with a disk plow. Soil was sampled three times during the first or third corn-growing season (V3, R1, and R6 stages). Soil samples were collected both near GM plants and between rows. ELISA was used to detect Cry3Bb1 residue in soil with the PALM buffer as the protein-extraction buffer. Only a very low level of Cry3Bb1 protein was detected (3.38–6.89 ng/g dry soil) in three near plant samples at one location, no detectable protein was found in any other samples. During the study, the abundance of surface and belowground nontarget arthropods in fields planted with YieldGard or non-Bt corn over 3 years was examined and no significant difference was observed (Ahmad et al., 2005).

Similarly, Cry1F maize soil accumulation was studied by planting Herculex I corn, which expresses Cry1F protein, at three different geographic locations in the United States for three consecutive seasons (Shan et al., 2008). At the last growing season, two types of soil samples were taken at tasseling and postharvest. Rhizosphere soil representing soil particles near roots and bulk soil was collected about 4 in. from plant side. Both ELISA and insect bioassay were used to detect Cry1F protein in soil samples. The SHAN buffer was used for protein extraction and the ELISA method had a validated detection limit of 4.5 ng/g dry weight soil. Tobacco budworm served as test insect for the bioassay, which has equivalent sensitivity as ELISA. No Cry1F protein was detected in any of soil samples collected after 3 years of cropping with both

ELISA and bioassay methods (Shan et al., 2008). Recently, a similar field accumulation study was conducted for Herculex RW hybrid corn, which expresses both Bt proteins Cry34Ab1 and Cry35Ab1. Validated ELISA methods were applied for quantitation and bioassay was used as verification. The SHAN buffer was used for Cry35Ab1 protein extraction and the Cry34Ab1 buffer was used to extract Cry34Ab1 protein from soil. Again, no detectable protein was found in any of the soil samples collected from fields after three consecutive cropping seasons (Shan, unpublished data).

In conclusion, effective protein extraction from soil matrices can be achieved by rational design and careful screening of four key components: buffer pH, cation salt, surfactant, protein, and other additives. Understanding of biochemical properties of a protein is extremely important for the buffer design and selection. Finally, one has to keep in mind that soil type is often one of the determining factors for protein extraction. An extract buffer that works well for a certain soil type may be not as effective for other soil types. An extraction validation is needed prior to formally using a buffer for a new soil type. With studies from numerous laboratories, immunoassay has been proven a sensitive and reliable analytical tool for soil matrices and can be used for multiple purposes to monitor proteins in ecosystems and conduct risk assessment studies.

REFERENCES

Accinelli, C., Koskinen, W. C., Becker, J. M., and Sadowsky, M. J. Mineralization of the *Bacillus thuringiensis* Cry1Ac endotoxin in soil *J. Agric. Food Chem.*, **2008**, *56*, 1025–1028.

Ahmad, A., Wilde, G. E., and Zhu, K. Y. Detectability of coleopteran-specific Cry3Bb1 protein in soil and its effect on non-target surface and below-ground arthropods. *Environ. Entomol.*, **2005**, *34*, 385–394.

Blackwood, C. B. and Buyer, J. S. Soil microbial communities associated with *Bt* and non-*Bt* corn in three soils, *J. Environ. Qual.*, **2004**, *33*, 832–836.

Brady, N. C. and Weil, R. R. *The Nature and Properties of Soils*, 13th edition, Prentice-Hall, Upper Saddle River, NJ, 2002.

Clark, B. W., Phillips, T. A., and Coats, J. R. Environmental fate and effects of *Bacillus thuringiensis* (Bt) proteins from transgenic crops: a review. *J Agric. Food Chem.*, **2005**, *53*, 4643–4653.

Crecchio, C. and Stotzky, G. Insecticidal activity and biodegradation of the toxin from *Bacillus thuringiensis* subsp. *kurstaki* bound to humic acids from soil, *Soil Biol. Biochem.*, **1998**, *30*, 463–470.

Donegan, K. K., Palm, C. J., Fieland, V. J., Porteous, L. A., Ganio, L. M., Schaller D. L., Bucao, L. Q., and Seidler, R. J. Changes in levels, species, and DNA fingerprints of soil microorganisms associated with cotton expressing the *Bacillus thuringiensis* var *kurstaki* endotoxin. *Appl. Soil Ecol.*, **1995**, *2*, 111–124.

Dubelman, S., Ayden, B. R., Bader, B. M., Brown, C. R., Jiang, C., and Vlachos, D. Cry1Ab protein does not persist in soil after 3 years of sustained *Bt* corn use, *Environ. Entomol.*, **2005**, *34*, 915–921.

Fiorito, T. M., Icoz, I., and Stotzky, G. Adsorption and binding of the transgenic plant proteins, human serum albumin, β-glucuronidase, and Cry3Bb1, on montmorillonite and kaolinite: microbial utilization and enzymatic activity of free and clay-bound proteins. *Appl. Clay Sci.* **2008**, *39*, 142–150.

Gao, Y., Schafer, B. W., Collins, R. A., Herman, R. A., Xu, X., Gilbert, J. R., Ni, W., Langer, V. L., and Tagliani, L. A. Characterization of Cry34Ab1 and Cry35Ab1 insecticidal crystal proteins expressed in transgenic corn plants and *Pseudomonas fluorescens*. *J. Agric. Food Chem.*, **2004**, *52*(26), 8057–8065.

Grothaus, G. D., Bandla, M., Currier, T., Giroux, R., Jenkins, G. R., Lipp, M., Shan, G., Stave, J. W., and Pantella, V. Immunoassay as an analytical tool in agricultural biotechnology. *AOAC Int.* **2006**, *89*, 913–928.

Gruber, H., Paul, V., Meyer, H. H. D., and Müller. M. Validation of an enzyme immunoassay for monitoring Cry1Ab toxin in soils planted with Bt-maize (MON810) in a long-term field trial on four South German sites. *J. Verbr. Lebensm.*, **2008**, *3*(Suppl. 2), 22–25.

Head, G., Surber J. B. Jr., Watson, J. A., Martin, J. W., and Duan, J. J. No detection of Cry1Ac protein in soil after multiple years of transgenic Bt cotton (Bollgard) use. *Environ. Entomol.*, **2002**, *31*, 30–36.

Herman, R. A., Evans, S. L., Shanahan, D. M., Mihaliak, C. A., Bormett, G. A., Young, D. L., and Buehrer, J. Rapid degradation of Cry1F delta-endotoxin in soil. *Environ. Entomol.*, **2001**, *30*, 642–644.

Herman, R. A., Wolt, J. D., and Halliday, W. R. Rapid degradation of the Cry1F insecticidal crystal protein in soil. *J. Agric. Food Chem.*, **2002a**, *50*, 7076–7078.

Herman, R. A., Scherer, P. N., and Wolt, J. D. Rapid degradation of a binary PS149B1 δ-endotoxin of *Bacillus thuringiensis* in soil, and a novel mathematical model for fitting curve-linear decay. *Environ. Entomol.*, **2002b**, *31*, 208–214.

Hopkins, D. W. and Gregorich, E. G. Detection and decay of the Bt endotoxin in soil from a field trial with genetically modified maize. *Eur. J. Soil Sci.*, **2003**, *54*, 793–800.

Icoz, I. and Stotzky, G. Cry3Bb1 protein from Bacillus thuringiensis in root exudates and biomass of transgenic corn does not persist in soil. Transgenic Research, **2007**. http://www.springerlink.com/content/pm447n1340n136t3/.

Icoz, I. and Stotzky, G. Cry3Bb1 protein from *Bacillus thuringiensis* in root exudates and biomass of transgenic corn does not persist in soil. *Transgenic Res.*, **2008a**, *17*, 609–620.

Icoz, I. and Stotzky, G. Review: fate and effects of insect-resistant Bt crops in soil ecosystems. *Soil Biol. Biochem.*, **2008b**, *40*, 559–586.

James, C. Executive summary: Global status of commercialized biotech/GM crops, *ISAAA Briefs* 41-2009, ISAAA, Ithaca, NY, 2009.

Kosaki, H., Coats, J. R., Wang, K., and Wolt, J. D. Persistence and degradation of maize-expressed vaccine protein, *Escherichia coli* heat-labile enterotoxin subunit B, in soil and water. *Environ. Toxicol. Chem.*, **2008**, *27*, 1227–1236.

Lipton, C. R., Dautlick, J. X., Grothaus, G. D., Hunst, P. L., Magin, K. M., Mihaliak, C. A., Rubio, F. M., and Stave, J. W. Guidelines for the validation and use of immunoassays for determination of introduced proteins in biotechnology enhanced crops and derived food ingredients, *Food Agric. Immunol.*, **2000**, *12*, 156–164.

Marchetti, E., Accinelli, C., Talame, V., and Epifani, R. Persistence of Cry toxins and *cry* genes from genetically modified plants in two agricultural soils. *Agronomy for Sustainable Development*, **2007**, *27*, 231–236.

Palm, C. J., Donegan, K., Harris, D., and Seidler, R. J. Quantification in soil of *Bacillus thuringiensis* var *kurstaki* δ-endotoxin from transgenic plants. *Mol. Ecol.*, **1994**, *3*, 145–151.

Palm, C. J., Schaller, K. K., Donegan, K. K., and Seidler, R. J. Persistence in soil of transgenic plant produced *Bacillus thuringiensis* var. *kurstaki* δ-endotoxin. *Can. J. Microbiol.*, **1996**, *42*, 1258–1262.

Parsons, T. R., Maita, Y., and Lalli, C. M. *A Manual of Chemical and Biological Methods for Seawater Analysis*, Pergamon Press, Oxford, UK, 1984.

Saxena, D. and Stotzky, G. *Bt* toxin is not taken up from soil or hydroponic culture by corn, carrot, radish, or turnip. *Plant Soil*, **2002**, *239*, 165–172.

Saxena, D., Folres, S., and Stotzky, G. Vertical movement in soil of insecticidal Cry1Ab protein from *Bacillus thuringiensis*. *Soil Biol. Biochem.*, **2002**, *34*, 111–120.

Shan, G., Lipton, C., Gee, S. J., and Hammock, B. D. Immunoassay, biosensors and other nonchromatographic methods. In Lee, P. W. (Ed.), *Handbook of Residue Analytical Methods for Agrochemicals*, John Wiley & Sons, West Sussex, England, 2003, pp. 623–679.

Shan, G., Embrey, S. K., Herman, R. A., and McCormick, R. Cry1F protein not detected in soil after three years of transgenic Bt corn (1507 Corn) use. *Environ. Entomol.*, **2008**, *37*, 255–262.

Sims, S. R. and Holden, L. R. Insect bioassay for determining soil degradation of *Bacillus thuringiensis* subsp. *kurstaki* Cry1Ab protein in corn tissue. *Environ. Entomol.*, **1996**, *25*, 659–664.

Sims, S. R. and Ream J. E. Soil inactivation of the *Bacillus thuringiensis* subsp. *kurstaki* CryIIA insecticidal protein within transgenic cotton tissue: laboratory microcosm and field studies. *J. Agric. Food Chem.*, **1997**, *45*, 1502–1505.

Sims, S. R., Berberich, S. A., Nidia, D. L., Segalini, L. L., Leach, J. N., Ebert, C. C., and Fuchs, R. L. Analysis of expressed proteins in fiber fractions from insect-protected and glyphosate-tolerant cotton varieties. *Crop Sci.*, **1996**, *36*, 1212–1216.

Soil Survey Division Staff. *Soil Survey Manual, Handbook 18*, Soil Conservation Service, U.S. Department of Agriculture, 1993.

Stotzky, G. Persistence and biological activity in soil of insecticidal proteins from *Bacillus thuringiensis* and of bacterial DNA bound on clays and humic acids. *J. Environ. Qual.*, **2000**, *29*, 691–705.

Stotzky, G. Persistence and biological activity in soil of the insecticidal proteins from *Bacillus thuringiensis*, especially from transgenic plants. *Plant Soil*, **2004**, *266*, 77–89.

Tapp, H., Calamai, L., and Stotzky, G. Adsorption and binding of the insecticidal proteins from *Bacillus thuringiensis* subsp. *kurstaki* and subsp. *tenebrionis* on clay minerals. *Soil Biol. Biochem.*, **1994**, *26*, 663–679.

Tapp, H. and Stotzky, G. Persistence of the insecticidal toxin from *Bacillus thuringiensis* subsp. *kurstaki* in soil. *Soil Biol. Biochem.*, **1998**, *30*, 471–476.

Venkateswerlu G. and Stotzky G. Binding of the protoxin and toxin proteins of *Bacillus rhuringiensis* subsp. *kurstaki* on clay minerals. *Current Microbiol.*, **1992**, *25*, 1–9.

Voparil, I. M. and Mayer, L. M. Commercially available chemicals that mimic a deposit feeder's (*Arenicola marina*) digestive solubilization of lipids. *Environ. Sci. Technol.*, **2004**, *38*, 4334–4339.

Wang, H., Ye, Q., Wang, W., Wu L., and Wu, W. Cry1Ab protein from *Bt* transgenic rice does not residue in rhizosphere soil, *Environ. Pollut.*, **2006**, *143*, 449–455.

Wang, H., Ye, Q., Gan, J., and Wu, L. Biodegradation of Cry1Ab protein from Bt transgenic rice in aerobic and flooded paddy soils. *J. Agric. Food Chem.*, **2007**, *55*, 1900–1904.

West, A. W., Burges, H. D., White, R. J., and Wyborn, C. H. Persistence of *Bacillus thuringiensis* parasporal crystal insecticidal activity in soil. *J. Invertebr. Pathol.*, **1984**, *44*, 128–133.

Zwahlen, C., Hilbeck, A., Gugerli, P., and Nentwig, W. Degradation of the Cry1Ab protein within transgenic *Bacillus thuringiensis* corn tissue in the field. *Mol. Ecol.*, **2003**, *12*, 765–775.

CHAPTER 13

IMMUNOASSAY APPLICATIONS IN PLANT-BASED BIOPHARMA

Thomas Patterson
Gregory Gilles

13.1   INTRODUCTION TO PLANT-BASED BIOPHARMA

Although using plants as a production platform for pharmaceuticals is often thought of as a modern development made possible by the advent of biotechnology, in reality, plants have been sources of therapeutic compounds for thousands of years. Many early cultures used various medicinal herbs for the treatment of ailments. Renewed interest in "natural" or homeopathic remedies will likely increase the demand for novel pharmaceutical actives from plants. The identification of the causative agent for pharmacological responses has been a fruitful area of research for natural-product chemists. Indeed, one can pick up a random copy of a journal such as Phytochemistry© and a quick perusal of the table of contents will likely turn up at

Immunoassays in Agricultural Biotechnology, edited by Guomin Shan
Copyright © 2011 John Wiley & Sons, Inc.

least several articles related to the chemistry of plant-based compounds with biological activities.

In addition to known pharmaceutically active compounds, there is considerable interest in "bioactives" from plants that confer a positive health benefit (Kris-Etherton et al., 2002). These can be compounds such as bioactive peptides released by the digestion of plant proteins. Small molecules such as glucosinolates in *Brassica* vegetables (broccoli, brussel sprouts) have been associated with positive health benefits. Beyond human health benefits, veterinary benefits of plant bioactives are receiving attention as well (Rochfort et al., 2008).

Through modern science, today many diseases are now preventable or treatable by therapeutic proteins; however, the cost to produce and the quantity of these proteins needed in order for the public to receive these treatments remains a limiting factor. Immunization is among the most successful and the most cost-effective public health interventions (Miller and Hinman, 2004; Hadler, 2004). In certain regions of the world, immunization has eradicated smallpox, measles, and poliomyelitis, yet 1.4 million children died from vaccine preventable disease around the world in 2002 (WHO, 2004). A newer approach to therapeutic proteins are antibody-based therapies (immunotherapies) that have been developed in recent years to treat immune system disorders, inflammatory diseases, cancers, disorders of the central nervous system, and infectious diseases (Gomond et al., 2004). However, the cost associated with production and the physical space and time associated with generating antibodies and vaccines restricts their use to certain segments and classes in the world.

Plant biotechnology may provide a suitable answer to the proposed problem. Plants have provided humans with a host of useful and therapeutic molecules for many years, but in the past 20 years, the development of genetic modification has enabled the production of specific heterologous proteins (Ma et al., 2003). The first example of producing vaccines in plants was carried out by Merck in 1986 (Barta et al., 1986). In 1989, the first antibody was expressed in tobacco (Hiatt et al., 1992). Molecular farming and biopharming are the latest catch phrases used to describe the production of recombinant proteins in plants. Molecular farming offers several advantages for the production of biopharmaceuticals. Plants are economical by being able to produce in larger-scale-up capacity and lower cost of production compared with fermentation of bacteria, yeast, or mammalian cell lines. Plants are able to provide posttranslational modifications similar to mammalian lines at a reduced risk of contamination from potential human pathogens. If the protein is delivered via edible tissue, then the expensive step of purifying the recombinant protein could be eliminated as well as the need of health professionals for their delivery. However, the uptake of the protein in the digestive tract into the blood stream is a potential limiting issue and possible degradation of the protein by digestive enzymes.

There are some hurdles, however, associated with the production of therapeutic proteins in plants. First and foremost is obtaining a sufficient level of protein expression in the plant tissue to achieve either a therapeutic dose (for direct oral consumption) or an economically relevant concentration for extraction and purification on a commercial scale. Currently, there are three ways to express transgenic proteins in plant production systems; nuclear transgenic plant, chloroplast transgenic plant, and plant virus-mediated transient expression.

Stable nuclear transformation results in the integration of the recombinant DNA in the nuclear genome of the plant via either a plant pathogen, *Agrobacterium tumefaciens* or by biolistic micro projectile bombardment. These methods result in the random integration of the transferred DNA in the nucleus genome. Multiple locus insertions can occur as well as the inserted DNA being subject to positional effects in the genome. This increases the probability of unstable gene expression, gene silencing, and complex patterns of inheritance of the transgene. Transgene containment due to unintended outcrossing is also an issue with field planting. With nuclear transgenic plants, the expression level also tends to be low, about 0.01–7.00% of total soluble protein (Thanavala et al., 2006). However, nuclear transgenic plants have a high-scale-up capacity by planting more acres of transformed crops.

Other advantages are that the transgene lies in the nuclear genome, the expressed protein can have posttranslational modifications. A possible advantage may also be the expression of the protein in an edible portion of the plant eliminating the need to purify the therapeutic protein. Chloroplast transformation may overcome the problem of low expressing transgenic plants. Chloroplast transgenic plants have a very high level of expression, 4.1–31.1% of total soluble protein due to the high chloroplast genome copy number (Thanavala et al., 2006). Because of the hyper-expression of chloroplast transgenic plants, one acre can produce up to 360 million doses of clean, safe, and fully functional anthrax vaccine antigen (Kumar and Dufourmantel, 2005; Grevich and Daniel, 2005). In this technique, the transgene is inserted into gene expression cassettes and introduced into the chloroplast by biolistic bombardment and integrates into the chloroplast genome by homologous recombination (Mason et al., 1998). Gene position effects that are seen in nuclear transgenic plants are eliminated due to site-specific recombination. Furthermore, gene silencing has not been observed in plants with transgenic chloroplast integration. Another advantage is that the chloroplast genome is inherited strictly maternally, which minimizes the possibility of unintended outcrossing of transgenes by pollen (Svab and Maliga, 1993). There has also been evidence of posttranslational modification in chloroplast-transformed proteins. Fully functional therapeutic proteins with disulfide bonds and appropriate posttranslational modifications have already been produced using transgenic chloroplast (Staub et al., 2000; Fernandez-San Millan et al., 2003; Leelavathi and Reddy, 2003; Glenz et al., 2006). Fully functional vaccine antigens from chloroplast-transformed plants have also been shown in the literature (Koya et al., 2005; Watson et al., 2004; Daniell et al., 2001, 2005; Tregoning et al., 2003; Molina et al., 2004).

Transient expression of the transgene is the third approach to plant expression systems. With this technique, plants are transformed with a plant virus that infects the plant cells and replicate autonomously to very high copy numbers. Plant virus expression systems are mainly developed from RNA viruses (Canizares et al., 2005). The transgenic viruses can be introduced into the host plant cells by *Agrobacterium*-mediated infection (Turpen et al., 1993) or by spraying the plants with a mixture of viral particles and an abrasive (Gleba et al., 2007). The major advantages of the viral vector expression system are speed and that the yield can be up to 5% of the plant protein (Kumagai et al., 1993). Until recently, one of the drawbacks of this system is that there is a size constraint for the genes that the viral vectors can successfully express at high yields. Vectors that fuse the protein of interest to the coat protein can be used to express long polypeptides that are at least 133 amino acids long (Werner et al., 2006).

While plants have the ability to fold proteins correctly and provide posttranslational modifications, plant-produced proteins differ subtly from mammalian-produced proteins. Both hosts use similar signals for the addition of N-linked glycans, but the nature of the sugars that are added differ. Plants add a xylose that mammals lack.

Delivering vaccines orally also has its own inherent problems. The immunogen must maintain its native structure in the stomach environment, which contains proteolitic enzymes and acidic pH. There have been several different attempts to enhance the oral immunogenicity of the antigen. The common approach has been to protect the antigens by encapsulating them by liposomes (Wachsmann et al., 1985), gelatin capsules (Filgueira et al., 1995; Mestecky and McGhee, 1992), and microspheres (Khoury et al., 1995; Klipstein et al., 1983; Moldoveanu et al., 1993; O'Hagan et al., 1991).

The most promising approach has been the use of virus-like particles (VLPs). VLPs consist of proteins that form a virus' outer shell and the surface proteins, without the RNA required for replication. VLPs are able to survive the harsh stomach environment and are taken up by M cells and activate the common mucosal immune system (Ball et al., 1998). Norwalk virus-like particles produced in tobacco has been shown to be eliciting a mucosal immunogenic response in mice (Santi et al., 2008).

There are also several regulatory issues when attempting to use molecular farming. Government approval and regulation of plant-made pharmaceuticals is overseen by not only the Food and Drug Administration (FDA), but also the U.S. Department of Agriculture (USDA) due to the possible risk to human health and the environmental risk associated with transgenic plants. WHO has published guidelines on good agricultural practices for medicinal plants (WHO, 2003). The use of contained facilities such as greenhouses or bioreactors, the incorporation of traceable markers, and the use of inducible promoters could insure a reduction in the safety risk to the environment.

13.2 IMMUNOASSAYS USED IN PLANT-BASED BIOPHARMA

Immunoassays typically are applied throughout the product development cycle for any transgenic-based pharmaceutical product. In general, the product development process can be defined by the following stages.

13.2.1 Initial Product Development

Product development steps begin with the earliest phase of project discovery and development. In the case of plant-based biopharma, this may be initial crude types of assays to verify the presence of transgenic protein expression in a plant callus or early-stage regenerated plant. Researchers often rely on polyclonal rabbit antibodies at this stage as a basic tool for simple sandwich ELISA's and Western blotting applications. Often, multiple antibodies are produced in small quantities if multiple antigens are being developed and tested. At this stage of the R&D development process, the goal is often to verify transgenic protein expression and select the best producing plants or cell cultures to advance to further development.

In the case of small molecule therapeutics, the early stage development focus is somewhat different. Depending on the analyte, it may be necessary to chemically synthesize a hapten molecule for use as an antigen. While not reviewed here, this can be the most challenging aspect of developing small molecule immunoassays. The success or failure of the immunoassay development hinges on the identification of a suitably specific antibody that reacts with the small molecule with sufficient specificity and sensitivity.

13.2.2   Product Production and Quality Assurance

As the product moves from early stage discovery and development, there is an increased volume demand for immunoassays. At these stages, validation experiments are initiated to characterize the performance of the assay (see Chapter 6). Also, at this stage of the development, production of a monoclonal-antibody-based assay or a polyclonal antibody using a larger animal, such as goats, will be initiated to generate a large supply of antibodies for future product support needs. Development of assay kits may be contracted out to kit manufacture companies for routine manufacture and supply of assays if the anticipated demand for the assay is likely to be large, or if it proves more cost-effective for the developer versus generating reagents internally.

13.2.3   Product Registration

At some point in the development pipeline, the decision is made to initiate product registration with the FDA. At this point, a full validation will be initiated in accordance with the guidelines of the regulatory agencies (see Chapter 6). In the case of biopharmaceuticals, the application of these assays could also extend to therapeutic drug monitoring in test patients. There are extensive requirements for product registration studies and the use of fully validated assays, which will only be mentioned here.

13.3   CONSIDERATIONS IN DEVELOPING PLANT BIOPHARMA IMMUNOASSAYS

13.3.1   Target Analytes in Plant Biopharma

A number of factors must be considered when immunochemical approaches are to be applied to the detection of pharmaceutical proteins. In the experience of the authors, careful up-front analysis of the analytical problems to be addressed will facilitate the design of the assays and their eventual validation. As with any assay, careful definition of the performance requirements and applications of the assay will aid in the development process and assist in achieving the goals with minimal rework or redesign. The following are a list of considerations applied in the author's laboratory that have proven to be useful.

13.3.1.1 Definition of the Analyte and its Characteristics

A clear understanding of the nature of the analyte protein that will be measured is the starting place for immunoassay design. The protein source and the transgene construction (animal, human, microorganism, etc.) will dictate many aspects of the assay design. For instance, will the protein be soluble or membrane bound? Is the protein monomeric or multimeric (e.g., scFV's or IgG's)? Does the extracted protein need to be functional (as enzymes, or can be reconstituted with co-enzymes and other co-factors?), and do functional assays need to be included as part of the evaluation process? Is the protein sequence likely to be distinct from the endogenous matrix proteins, or does the expression hosts containing homologous proteins that may cross react?

In some cases, such as the expression of humanized immunoglobulins in plants, specific antibodies to both the whole IgG and its subunits are commercially available, requiring only the evaluation of antibody pairs for the specific application. Other more exotic proteins such as viral antigens may prove to be more challenging, and require the development of reagents. The production of antibodies is discussed elsewhere in Chapters 2 and 3 and will not be covered here.

13.3.1.2 Assay Requirements

When designing an assay for a plant biopharm project, the requirements of the particular process development stage need to be considered. In the early research and development stages of the product development, plants may be transformed with different vector constructs in an effort to identify the best combination of structural genes, promoters, and other gene elements that will yield the highest commercial yield of product. Each gene construct typically produces a number of transformation events that may vary in number of gene copies, insertion location, orientation, completeness of the insertion, and ultimate expression level. Each event will require evaluation to determine the expression of the transgenes. The assay requirements may also vary depending on the complexity of the transgenic protein being expressed in the plant tissue. For instance, a monomeric protein may only require a single assay to determine the concentration of the protein in tissue extracts. For a multimeric protein, the assay requirement may become more complex and may require multiple assays to address specific fundamental questions.

Using the example of expressing IgG in plant tissue, where transgenes encoding the heavy and light chains of the antibody are transformed into the plant, a series of questions related to protein expression arise, which require different assays. Examples of these are as follows:

1. Are the heavy and light chains expressed in a 1:1 stoichiometric ratio or is one chain over-expressed relative to the other?

2. Do the heavy and light chains fully assemble into the whole IgG or is there an accumulation of a partially assembled IgG?

3. What is the ratio of fully assembled IgG to unassembled heavy and light chain subunits?

4. Does the plant accumulate truncated forms of the IgG fragments (such as Fab_2, Fab, Fc)?

5. Is the expressed IgG functionally active?

In the earlier stages of product development, all of these analytical questions need to be addressed as part of identifying the most commercially relevant event for product development. As the product moves through the development phase and less efficacious events are eliminated, the requirement for all of these assays will diminish once it has been verified that the best events are expressing the desired assembled product. At the end of product development, it is likely that a single immunoassay to quantitate the concentration of the product in the tissue is sufficient both for routine quality control applications as well as product-purification/process-monitoring applications.

13.3.1.3   Differences in Plant Versus Microbial Versus Mammalian Protein Expression   As reviewed by Chen et al. (2005) and Kermode (2006), differences exist between transgenic protein-expression systems, with no one-expression system necessarily being considered as "universal." Transgenic plants have been suggested to offer advantages in terms of easily scalable production, low risk of contamination from endotoxins, prions or mammalian viruses, and relatively high product quality. These advantages can be offset by the seasonal nature of production, especially for field-based production systems. Early in the development of "Pharming" systems, it was envisioned that pharmaceutical production would take place in existing agricultural production systems. More recently, concerns about food crops containing Pharma products and potential contamination of the food supply and the subsequent regulation of BioPharma production have significantly impacted the ease of production. Production of BioPharma products is now limited to controlled greenhouse production or carefully regulated field production. Not withstanding, to date the "ideal" plant BioPharma production platform still has not been identified.

Protein Glycosylation   A major difference between mammalian and plant-based expression systems is the posttranslational processing and glycosylation. While plants have the ability to fold proteins correctly and provide posttranslational modifications, plant-produced proteins differ subtly from mammalian-produced proteins. Both hosts use similar signals for the addition of N-linked glycans, but the nature the sugars that are added differ. Plants add a xylose that mammals lack, to the β-1,2 position and adds a fucose to α-1,6 instead of an α-1,3 position found in mammals. Plants also lack terminal galactose and sialic acid, which may result in an unwanted or redirected response from the therapeutic protein (Van Ree et al., 2000). Additionally, some types of plant glycosylation structures are thought to be allergenic in humans, although this is still an area of controversy. However, it is possible to alter plants to produce more mammalian-like sugars through genetically engineering the plants (Bakker et al., 2001; Palacpac et al., 1999).

Modification of Expressed Proteins   Probably, the most common form of modification of transgenic proteins in plants is truncation. Presumably, this arises from the action of endogenous proteases. Truncation can be identified by shifting migration on SDS-PAGE gels with Western blotting, although the detection of the removal of only a few amino acids may require other techniques such as MALDI-TOF mass spectroscopy. Other modification, such as the coupling of phenolic compounds to exposed amino or sulfhydryl groups by either enzyme-mediated reactions (poly-phenol oxidase) or the reaction of oxidized phenols, can be random and difficult to detect without extensive structural characterization. The potential for modification exists in the method of extracting the proteins from the plant tissue, and is a consideration in process design for plant biopharma production.

13.3.2   Tissue Matrix and Genetic Considerations

Transgenic plant biopharma products can be produced in a number of different plant species, tissues, or even cell cultures. Naturally occurring plant therapeutics can occur in even more diverse species and challenging tissues. Each plant matrix material presents unique challenges and consideration for the extraction and assay of the biopharma analyte. For transgenic plant, the expression of the gene in the tissue is intimately associated with the underlying plant genetics.

In the case of plant cell cultures or clonally propagated whole plants, once the desired transgenic event has been identified, the material can be propagated without consideration of sexual reproduction. However, unlike bacterial or yeast production systems, plant cell cultures may change in expression competency over time.

For biopharma products produced in whole plants that require sexual propa-gation, the event selection process typically runs in parallel with a plant breeding process to select an elite plant line as a production platform. This process is somewhat species specific and depends on whether or not an elite plant variety can be directly transformed. When the plant that is transformed is not an elite line, then the transgenes must be backcrossed into an elite variety. The backcrossing and selection occurs in parallel with the transgene event selection process. In the case of a single gene insertion event located on a single chromosome in the plant genome, the inheritance of the transgene will typically follow a simple Mendelian segregation ratio. Progeny in an F2 generation should segregate in a ratio of 25% homozygous positive, 50% heterozygous, and 25% homozygous null, if there are no unusual circumstances influencing the integration of the transgene in the genome nor linkage to gamete-fitness traits. Events with multiple gene insertions or linkage to gamete-fitness traits may give rise to distorted segregation ratios. When measuring expression levels of the transgene, the segregation of the material and the impact of homozygous versus heterozygous gene dose effects must be considered by the investigator.

The tissue the transgenic BioPharma product is designed to accumulate in is also a consideration. Leaf or other vegetative tissue expression allows the sampling of tissue for assay over the life cycle of the plant, often at very early development stages. Conversely, seed-specific expression is limited to the end of the growing cycle when the plant has reached maturity. Each tissue system offers different challenges and opportunities in terms of product assays, some of which are outlined below.

13.3.2.1 *Vegetative Tissue Expression*

a. Tissue can be sampled soon after the seed germinates and sufficient leaf material has developed to allow sampling.

b. Null plants can quickly be identified and discarded, reducing the population size.

c. Concentrations of the BioPharma product may be subject to temporal variability, impacted by the age of the tissue.

d. Concentration gradients may exist in the plant, with levels being either higher or lower in young/older tissue sampled from the same plant.

e. Older tissue may have increased truncation of the desired protein product.

f. The stability of the BioPharma product in fresh tissue must be determined and also be determined the appropriate sampling/tissue storage conditions.

13.3.2.2 *Seed Tissue Expression*

a. Tissue can only be sampled at maturity.

b. Seed tissue typically is stable and easily stored.

c. Processing seed for analysis can be destructive (grinding whole seed or seed bulks) or semidestructive (seed slices, half seeds).

d. Identification of nulls can be done prior to planting.

e. Seeds represent the next generation in the breeding cycle (i.e., an F1 plant produces F2 seed, etc.)

f. Expression levels can change with growing conditions, environment, fertility, and stress.

g. Event sorting may require rapid analysis of segregating seeds in a short period of time to allow planting of the next generation for product development.

13.3.3 Extraction of Plant Tissue

The sampling and extraction of plant tissue can be quite diverse, with numerous variations in sample collection, preservation, and processing. Some basic principles can be applied as guidelines, but the investigator must be aware that the nature of the analyte will dictate how the tissue is processed. Very stable transgenic proteins that are readily soluble may tolerate aggressive tissue homogenization techniques and require relatively simple buffer systems. Other, more labile or difficult to solubilize proteins may require gentle techniques and/or novel extraction buffer systems. Once solubilized, the stability of the protein in the extract must be determined. All these variables must be considered and factored into the design of the immunoassay development.

In addition to the solubilization of the transgenic protein, the completeness of extraction needs to be considered. A practice the authors have found useful is to sequentially extract the sample with the extraction buffer and assay the concentration of analyte in each sequential extract. Once no more detectable analyte is found in the extract solution, the remaining tissue pellet is treated with SDS-PAGE extraction buffer to solubilize any remaining protein. The SDS extract is analyzed by SDS-PAGE and Western blotting to verify that no un-extracted analyte protein is remaining in the tissue. If a significant amount of protein is still found in the tissue pellet, then alternative extraction procedures need to be investigated. These can include other homogenization techniques, different buffers, pH, ionic strength, additives such as chelators or detergents, and so on. In some cases, it may be necessary to treat the tissue homogenate with cell-wall degrading enzymes to aid in releasing the protein into solution.

13.3.3.1 Seed
Seed tissue can be either relatively simple to process or extremely challenging, depending on the tissue and the assay requirements. Processing of bulk seed often is amenable to bulk seed grinding methods using various laboratory mills. Roller mills, knife mills, and hammer mills are common techniques used for seed grinding on a larger scale. Avoiding excessive heating of the grain during grinding is desirable to avoid protein denaturation. Smaller quantities of seed can be easily processed in domestic coffee mills. Individual seeds can be ground in ball mills, crushed in a mortar and pestle, or bead mills.

Depending on the seed type and location of the expression within the seed (embryo versus endosperm or cotyledons), in many cases it will be possible to obtain a subsample of the seed for extraction while maintaining the rest of the seed for grow-out. Dicot plants can be partially sprouted and a cotyledon slide taken by excision with a scalpel. Dried corn kernels can be chipped to obtain an endosperm slice. In the author's laboratory, a commercial tablet splitter obtained from a local pharmacy has proven useful for this technique. Small seed fragments typically can be homogenized in a bead mill such as a GenoGrinder™ (Spex CertiPrep Group).

Extracting seed tissue can involve aggressive homogenization techniques such as bead milling, homogenization, sonication, or gentler processes such as stirring. Buffer systems can be quite variable depending on the properties of the analyte being extracted. Additives such as EDTA, PMSF, ascorbic acid–have all been described. For immunoglobulin extraction, PBS buffer systems often are sufficient to achieve solubilization of the IgG. Less soluble proteins may require the addition of surfactants such as Triton X-100 or Tween 20. Membrane-bound proteins may even need strong detergents such as LDS or SDS for sufficient extraction and solubilization. The amount of ground seed to extract, and the seed:buffer ratio may need to be determined experimentally. In the author's laboratory, large-scale screenings of samples was achieved using 20 mg of corn seed flour in 500 μL of buffer, using a GenoGrinder, and a 96-cluster tube array for homogenization.

13.3.3.2 Green or Fresh Tissue
Sampling techniques for fresh tissue are determined by the specific tissue type being sampled. For leaf tissue, it is convenient to take tissue samples on a leaf area basis by using a punch of defined area to quickly collect leaf discs. The size of the punch depends on the dimensions of the leaf being sampled and the dimensions of the sample collection/extraction. In the author's laboratory, it has proven convenient to use a standard hand paper punch to collect leaf samples. Typically, four punches with a total leaf area of ~ 1.1 cm^2 are placed in a

cluster tube and extracted with 500 µL buffer using a GenoGrinder. If it is desired to express the protein concentration on a dry weight tissue basis, a parallel set of leaf discs can be collected in a tared tube, dried, and the tissue dry weight is determined. If tissue fresh weights are collected, prior to drying, the data can be expressed on either a fresh or dry weight basis. If only fresh weight data were collected, the expression level on a dry weight basis often can be approximated by assuming fresh tissue has ~80% moisture content.

When sampling other nonleaf tissues with uneven geometry, expressing data on an area basis likely will have little meaning. In these cases, expression on a fresh weight or dry weight basis is more relevant. Tissue such as potato tuber can be sampled by slicing or with a device like a cork borer to remove a plug of tissue. Other tissue types such as flower petals, fruiting structures, pollen, stems, and roots can likely only be samples on a weight basis. In many cases, it may prove difficult to obtain a uniform subsample of these tissues. In that case a larger quantity of tissue should be collected and homogenized for sampling. Freeze drying fresh tissue is often advisable for sample storage. Dried tissue can also be ground to obtain homogeneous tissue samples. Similar results can be achieved by freezing fresh tissue on dry ice or liquid nitrogen and grinding in a frozen state to achieve a powder.

13.3.3.3 *Plant Cell Cultures*

Plant cell suspension cultures or tissue cultures are essentially a fresh tissue sample that can be processed by the previously mentioned techniques for whole fresh tissue. The challenges presented by these tissues are collecting a representative subsample while maintaining sterile conditions in the culture system. Callus tissue can be particularly challenging, in which calli are often variable in size and shape. Also, expression in callus tissue may not be homogeneous, creating further sources of analytical variability. Suspension cultures may be dilute, and need to be concentrated by centrifugation to remove culture media.

13.3.3.4 *Sample Dilution and Expression of Results*

After a sample has been homogenized, the sample will likely need to be diluted to (1) reduce matrix interference effects in the assay; (2) bring the analyte concentration in the extract to a concentration that will be in range of the immunoassay. Matrix interference must be determined specifically for each assay and analyte and the appropriate dilution factor established, which avoids or at least minimizes matrix interference with the assay. In some cases, it is desirable to dilute the sample extract with a buffer containing a carrier protein such as bovine serum albumin to help reduce interferences caused by nonspecific protein interactions.

The sample working dilution for an immunoassay analysis is a function of the concentration of the analyte in the tissue extract, and the working range of the immunoassay. Typically, an ELISA assay will have a working range of 10×, (e.g., 1–10, 2–20, 10–100, 100–1000). If the concentration range of the samples to be analyzed varies widely (as often is the case in early stage event sorting), then multiple dilutions of the sample will be required to achieve analytical results that remain in range in the assay. While serial dilutions (1:2 or 1:3) traditionally have been used for this purpose, when large numbers of samples need to be assayed, this can

TABLE 13.1 Example of Bracketed Sample Dilutions to Cover a Broad Sample Concentration Range

Dilution	Analyte concentration in extract (ng/mL)												
	10	20	40	80	160	320	640	4080	240	5120	10,240	20,480	40,960
1:10	1.0	2.0	4.0	8	16	32							
1:70	0.14	0.29	0.6	1.1	2.3	4.6	9.1	18.3					
1:610	0.02	0.0	0.1	0.1	0.3	0.5	1.0	2.1	4	8	17		
1:4270	0.00	0.00	0.01	0.02	0.04	0.07	0.15	0.30	0.6	1.2	2	5	10

become prohibitive in terms of the number of assays to be performed. The authors have used a simple calculation model to design dilution series for samples with widely varying ranges of concentrations to maximize the number of samples analyzed using the smallest number of analytical assays (Table 13.1).

In the above example, assuming the assay has a working range of 1–10 ng/mL, then a series of four dilutions will efficiently bracket a potential analyte concentration range from 10 to 40,960 ng/mL in only four dilution steps, with overlap between the upper and lower ranges of each assay plate. This technique has worked efficiently in cases of early event sorting when expression levels can vary widely. As transgenic events move forward in the development process, expression levels eventually become more predictable, and the range of extract concentrations will narrow, reducing the number of dilutions necessary.

13.3.3.5 *Expression of Analyte Concentration Results* Transgenic protein expression is typically expressed on a mass basis (ng protein per mg tissue dry weight, or ppm). Alternatively, the protein concentration can be expressed on a tissue fresh weight basis or on a leaf area basis. In some cases, investigators have attempted to normalize transgenic protein expression on a mass of total sample protein basis (ng transgenic protein/mg matrix protein) to correct for variation in tissue protein contents. The appropriate basis for reporting data will depend on the use of the information. Regulatory agencies may require the data to be reported on a ppm basis, while processing engineers may want the data reported on a basis that is relevant to the process feedstock that will be processed (fresh material, dry seeds, etc.).

13.4 EXAMPLES OF IMMUNOASSAYS

One simple and quick method for screening transgenic tissue cultures for the protein of interest is by a dot blot (or slot blot). Dot blotting is a method of placing nondenatured proteins on nitrocellulose or PVDF membranes for imaging with antibodies similar to an immunoblot. This technique offers a medium throughput semiquantitative way to choose which tissue culture lines you wish to advance. Significant timesavings is obtained over a traditional immunoblot since separation of the proteins is not required; however, no information on the size of the target protein can be obtained.

Apparatuses such as the Bio-Rad Bio-Dot Microfiltration Apparatus can provide reproducible methods for binding proteins and come in a 96-well and 48-well formats. Special precautions have to be taken when performing dot bots since there is no denaturation of the sample native peroxides in the plants, which can present false positives when using a horseradish peroxidase secondary antibody and substrate. The authors have been able to circumvent this by placing samples at 65°C for an hour to inactivate the peroxidases. The authors have also performed a semiquantitative dot blot by adding a bovine serum albumin (BSA) standard curve to the blot. Samples should be serial diluted to ensure that they will fall within the BSA standard curve. Imagining was performed in a BioRad Fluor S imager using Quality One software; the authors were able to determine peak density of the samples to get a rough estimation of the amount of the protein of interest.

13.4.1 Plant-Expressed Antibodies

The production of monoclonal antibodies in plants has been described in a number of patents (Greco et al., 2007; Li et al., 2006); however, at the time of writing of this chapter, most recombinant antibody approved by the FDA are produced in Chinese hamster ovary (CHO) cell lines. With plants, genes can be placed into plants encoding heavy and light chain of the desired antibody, and when expressed, will assemble into a full-length antibody via a disulphide bridge, which so far can only be obtained from cultured mammalian cells or plant lines. Even though proteins are produced similarly in plants and animals resulting in correctly folded proteins by the chaperones and processing enzymes, glycosylation still differs and may be determinant of plant allergens.

Leyva et al. (2007) describe the development of an ELISA to quantify an anti-HBsAG antibody produced transgenic tobacco plants. The monoclonal-based immunoassay was tested and found to be specific for the desired antibody and not subject to interference from plant-matrix proteins. The authors determined that the immunoassay could be useful for process monitoring of the large-scale purification of this antibody (Valdes et al., 2003).

The production of an anti-HIV–ELP fusion protein was studied by Floss et al. (2008). In this study, the monoclonal 2F5 antibody was designed as a fusion protein with an elastin-like peptide (ELP) and expressed in tobacco plants. The recovery and characterization of the antibody fusion was studied in detail, using multiple techniques such as Western blotting, functional assays, and glycosylation analysis. Generation of the ELP fusion protein was found to improve both the stability and yield of the transgenic antibody from plants.

13.4.1.1 Vaccines Production of various vaccines in plant requires both the quantification of the antigen by immunoassay and also the evaluation of its efficacy in developing an immune response in the treated subject. Antigens reported to be expressed in plants have included antigens for an influenza vaccines (Mett et al., 2008; Shoji et al., 2008), anthrax toxin (Koya et al., 2005), dust mite allergens (Yang et al., 2008), an anti-cancer HPV antigen (Massa et al., 2007), and HIV-1/HBV (Greco et al., 2007). Each of these references gives excellent examples of the use of immunoassays to monitor both the production and efficacy of the transgenically produced antigens.

13.4.1.2 Small Molecules Small plant molecules can also have therapeutic effects. For example, Vincristine (VCR) and vinblastine (VLB) are dimeric indole alkaloids from *Catharanthus roseus*. These drugs were marketed by Eli Lilly and Company as mitotic inhibitors for use in cancer chemotherapy. VCR has rather low content in the raw plant material and therefore is obtained by oxidation of VLB. To investigate the conversion, pharmacokinetics of these drugs, or the drug levels in biological samples for the specific dimmers, numerous immunoassays have been developed (Hacker et al., 1984; Lapinjoki et al., 1986; Volkov, 1995).

Paclitaxel, a taxol from *Taxus brevifolia*, is also used as mitotic inhibitors in cancer chemotherapy. This drug was manufactured by Bristol-Myers Squibb. Numerous assays also exist for the identification of various taxols; however, many of these assays are patented.

13.5 VALIDATION OF BIOPHARMA ASSAYS

As with any assay development process, the general principles of method validation apply, whether they are based on chromatography or immunochemistry. The purpose of method validation is to verify the sensitivity, linearity, stability, and repeatability of the assay procedure. The reader is referred to Chapter 6 for a more extensive discussion on this topic.

Geng et al. (2005) outline an efficient process for validating immunoassays that covers the specific experiments required in method validation.

Typical validation characteristics, which should be considered are as follows:

1. accuracy,
2. precision,
3. specificity,
4. detection limit,
5. limit of quantitation,
6. linearity,
7. range, and
8. ruggedness and robustness.

Detailed explanation of these characteristics can be found at the referenced citation (Valdes et al., 2003).

13.5.1 FDA Guidelines

The FDA has published guidelines in the "Guidance for Industry: Bioanalytical Method Validation" (CVM, 2001), which describe the agency expectations for full, partial, and cross-validation of analytical procedures, with a special discussion on ligand binding assays. While this guidance covers all analytical methods such as chromatography and mass spectroscopy, it is also intended for immunological assays. The fundamental principles of method validation apply to all analytical techniques and, as such, are excellent guidelines to follow when developing a new immunological technique.

13.5.2 Designing Validation Studies

The above-cited references give an excellent outline of each aspect of method validation. In general, careful up-front planning of experiments will guide the developer to an efficient completion of the validation process with a minimum amount of rework. Fundamental questions such as matrix interference, antibody specificity, potential cross-reactivity, reagent stability, and sample extraction should be studied and understood before embarking on extensive validation studies.

While much focus typically is applied to the assay performance, equal consideration should be applied to the analytical standard. This can be particularly challenging when working with protein-based biopharma products. In some cases, the specific protein being studied may only be available in limited amounts, or impure protein preparations. Since the calibration of any immunoassay is dependent on knowing the specific concentration of the calibration standards, any inaccuracies in calibration standards will be propagated throughout the immunoassay. Since proteins vary widely in stability, solubility, and structure, it is difficult to describe one "best practice" for establishing purity and stability values. General concepts such as freeze-thaw stability, short- and long-term stability, stock-solution stability, and even post-preparation stability need to be evaluated (Valdes et al., 2003).

13.6 SUMMARY

Plant biopharma is an intriguing but rapidly changing field that is difficult to summarize. Much of the early enthusiasm for the "pharming" field has been tempered by difficulties in the cultivation and containment of transgenic biopharma crops, especially when the crop species is also a major food source. The trend at the writing of this chapter has been to move toward contained productions under strict greenhouse conditions, or to plant cell culture-based systems. Although numerous investigators continue to study plant-based production of transgenic therapeutics, routine commercialization of these products has yet to be achieved. On the other hand, the ongoing quest for endogenous bioactive in plants is an area that will likely to undergo continued study and growth and may ultimately yield a number of commercially produced products.

REFERENCES

Ball, J. M., Hardy, M. E., Atmar, R. L., Conner, M. E., and Estes, M. K. Oral immunization with recombinant Norwalk virus-like particles induces a systemic and mucosal immune response in mice. *J. Virol.*, **1998**, 72, 1345–1353.

Bakker, H., Bardor, M., Molthoff, J. W., Gomord, V., Elbers, I., Stevens, L. H., Jordi, W., Lommen, A., Faye, L., Lerouge, P., and Bosch, D. Galactose extended glycans of antibodies produced by transgenic plants. *Proc. Natl. Acad. Sci. USA*, **2001**, *98*, 2899–2904.

Barta, A., Sommergruber, K., Thompson, D., Hartmuth, K., Matzke, M. A., and Matzke, A. J. M. The expression of nopaline synthase human growth hormones chimeric gene in transformed tobacco and sunflower callus tissue. *Plant Mol. Biol.*, **1986**, *6*, 347–357.

Chen, M., Liu, X., Wang, Z., Song, J., Qi, Q., and Wang, P. G. Modification of plant N-glycans processing: the future of producing therapeutic protein by transgenic plants. *Med. Res. Rev.*, **2005**, *25*, 343–360.

Canizares, M. C., Nicholson, L., and Lomonossof, G. P. Use of viral vectors for vaccine production in plants. *Immunol. Cell Biol.*, **2005**, *83*, 263–270.

CVM. Bioanalytical method validation U.S. Department of Health and Human Services Food and Drug Administration Center for Drug Evaluation and Research (CDER) Center for Veterinary Medicine (CVM): guidance for industry: 2001, available at http://www.fda.gov/downloads/Drugs/GuidanceComplianceRegulatoryInformation/Guidances/ucm070107, (accessed June 10, 2009), **2001**.

Daniell, H., Lee, S. B., Panchal, T., and Wiebe, P. O. Expression of the native cholera toxin B subunit gene and assembly as functional oligomers in transgenic tobacco chloroplast. *J. Mol. Biol.*, **2001**, *311*, 1001–1009.

Daniell, H., Cheboulu, S., Kumar, S., Singleton, M., and Falconer, R. Chloroplast-derived vaccine antigens and other therapeutic proteins. *Vaccine*, **2005**, *23*, 1779–1783.

Fernandez-San Millan, A., Mingo-Castel, A., Miller, M., and Daniel, H. A chloroplast transgenic approach to hyper-express and purify human serum albumin, a protein highly susceptible to proteolytic degradation. *Plant Biotechnol. J.*, **2003**, *1*, 71–79.

Filgueira, D. M. P., Berinstein, A., Smitsaart, E., Borca, M. V., and Sadir, A. M. Isotype profiles induced in Balb/c mice during foot and mouth disease (FMD) virus infection or immunization with different FMD vaccine formulations. *Vaccine*, **1995**, *13*, 953–960.

Floss, D. M., Sack, J., Stadlmann, J., Rademacher, T., Scheller, J., Stoger, E., Fischer, R., and Conrad, U. Biochemical and functional characterization of anti-HIV antibody-ELP fusion proteins from transgenic plants. *Plant Biotechnol. J.*, **2008**, *6*, 379–391.

Geng, D., Shankar, G., Schantz, A., Rajadhyaksha, M., Davis, H., and Wagner, C. Validation of immunoassays used to assess immunogenicity to therapeutic monoclonal antibodies. *J. Pharm. Biomed. Anal.*, **2005**, *39*, 364–375.

Gleba, Y., Klimyuk, V., and Marillonnet, S. Viral vectors for the expression of proteins in plants. *Curr. Opin. Biotechnol.*, **2007**, *18*, 134–141.

Glenz, K., Bouchon, B., Stehle, T., Wallich, R., Simon, M. M., and Warzecha, H. Production of a recombinant bacterial lipoprotein in higher plant chloroplast. *Nat. Biotechnol.*, **2006**, *24*, 66–76.

Gomond, V., Sourrouille, C., Fitchette, A.-C., Bardor, M., Pagny, S., Lerouge, P., and Faye, L. Production and glycosylation of plant-made pharmaceuticals: the antibodies as a challenge. *Plant Biotechnol. J.*, **2004**, *2*, 83–100.

Greco, R., Michel, M., Guetard, D., Cervantes-Gonzalez, M., Pelucchi, N., Wain-Hobson, S., Sala, F., and Sala, M. Production of recombinant HIV-1/HBV virus-like particles in *Nicotiana tabacum* and *Arabidopsis thaliana* plants for a bivalent plant-based vaccine. *Vaccine*, **2007**, *25*, 8228–8240.

Grevich, J. J. and Daniel, H. Chloroplast genetic engineering: recent advances and future perspectives. *Crit. Rev. Plant Sci.*, **2005**, *24*, 83–107.

Hacker, M. P., Dank, J. R., and Ershler, W. B. Vinblastine pharmacokinetics measured by a sensitive enzyme-linked immunosorbent assay. *Cancer Res.*, **1984**, *44*, 478–481.

Hadler, S. C. Vaccination programs in developing countries. In Plotkin S. A. and Orenstein, W. A. (Eds.), *Vaccines*, 4th edition, Elsevier Inc., Philadelphia, PA, 2004.

Hiatt, A., Cafferkey, R., and Bowdish, K. Production of antibodies in transgenic plants. *Proc. Natl. Acad. Sci. USA*, **1992**, *89*, 11745–11749.

Kermode, A. Plants as factories for production of biopharmaceutical and bioindustrial proteins: lessons from cell biology. *Can. J. Bot.*, **2006**, *84*, 679–694.

Khoury, C. A., Moser, C. A., Speaker, T. J., and Offit, P. A. Oral inoculation of mice with low doses of microencapsulated noninfectious rotavirus induces virus-specific antibodies in gut-associated lymphoid tissue. *J. Infect. Dis.*, **1995**, *172*, 870–874.

Klipstein, F. A., Engert, R. F., and Sherman, W. T. Peroral immunization of rats with *Escherichia coli* heat-labile enterotoxin delivered by microspheres. *Infect. Immun.*, **1983**, *39*, 1000–1003.

Koya, V., Moayeri, M., Leppla, S. H., and Daniell, H. Plant based vaccine: mice immunized with chloroplast-derived anthrax protective antigen survive anthrax lethal toxin challenge. *Infect. Immun.*, **2005**, *73*, 8266–8274.

Kris-Etherton, P. M., Hecker, K. D., Bonanome, A., Coval, S. M., Binkoski, A. E., Hilpert, K. F., Griel, A. E., and Etherton, T. D. Bioactive compounds in foods: their role in the prevention of cardiovascular disease and cancer. *Am. J. Med.* **2002**, *113*, 71–88.

Kumagai, M. H., Turpen, T. H., Weinzettl, N., Della-Cioppa, G., Turpen, A. M., Donson, J., Hilf, M. E., Grantham, G. L., Dawson, W. O., Chow, T. P., Piatak, M., and Grill, L. K. Rapid, high-level expression of biologically active a-techosanthin in transfected plants by an RNA viral vector. *Proc. Natl. Acad. Sci. USA*, **1993**, *90*, 427–430.

Kumar, D. H. and Dufourmantel, N. Breakthrough in chloroplast genetic engineering of agronomically important crops. *Trends Biotechnol.*, **2005**, *23*, 238–245.

Lapinjoki, S. P., Verajankorva, H. M., Huhtikangas, A. E., Lehtola, T. J., and Lounasmaa, M. An enzyme-linked immunosorbent assay for the antineoplastic agent vincristine. *J. Immunoass.*, **1986**, *7*, 113–128.

Leelavathi, S. and Reddy, V. S. Chloroplast expression of His-tagged GUS-fusions: a general strategy to overproduce and purify foreign proteins using transplastomic plants as bioreactors. *Mol. Breed.*, **2003**, *11*, 49–58.

Leyva, A., Franco, A., Gonzalez, T., Sanchez, J. C., Lopez, I., Geada, D., Hernandez, N., Montanes, M., Delgado, I., and Valdes, R. A rapid and sensitive ELISA to quantify an HBsAg specific monoclonal antibody and a plant-derived antibody during their downstream purification process. *Biologicals*, **2007**, *35*, 19–25.

Li, J.-T., Fei, L., Mou, Z.-R., Wei, J., Tang, Y., He, H.-Y., Wang, L., and Wu, Y.-Z. Immunogenicity of a plant-derived edible rotavirus subunit vaccine transformed over fifty generations. *Virology*, **2006**, *356*, 171–178.

Ma, J. K.-C., Drake, P. M. W., and Christou, P. The production of recombinant pharmaceutical proteins in plants. *Nature*, **2003**, *4*, 794–804.

Mason, H. S., Haq, T. A., Clements, J. D., and Arntzen, C. J. Edible vaccine protects mice against *Escherichia coli* heat-labile enterotoxin (LT): potatoes expressing a synthetic LT-B gene. *Vaccine*, **1998**, *16*(13), 1336–1343.

Massa, S., Franconi, R., Brandi, R., Muller, A., Mett, V., Yusibov, V., and Venuti, A. Anti-cancer activity of plant-produced HPV16 E7 vaccine. *Vaccine*, **2007**, *25*, 3018–3021.

Mestecky, J. and McGhee, J. R. Prospects for human mucosal vaccines. *Adv. Exp. Med. Biol.*, **1992**, *327*, 13–23.

Mett, V., Musiychuk, K., Bi, H., Farrance, C. E., Horsey, A., Ugulava, N., Shoji, Y., de la Rosa, P., Palmer, G. A., Rabindran, S., Streatfied, S. J., Boyers, A., Russell, M., Mann, A., Lambkin, R., Oxford, J. S., Schild, G. C., and Yusibov, V. A plant-produced influenza subunit vaccine protects ferrets against virus challenge. *Influenza Other Respir. Viruses*, **2008**, *2*, 33–40.

Miller, M. A. and Hinman, A. P. Economic analyses of vaccine policies. In Plotkin S. A., Orenstein W. A., (Eds.), *Vaccines*, 4th edition, Elsevier Inc., Philadelphia, PA, 2004.

Moldoveanu, A. M., Novak, M., Huang, W., Gilley, R. M., Staas, J. K., Shafer, D., Compans, R. W., and Mestecky, J. *Oral Immunization Influenza Virus Biodegrad. Microspheres.*, **1993**, *167*, 84–90.

Molina, A., Hervas-Stubbs, S., Daniel, H., Mingo-Castel, A. M., and Veramendi, J. High-yield expression of a viral peptide animal vaccine in transgenic tobacco chloroplast. *Plant Biotechnol. J.*, **2004**, *2*, 141–153.

O'Hagan, D. T., Jeffery, H., Roberts, M. J. J., McGee, J. P., and Davis, S. S. Controlled release microparticles for vaccine development. *Vaccine*, **1991**, *9*, 768–771.

Palacpac, N. Q., Yoshida, S., Sakai, H., Kimura, Y., Fujiyama, K., Yoshida, T., and Seki, T. Stable expression of human beta 1,4-falactosyltransferase in plant cells modifies N-linked glycosylation patterns. *Proc. Natl. Acad. Sci. USA*, **1999**, *96*, 4692–4697.

Rochfort, S., Parker, A. J., and Dunshea, F. R. Plant bioactives for ruminant health and productivity. *Phytochemistry*, **2008**, *69*, 299–332.

Santi, L., Batchelor, L., Huang, Z., Hjelm, B., Kilbourne, J., Arntzen, C. J., Chen, Q., and Mason, H. S. An efficient plant viral expression system generating orally immunogenic Norwalk virus-like particles. *Vaccine*, **2008**, *26*, 1846–1854.

Shoji, Y., Chichester, J. A., Bi, H., Musiychuk, K., de la Rosa, P., Goldschmidt, L., Horsey, A., Ugulava, N., Palmer, G. A., Mett, V., and Yusibov, V. Plant-expressed HA as a seasonal influenza vaccine candidate. *Vaccine*, **2008**, *26*, 2930–2934.

Staub, J. M., Garcia, B., Graves, J., Hajdukiewicz, P. T. J., Hunter, P., Nehra, N., Paradkar, V., Schlittler, M., Carroll, J. A., Spatola, L., Ward, D., Ye, G., and Russell, D. High yield production of a human therapeutic protein in tobacco chloroplast. *Nat. Biotechnol.*, **2000**, *18*, 333–338.

Svab, Z. and Maliga, P. High-frequency plastid transformation in tobacco by selection for a chimeric aadA gene. *Proc. Natl. Acad. Sci. USA*, **1993**, *90*, 913–917.

Thanavala, Y., Huang, Z., and Mason, H. S. Plant-derived vaccines: a look back at the highlights and a view to the challenges on the road ahead. *Expert Rev. Vaccines*, **2006**, *5*, 249–260.

Tregoning, J. S., Nixon, P., Kuroda, H., Svab, Z., Clare, S., Bowe, F., Fairweather, N., Ytterberg, J., van Wijk, K. J., Dougan, G., and Maliga, P. Expression of tetanus toxin fragment C in tobacco chloroplast. *Nucleic Acids Res.*, **2003**, *31*, 1174–1179.

Turpen, T. H., Turpen, A. M., Weinzettl, N., Kumagai, M. H., and Dawson, W. O. Transfection of whole plants from wounds inoculated with *Agrobacterium tumefaciens* containing cDNA of tobacco mosaic virus. *J. Virol. Methods*, **1993**, *42*, 227–239.

Valdes, R., Gomes, L., Padilla, S., Brito, J., Reyes, B., Alvarez, T., Mendoza, O., Herrera, O., Ferro, W., Pujol, M., Leal, V., Linares, M., Hevia, Y., Garcia, C., Mila, L., Garcia, O., Sanchez, R., Acosta, A., Geada, D., Paez, R., Vega. J. L., and Borroto, C. Large-scale purification of an antibody directed against hepatitis B surface antigen from transgenic tobacco plants. *Biochem. Biophys. Res. Commun.*, **2003**, *308*, 94–100.

Van Ree, R., Cabanes-Macheteau, M., Akkerdaas, J., Milazzo, J. P., Louthelier-Bourhis, C., Rayon, C., Villalba, M., Koppelman, S., Aalberse, R., Rodriguez, R., Faye, L., and Lerouge, P. Beta(1,2)-xylose and alpha(1,3)-fucose residues have a strong contribution in IgE binding to plant glycoallergens. *J. Biol. Chem.*, **2000**, *275*, 11451–11458.

Volkov, S. K. Investigation of the structure of chemical compounds, methods of analysis and process control. *Pharm. Chem. J.*, **1995**, *29*, 567–570.

Wachsmann, D., Klein, J. P., Scholler, M., and Frank, R. M. Local and systemic immune responses to orally administered liposome-associated soluble *S. mutans* cell wall antigens. *Immunology*, **1985**, *54*, 189–193.

Watson, J., Koya, V., Leppla, S. H., and Daniel, H. Expression of *Bacillus anthracis* protective antigen in transgenic chloroplast of tobacco, a non-food/feed crop. *Vaccine*, **2004**, *22*, 4374–4384.

Werner, S., Marillonnet, S., Hause, G., Klimyuk, V., and Gleba, Y. Immunoabsorbent nanoparticles based on a tobamovirus displaying protein A. *Proc. Natl. Acad. Sci. USA*, **2006**, *103*, 17678–17683.

WHO. Guidelines on good agricultural and collection practices (GACP) for medicinal plants. World Health Organization, Geneva, **2003**.

World Health Organization. Global immunization data, available at http://who.int/immunization_monitoring/data/GlobalImmunizationData.pdf, (accessed June 10, 2009). **2004**.

Yang, L., Kajiura, H., Suzuki, K., Hirose, S., Fukiyama, K., and Takaiwa, F. Generation of a transgenic rice seed-based edible vaccine against house dust mite allergy. *Biochem. Biophys. Res. Commun.*, **2008**, *365*, 334–339.

CHAPTER 14

IMMUNOASSAYS IN VETERINARY PLANT-MADE VACCINES

Giorgio De Guzman
Robert P. Shepherd
Amanda M. Walmsley

14.1 INTRODUCTION

Plant-made vaccines are a type of subunit vaccine where the reactor is the whole plant, or cultured plant cells, or organs. Plant-made vaccines can be produced and delivered in a number of ways; however, no matter the means, the potential of expressing and delivering a protective vaccine antigen in plants has sparked interest, wonderment,

Immunoassays in Agricultural Biotechnology, edited by Guomin Shan and incredulity, resulting in increasing research dedicated to advancing the fledgling biotechnology. A basic introduction to plant-made vaccines is given by Walmsley and Arntzen (2000), while a thorough review of plant-made vaccines for veterinary applications is given by Floss et al. (2007).

Producing a plant-made vaccine begins by selecting a suitable antigen. The corresponding gene is cloned into an expression cassette that contains plant regulatory sequences that drive gene expression. This cassette is then used in plant transformation. *Agrobacterium*-mediated transformation, stable or with assistance from plant viral elements for magnified transient transformation, is usually the preferred method for transformation of the plant cell nucleus. *Agrobacterium* is a plant pathogen that in the process of infection, transfers a segment of its DNA (T-DNA) into the nucleus of the host. Molecular biologists have taken advantage of this process to transfer genes of interest, in plant expression cassettes, into plant genomes. Transfer of the T-DNA from the bacterium into the host's genome occurs upon incubation of the transgenic *Agrobacterium* with plant cells. In transient expression systems, the infected plant tissues are harvested 2–9 days later and yield significant quantities of the protein of interest. For stable transformation, the transformed cells are positively selected during tissue culture using a marker or resistance gene and regenerated into transgenic plants or multiplied into plant cell lines. The time taken to regenerate a transgenic line is species dependent and ranges from 6 weeks to 18 months. Stably or transiently transformed, plant tissues are selected for further development based on authenticity of the plant-made subunit protein to the native form, and its concentration in plant tissues. The materials are amplified either in the greenhouse or fermentation reactors (plant cell and organ cultures), fully characterized and then tested for immunogenicity in animal trials.

In the years spanning the present day and the first mention of plant-made vaccines in peer-reviewed literature (Mason et al., 1992), the dogma of plant-made vaccine technology has evolved from being food items ("edible vaccines" in 1992) to prescribed fruit (1998) to plant-derived pharmaceuticals (2001). This change in dogma was brought about by the gradual realization that both agricultural and pharmaceutical regulations needed to be adhered to in order for this technology to produce a commodity. That is, not only is the growth and economical production of genetically modified organisms needing to be addressed, but also the adherence to good laboratory and manufacturing practices (GLP and GMP) and demonstration of consistency of vaccine batch production and performance.

On January 31, 2006, Dow AgroSciences LLC announced that in collaboration with Arizona State University and Benchmark Biolabs, Inc. (Lincoln, NE) they had received the world's first regulatory approval for a plant-made vaccine (USDA APHIS, 2006). The developed plant-made vaccine combating Newcastle Disease Virus (NDV) was made using a contained, plant-cell production system. The plant-made NDV vaccine is produced by a stably transformed plant cell line that is grown as a suspension culture in a conventional bioreactor system. The resulting plant cell cultures are harvested and partially purified before the antigen is formulated into the final vaccine. Chickens vaccinated subcutaneously with the plant-made, Newcastle Disease Virus vaccine, proved to be protected against lethal challenge by a highly virulent NDV strain. The dose response capable of greater than 90% protection ranged between 3 and 33 µg/dose with overall protection of 95% (Mihaliak et al., 2005). A formulation was advanced through the United States of America Department of Agriculture (USDA) Center for Veterinary Biologics' (CVB) regulatory approval process in a feat that demonstrated plant-made vaccines could be developed within an existing regulatory framework.

Many antigens have been successfully expressed in plant systems so the question is not whether plants can act as vaccine productions systems but whether they are consistent and economically feasible vaccine expression systems. Many factors will play important roles in the quest for plant-made vaccine feasibility. However, development of dependable immunoassays will play no small role in this achievement by reliably characterizing and quantifying protective antigens in plant cells and also determining the type, strength, location, and duration of the antigen-specific immune response induced. Other chapters in this book have discussed various aspects regarding immunoassay development (Chapters 4 and 5), validation (Chapter 6), and applications in GE product development (Chapter 10) of plant-derived proteins. This chapter provides a series of conditions, protocols, and trouble-shooting exercises to enable development of dependable immunoassays for detecting plant-made antigens and resultant induced immune responses.

14.2   PLANT-MADE ANTIGEN EXTRACTION

Functional proteins (as opposed to structural) ultimately act due to their tertiary configuration enabling specific binding to different compounds, molecules, or proteins. To enable detection and quantification of functional plant-made antigens, they first need to be extracted from plant tissues in a manner that allows their tertiary structure to be maintained. This extraction process, though seemingly simplistic, is one of the most critical aspects of proteomic analysis and is often problematic in plants due to an abundance of proteases and other interfering structures and compounds such as the plant cell wall, storage polysaccharides, various secondary metabolites, such as phenolic compounds, often low vaccine antigen expression (Nanda et al., 1975). Therefore, there are several factors needing consideration for optimal extraction of authentic and functional plant-made antigen.

14.2.1   Factors Influencing Effective Antigen Extraction

The nature of the antigen itself plays a major role in extraction from plant tissues. Protein extraction should therefore not be a standard procedure but investigated on a case by case basis. Knowing the range of pH and temperatures at which the protein is stable and when and where it is expressed in the plant cell is invaluable. Does it have transmembrane domains and therefore require detergent dispersal? Is it secreted into the apoplast and perhaps present in the media? Variables that should be taken into consideration include: optimal harvest point, extraction buffer, and physical disruption method (Figure 14.1).

14.2.1.1   Optimal Harvest Point   Expression of heterologous proteins in plant cells is effected by a myriad of factors both environmental and genetic including temperature, water availability, age of plant material, location of transgene insertion

Plant-protein extraction

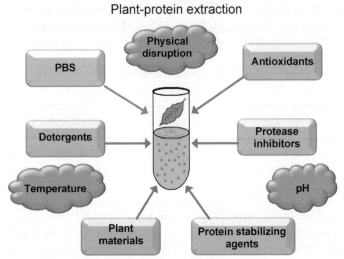

Figure 14.1 Plant protein extraction.

into the genome, and the promoter elements used. No matter the transformation method, investigations should be first made into what point of culture or plant development antigen expression is optimal. For example, our investigations in tomato have found that tomato fruit developmental stage accounts for 60.54% of total variance in antigen expression, and that harvest at the optimal fruit developmental stage resulted in up to 10-fold increase in detectable antigen.

14.2.1.2 Extraction Buffers Below are the buffer constituents and their roles:

Phosphate buffered saline (PBS) (137 mM NaCl, 8 mM Na_2HPO_4, 2 mM KH_2PO4, 2.7 mM KCl, and a pH of 7.4) is more effective than H_2O in maintaining the integrity of protein structures since it buffers changes in pH.

Nonfat dried skim milk powder (NFSM) can be added to act as an adsorbent of plant secondary metabolites such as phenolics that may react with and hamper the extraction of the protein of interest. NFSM also competes with the antigen for degradation by proteases and therefore increases the amount of antigen detected.

Protease inhibitors result in higher antigen detected (Dogan et al., 2000) since more intact antigen is present. Addition of leupeptin, aprotinin, E-64, pepstatin, and pefabloc to extraction buffers has resulted in increased detectable antigen. Commercial protease inhibitor cocktails are available, such as Complete protease inhibitor cocktail from Roche Pty. Ltd, and several plant-specific products from Sigma Aldrich. Phenylmethylsulfonyl fluoride (PMSF) has also been extensively used as a protease inhibitor in the past; however, the commercial cocktails are as effective without having the toxicity of PMSF.

Sodium ascorbate 1–20% (w/v) improves levels of monoclonal-reactive antigen 4–12-fold. Being an antioxidant, it prevents the extraction buffer and extracted proteins from reacting with atmospheric oxygen. Using sodium ascorbate rather than ascorbic acid is recommended because it does not change the pH of the buffer.

Glycerol at 10–20% (v/v) is often used to stabilize active proteins in solution (Smith et al., 2002). It also aids protein stability after samples are freeze-thawed.

14.2.1.3 *pH*   In addition to the constituents of the extraction buffers, the pH of the buffer can greatly affect not only the amount of proteins extracted, but also protein integrity. The pH affects the charge and therefore solubility of the protein of interest. For example, the pH of the extraction buffer should not be at or near the protein of interest's isoelectric point as proteins are least soluble in any given solvent at this point. Generally, it has been seen that an increase in pH of sample extractions leads to an increase in solubilization of proteins; however, it is recommended that the pH should not exceed 8 to avoid an adverse effect on the biological activity of the proteins (Nanda et al., 1975). A pH of 7–8 is generally desired for most extraction protocols.

14.2.1.4 *Temperature*   Extreme temperatures denature proteins. In general, a low temperature is desired to inhibit endogenous protease activity and slow the degradation of desired protein products. It is recommended to keep extraction procedures close to 4°C. However, certain antigens are more stable at higher temperatures, for example, 50°C gave optimal protein extraction yields of hepatitis B surface antigen (HbsAg) (Dogan et al., 2000).

14.2.2 Antigen Release

Since detergent helps to disrupt membranes, addition of detergent can often increase antigen extracted if the antigen is localized within a subcellular organelle. It is noteworthy that high detergent concentration may interfere with ELISA and that the amount of antigenic protein extracted is not affected by the concentration of detergent but the detergent to cell ratio (Smith et al., 2002).

Samples of callus, leaf or root are often physically disrupted to determine antigen production. However, care must be taken not to introduce extra variability by having vastly differing cell types in different samples. Different cell types not only produce different amounts of protein but may also differ in the ease of cell disruption and protein release (e.g., vein material versus lamina tissue). Therefore, taking samples from areas of the leaves/roots that contain similar cell types will greatly increase consistency in the amount of protein extracted.

There are two main tissue-grinding techniques for extraction of proteins of interest from plant tissues. These are manual and mechanical grinding. Manual grinding is generally performed with a pestle in either a mortar (large tissue mass) or in centrifuge tubes (small tissue mass). The main issue with manual grinding is consistency between sample extractions. In general, manual grinding is less consistent between extractions than mechanical grinding because of operator variability. Mechanical grinding involves addition of glass/ceramic/metallic beads alongside the sample and extraction buffer to the extraction vessel and using a vortex/mixer bead mill apparatus to consistently agitate the vessel (with respect to time and force applied). Other tissue grinding mills can use blades to cut up the material. The use of machines for grinding plant tissues not only eliminates operator variation but also greatly reduces time spent and labor involved performing extraction. The type of grinding technique performed is largely dependent on the mass and the number of samples that are to be extracted. For large masses of tissues or small sample sizes, it is more efficient to grind the tissue manually, whereas when working with large sample size or small tissue masses, mechanical grinding is recommended.

14.2.2.1 Examples of Plant-Made Antigen Extraction Protocol

General Extraction Buffer: Phosphate-buffered saline (PBS) (0.15% Na_2HPO_4, 0.04% KH_2PO_4, 0.61% $NaCl_2$ w/v, pH 7.2), 10 mM EDTA, and 0.1% Triton X-100.

Possible buffer additions to increase antigen authenticity:

Protease inhibitor

1–20% (w/v) Ascorbic acid

5% (w/v) Nonfat skim milk powder

10–20% (v/v) Glycerol

Example: Tomato Fruit Extraction Buffer consists of 50 mM sodium phosphate, pH 6.6, 100 mM NaCl, 1 mM EDTA, 0.1% Triton X-100, 10 µg/mL leupeptin, and 1 M phenylmethylsulfonyl fluoride.

14.2.2.2 Total Soluble Protein Extraction

A general protocol of total soluble protein extraction is listed in Table 14.1. Quantification of total soluble protein is often performed to enable comparison of the amount of antigen produced to units of total soluble protein and to gain insight into success of protein extraction. Protocols are usually described in Bradford reagent dye manuals such as Bio Rad's Protein Assay Dye Reagent Concentrate (Catalogue number 500-0006). After performing the required dilutions and reading absorbance. Plot and calculate the equation of the standard curve from the BSA readings (Figure 14.2). To ensure accurate protein content determination, the standard curve equation should contain at least five BSA dilution points, including a zero point, be in the linear region of the curve and have an r^2 value of 0.95 or higher. The linear region of standard curves gives the most accurate portrayal of concentration as absorbance is directly proportional to concentration and the assay is not restricted by biological or instrumental limitations. Using the equation of the BSA standard curve and the OD readings of the extracted samples enables calculation of the total soluble protein content.

TABLE 14.1 TSP Extraction Procedure

Step	Procedure
1	Snap freeze plant material and add ice cold extraction buffer
2	Homogenize sample
3	Centrifuge at 20,000 × g for 10–20 min
4	Store extract supernatant after adding 10–20% glycerol at -20°C

See Figure 14.3 for an example of a BSA standard curve and TSP concentration.

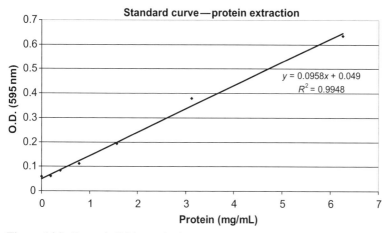

Figure 14.2 Example BSA standard curve.

14.3 ANTIGEN DETECTION AND QUANTIFICATION IN PLANT TISSUES

There are a number of methods that enable detection and quantification of a particular antigen in plant extracts. The technique selection relies on the information required such as qualitative (How big is the protein? Is it glycosylated, cleaved, or multimeric?) or quantitative; the number and type of antigen-specific antibodies available, whether a standard protein is available to act as a reference for quantification, and the required sensitivity of your assay to enable detection of your antigen of interest.

14.3.1 Dot Blots

Dot blotting in its simplest form is a qualitative assay where a 0.5–5 µL of extract is applied to a membrane and then immunohistochemically detected using appropriately derived antibodies. Dot blots are generally used for determining optimal conditions for Western blot analysis including blocking conditions, antibody dilutions and wash buffers. By comparing intensity of sample signals with a serial dilution of a standard antigen, dot blots can be semiquantitative. Since dot blots do not separate the samples according to protein size as accomplished through PolyAcrylamide Gel Electrophoresis (PAGE), less time is required. However, dot blots do not differentiate between specific and nonspecific binding. Therefore, negative controls should always be loaded.

14.3.2 Organ/Tissue Blots

Squash blots were initially used to detect virus infection in plant leaves. Recently, they have been used as a rapid method for detecting recombinant proteins in plant tissues with the advantage of indicating site of protein accumulation. However, organ/tissue

Immunoblotting Flow Diagram

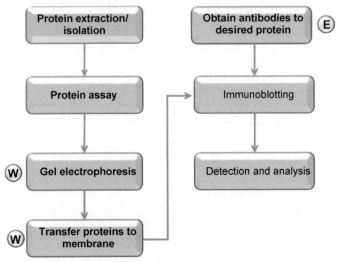

Figure 14.3 Immunoblotting flow diagram.

blots are not particularly sensitive. Therefore, the protein of interest must be produced abundantly.

14.3.3   Western Blot Analysis

Western blot or immunoblot analysis first involves separating extracted proteins according to size and charge using PAGE. The separated proteins are then transferred and immobilized onto a membrane and detected using antigen-specific antibodies (Figure 14.3; Table 14.2). Western analysis can detect native or denature proteins and determine if the protein of interest is the correct size, being degraded, is glycosylated or forms oligomers. Western analysis is usually qualitative, although if known amounts of standard protein are run along side the samples, an estimate of the amount of the protein of interest present can be made through comparing intensity of the signal. If protein extracts from a negative, wild-type sample also runs along side the samples, differentiation can be made between specific and nonspecific binding.

14.3.4   ELISA

Direct-bind ELISA can detect nanogram to picogram quantities of a protein of interest. In this assay, the protein of interest is directly adsorbed onto the surface of a polystyrene plate. The plate is then blocked with a blocking buffer to prevent unwanted binding proteins to the plastic surface. An antibody (primary or detector antibody) specific to the protein of interest is applied. This primary antibody may either be directly conjugated with a marker enzyme such as an alkaline phosphotase, or horseradish peroxidase, or another secondary antibody-conjugate may be used to

TABLE 14.2   General Protocol for Western Analysis

Step	Procedure
1	Determine total soluble protein content and dilute samples to contain same amount of TSP
2	Add protein-loading buffer and denature by boiling for 10 min, then snap cool on ice
3	Load into an acrylamide gel, along with molecular mass ladder. Run at 30 milliamps per gel until the gel front is approximately 5–10 mm from the end of the gel
4	Equilibrate the gel and transfer to membrane
5	Block membrane in PBST + NFSM for 1 h at 37°C or overnight at 4°C
6	Rinse the membrane with PBST then incubate the membrane with the optimal dilution of primary antibody in PBST + NFSM for 1 h at room temperature
7	Rinse membrane with PBST then incubate with the optimal dilution of your secondary antibody (with conjugate) in 10 mL PBST + NFDM at room temperature for 1 h
8	Rinse the membrane then visualize using film and ECL + kit as per manufacturer's directions

For detailed procedure see Walmsley et al. (2003).

detect the primary antibody. A quantitative analysis of the total number of antibody–enzyme conjugate molecules present on the plate surface is performed using a chromogenic or fluorescent substrate.

While direct-bind ELISA is sufficient to detect high abundance of proteins of interest in samples, low abundance molecules may not bind to the plate in sufficient numbers or may be sterically hindered in complex sample mixtures such that binding to the plastic surface is not enabled. Capture ELISA (or indirect, sandwich ELISA) can often detect picogram to microgram quantities of molecules in more complex mixtures (Figure 14.4).

In this assay, a monoclonal- or polyclonal-antibody specific to the antigen of interest is adsorbed onto the surface of the polystyrene plate. The plate is then blocked with a blocking buffer to prevent unwanted binding of molecules to the plastic surface. Samples are then added to the plate in a series of dilutions with buffer as with the direct-bind ELISA. The antigen in the samples specifically binds to the coating antibody. Other nonspecific molecules in the samples are washed away. All subsequent steps are performed in a similar manner to direct-bind ELISA.

A variety of factors may influence the choice of direct-bind or capture ELISA. These include whether or not whole serum or purified antibodies from a heterologous species are available to use as the capture antibody, the level of background noise associated with the specific interaction of different samples with the plastic adsorption surface, the sensitivity of the assay required, and the time limitations of the assay procedure (as the addition or the capture antibody requires an additional binding step).

14.3.4.1 ELISA Protocol Examples of general protocols for direct and indirect ELISAs are listed in Tables 14.3 and 14.4.

Coating Buffer Coating buffers are used in ELISA analysis to stabilize protein tertiary structure and allow strong adsorption to the polystyrene plate matrix.

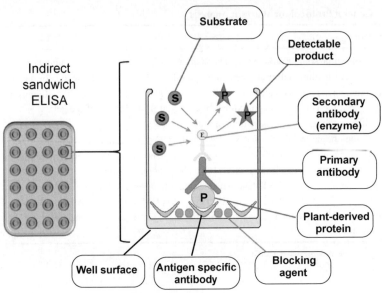

Figure 14.4 Indirect sandwich ELISA diagram.

TABLE 14.3 Direct ELISA Protocol

Step	Procedure
1	Coat a 96-well plate with 1–10 μL per well of plant extract diluted in PBS/coating buffer
2	Perform twofold serial dilutions of each sample down the plate. PBS + 0.05% Tween (PBST) + 1% NFSM
3	Make the standard curve by diluting the antigen stock to 50 ng/mL, then perform twofold dilutions down the plate in PBS + 0.05% Tween (PBST) + 1% NFSM
4	Incubate for 1 h at 37°C of overnight at 4°C
5	Wash three times with PBST
6	Block wells with 200 μL/well of 5% NFSM in PBST at 37°C for 1 h
7	Wash three times with PBST
8	Add 50 μL/well of primary antibody diluted in 1% PBST + NFSM. Incubate at 37°C for 1 h
9	Wash three times with PBST
10	Add 50 μL/well of secondary antibody diluted in 1% PBST + NFSM. Incubate at 37°C for 1 h
11	Wash four times with PBST
12	Add 50 μL/well of TMB substrate. Incubate at room temperature for 5 min
13	Add 50 μL/well of 1 N H_2SO_4
14	Read absorbance at 450 nm
15	Construct standard curve and calculate sample antigen concentrations

TABLE 14.4   Indirect ELISA Protocol

Step	Procedure
1	Coat plates with 50 μL/well of diluted capture antibody (10–100 μg/mL) in coating buffer. Incubate overnight at room temperature
2	Wash three times with PBST
3	For the standard curve, dilute stock to 50 ng/mL
4	Perform twofold dilutions down the plate in PBST/1% NFSM
5	Coat plates with 1–10 μL/well of plant extract and make up to 100 μL with PBS
6	Perform twofold serial dilutions down the plate
7	Incubate shaking for 2 h at room temperature or overnight at 4°C
8	Wash plate three times with PBST (0.05% Tween)
9	Block wells with 200 μL/well of 5% NFSM in PBS (0.05% Tween-20), 37°C for 1 h
10	Wash three times with PBST
11	Add 50 μL/well of primary antibody diluted in 1% NFSM in PBS. Incubate at 37°C for 1 h
12	Wash three times with PBST
13	Add 50 μL/well of secondary antibody diluted in 1% NFSM in PBS. Incubate at 37°C for 1 h
14	Wash four times with PBST
15	Add 50 μL/well of TMB substrate. Incubate at room temperature 5 min
16	Add 50 μL/well of 1 N H_2SO_4
17	Read absorbance at 450 nm
18	Construct standard curve and calculate sample antigen concentrations

A general schematic for an indirect ELISA can be found in Figure 14.8.

A good starting point for determining the optimal binding conditions for antibody or antigen adsorption onto the solid-phase of the 96-well capture is to simply use PBS as the coating buffer. Several other buffers are reported in the literature. These include carbonate buffers (0.1 M carbonate or bicarbonate in H_2O at pH 9.6) or buffers containing protamine sulfate. The coating buffer used can significantly affect the amount of specific protein ultimately detected (Figure 14.4). It is therefore important to ensure that the buffer is optimized.

Blocking Agents   Many reagents can be used to block surfaces of plates or membranes to which proteins of interest are bound. This blocking step is required to minimize the nonspecific binding of antigens and antibodies in subsequent steps, and to reduce the background of the assay. Nonfat dried skim milk powder made up in PBS works very well and is relatively inexpensive. However, caution should be used in situations where blocking agents are derived from the same species as will be subsequently tested, that is, do not use NFSM if you will be testing the IgM response to vaccination in cattle. Other common blocking agents include bovine serum albumin at 1–3% (w/v) in PBS (He et al., 2008).

Pipette Tips and Multichannel Pipette   Although ELISA analysis can be performed using a single channel pipette, a multichannel pipette (when properly used) reduces pipetting error, time-spent pipetting, and operator fatigue. Good pipetting technique is essential for the success of ELISA analysis. Due to the lack of internal reference, small errors introduced by inaccurate or imprecise technique may introduce significant noise in the final analysis. Always inspect the pipette and tips for correct seal, and ensure that consistent dispensing technique is used. Electronic multichannel pipettes are ideal for this purpose as they generally have a higher precision of dispensed volume than can be achieved by hand. Fully read the manufacturer's instructions to determine the range of volume to be dispensed.

Shaking Platform and/or Incubator   While not essential, increased temperature during reactions and/or shaking will help to reach reaction equilibrium conditions faster. For example, a typical incubation step may be to place plate with reagents at room temperature for 2 h. Typically, with shaking at 90–120 rpm or incubation at 37°C, this can be decreased to 30–60 min. When designing a protocol for immune-response detection, these variables need to be standardized early during protocol development and followed strictly. Large variation between assay runs is possible when these conditions are not adhered to.

Adsorbent Polystyrene 96-Well Flat-Bottom Plates   These plates can be procured from any manufacturer, but it is important that they are specifically formulated for protein adsorption and are generally sold as being specific to ELISA. It is best to standardize protocols for a specific plate model as differences in binding do occur. This will allow confidence in later comparison of assay results between samples, and will provide a consistent background signal between experiments. Also remember that vinyl plate covers will prevent evaporation of samples and may avoid spills.

Monoclonal or Polyclonal Antibody   In ELISA, it is important to ensure that the animal species from which the antibodies are raised for capture, primary or secondary binding are all different. This will minimize the chance of cross-reactivity between already adsorbed but still exposed antibody motifs (e.g., to investigate the immune response of chickens to an influenza vaccine, antibodies specific to the influenza antigen should be produced in guinea pig, for example and guinea pig capture or conjugate antibody is not subsequently used in the detection for the antibody isotypes). Commercial, recombinant, primary antibodies should be made up into a range of 1 mg/mL of buffer (PBS), and a starting solution of 0.1–0.5 µg/mL (1:2000–1:10,000) is recommended. A starting dilution is often suggested by the manufacturer. It is also important for antibody dilutions used in Western blot and ELISA to be optimized. However, for whole serum, good starting dilutions to try for a primary antibody are 1:1000 or 1:5000 and for a secondary antibody 1:10,000–1:15,000.

Some commercially available primary antibodies are tailored for use in either ELISA or Western blot. Ensure that the primary antibody you are using was raised in a manner appropriate to your method. For example, in ELISA, the antibody should be raised using the native antigen and not raised from denatured linear protein as is commonly used in nonnative Western blot and immunoblot procedures. This information should be available in the antibody product sheet or upon request from the manufacturer.

Purified or Recombinant Reference Antigen   Antigens may be purified from their host source, or manufactured in a heterologous recombinant system. Antigens are recommended to be stored at concentrations of 1 mg/mL and should be diluted to a range of 1–0.1 µg/mL (1:1000–1:10,000 dilution). Best-practice ELISA strategies use a reference antigen from the pathogen or viral host rather than purified from the tested system. For example, after vaccinating chickens with recombinant plant-produced proteins, the reference antigen should ideally be produced in a host system (e.g., avian cell line) and not in plant cells. This gives a clear indication that the antibodies raised during vaccination are able to bind and potentially neutralize the native or near-native antigens.

Commercial Colorimetric Substrate   This is generally purchased as a pre-mixed solution from a commercial supplier. Due to variation in propriety solutions, it is best practice to standardize this reagent during assay development to minimize variation in subsequent assays. Commonly used substrates include: 0.004–0.02% H_2O_2 plus *ortho*-phenylene diamine (OPD), tetra-methylbenzidine (TMB), 2,2′-azino di-ethylbenzonthiazoline-sulfonic acid (ABTS), 5-aminosialic acid (5AS), or di-aminobenzidine (DAB) for use with horseradish peroxidase enzyme labels, or 2.5 mM *para*-nitrophenyl phosphate (pnpp) for alkaline phosophotase enzyme labels, or 3 mM *o*-nitrophenyl beta-D-galactopyranoside (ONPG) or urea and bromocresol dyes for urease enzymatic markers.

Spectrophotometer   There are many models of spectophotometers capable of reading fluorometric or colorimetric results from Bradford and ELISA. The complexity of the model you use is reliant on what other research is performed within your laboratory. If the spectrophotometer is to be used only for reading Bradford's and ELISA plates then it should be compatible with 96-well plates and fitted with filters specific to your preferred substrate. There are also many software packages available to read and analyze results and these usually come with the spectrophotometer.

14.3.5   Optimizing Antibody Dilutions for Western Analysis

Reduced wastage of reagents, reduced background, and increased sensitivity are three desirable advantages of optimizing antibody dilutions used in Western analysis. Dot blots using a range of antigen concentrations and different primary and secondary antibody dilutions and permutations should be used to determine the antibody dilution combination that is able to detect the least amount of antigen using the highest dilution of antibodies. Manufacturers of reagents kits often used to detect Western analysis (e.g., Stratagene's ECL + Kit), often include a general protocol in their instruction booklet. Good starting points for antibody dilutions for Western optimization are three primary antibody dilutions (1 in 1000, 1 in 2500, 1 in 5000) and three secondary antibody dilutions (1 in 10,000; 1 in 25,000; 1 in 100,000). Remember to treat the antigen of interest during dot blot optimization the same way as you would during Western analysis, that is, in its native or denatured state. A general protocol for Western analysis is given in Table 14.2.

14.3.6 Optimizing Antibody Dilutions for ELISA Analysis

As per Western analysis, antibody dilutions used in ELISA analysis should be optimized to decrease the amount of reagents used, decrease background, and increase sensitivity. Depending on how optimization is performed, this important step can also flag possible undesirable interactions of the antibodies used (interactions not based on antigen presence). This possibility is accounted for in the optimization protocol described by Bruyns et al. (1998). Good starting points for antibody dilutions for ELISA optimization are four primary antibody dilutions (1 in 500, 1 in 1000, 1 in 2500, 1 in 5000) and four secondary antibody dilutions (1 in 5000; 1 in 10,000; 1 in 25,000; 1 in 100,000). Make sure the amount of secondary antibody is in excess to antigen to ensure the assay is quantitative and that each blank only gives background signal. A general schematic of indirect ELISA can be found in Figure 14.4 and a protocol in Table 14.4. An example of coating buffer: in every 300 mL of buffer, it contains 0.48 g $NaHCO_3$, 0.88 g $NaHCO_3$, and 0.06 g NaN_3 (pH 9.6). It is usually stored at 2–8°C.

14.4 IMMUNOASSAYS FOR DETERMINING RESPONSE TO VETERINARIAN PLANT-MADE VACCINES

The overall aim of vaccination is the development of an immunological memory toward a particular antigen/pathogen to enable a rapid immunological response against the antigen/pathogen upon future exposure. Determining if a vaccination trial is efficacious ultimately relies upon the survival of animals when challenged with the live pathogen. Due to the cost, complexity, and ethical implications of challenge trials (as well as the difficulty in obtaining approval for human trials), proxy markers of a protective immune response are often developed and utilized. It is important to note that these markers are unable to ultimately indicate if vaccination is effective, but instead indicate whether an avenue of vaccine development is worth pursuing, or if there is an advantage to existing vaccines. These markers are often the type of cytokines or antibodies induced following effective vaccination. Cell-mediated responses to pathogens are usually indicated by increased cytokine and chemokine production, while humoral responses to intercellular pathogens are usually indicated by specific antibodies. For example, an inflammatory (Th1) response is usually required for an intracellular pathogen, such as a virus, and this response can be indicated by increase production of interferon gamma (INF-γ). A Th2 response is usually required by an intercellular pathogen such as bacteria that produce toxins. A Th2 response is often indicated by chemokines interleukin-10 (IL-10) and IL-4 and antibody specifically produced toward the pathogen.

This section will focus on characterization of Th2 responses and antigen-specific antibody analysis post-vaccination with protein antigens. These assays require only minimal experimental apparatus, and do not require the expensive and technically demanding cell culture techniques used for characterization of cellular response. All that is required for determination of the humoral response is samples of blood, fecal and/or mucosal surface washes collected over the vaccination schedule.

Antibodies secreted from lymphocytes are retained in the serum, or are secreted to the mucosal surfaces of the gastrointestinal, respiratory or genital tract where they act as the first line of immune defense in neutralizing toxins and pathogens. The total antibody titer (how many individual antibodies molecules are present), and the specific antibody structural isotypes present in the serum or mucosal surface may provide a marker of the intensity and specificity of the immune response to vaccination with the plant-made protein. To quantify the concentration of antibodies in the blood or fecal samples raised by vaccination, ELISA is performed to measure the total number of dilutions required to reduce the signal to the background for the assay. This measurement does not rely on an external standard, and as such indicates the ultimate intensity of immune response. This antigen-specific antibody response is determined using either a direct noncovalent adsorption of the antigen to the plastic substrate of a 96-well plate, or as a capture ELISA as where an antigen-specific antibody is bound to the plate and subsequently used to capture and present the antigen to the sample serum.

14.4.1 Endpoint Determination

The goal of the end point ELISA is to have the total antibody titer to be exhausted within the range of the plate, while at the same time maximizing the resolution of response detection. This is a balancing act that requires careful optimization of the starting dilution of serum or fecal matter, and will be dependent on the dynamic range of immune responses expected. It is suggested to either have the dilution series running down or across an entire column or row of a 96-well plate. This will give the highest chance of being able to assess the end point across the high dynamic range of results that are to be expected for "real-world" biological data sets.

14.4.2 Blood Sample

Whole blood should be taken from the animal in accordance with prescribed animal ethics. After collection, the blood should be transferred to a suitable centrifuge tube and kept at room temperature for an hour to allow coagulation. As soon as feasible (within hours), centrifuge collected blood at $6000 \times g$ for 5 min until the erythrocytes and lymphocytes have separated from the serum. Longer spinning of cells may be required for larger quantities of blood. Make sure not to spin at too high a force as shearing of erythrocytes will release the hemoglobin into the supernatant. A maximal force of around $8000 \times g$ is recommended. Release of the cellular contents into the supernatant dramatically increases extraneous binding of primary and/or secondary antibodies, and increases the ELISA signal regardless of antigen-specific antibody concentration. Aspirate the upper supernatant and store at -20 or $-80°C$ for long-term storage. Serum can be stored at $4°C$ for short periods of time. Typically, only $100–200\,\mu L$ of blood is required for a series of end point analysis assays. However, if replicates or repeated measurements are required for the assay, greater than 1 mL of serum may be required (particularly, if low antibody titers are observed). This is not commonly a problem in large animals, but if small animal such as mice are assayed, it may be an important consideration when planning experimental protocols for control and middle time point bleeds.

14.4.3   Fecal Sample

Antibodies, particularly secreted gut IgAs can be detected in fecal matter by resuspending the dry or wet matter in PBS, and extracting the total soluble protein fraction. For dry fecal matter, start by making a 1:10 (w/v) suspension of fecal matter in PBST containing 0.1% Tween-20 and protease inhibitors. For hydrated fecal matter, begin extraction with a 1:5 (w/v) suspension. Homogenize samples using a mechanical grinding device or by adding a ceramic or tungsten bead to a flat bottom tube and using a lab vortex or shaker plate for 30–60 s. Centrifuge suspensions for 10 min at $12,000 \times g$ at 4°C, collect the supernatant and centrifuge the suspensions again for 10 min at $12,000 \times g$ at 4°C. Quantify the total soluble protein component, and store samples at -20°C until assay. Load approximately 500 μg of total soluble protein onto plate when performing ELISA.

14.4.4   Detection with Direct and Indirect ELISAs

Direct coating plates may require more antigen to achieve similar signals to that received from indirect sandwich ELISAs and should always be applied to the plate in coating buffer without a blocking agent to maximize the number of bound motifs (Kulkarni et al., 2008).

A Primary antibody specific to the subject animal and isotype to be detected is required. For example, to determine the response of antigen-specific IgG1 antibodies in a sheep trial, an antisheep IgG1 produced in goat, mouse, or rabbit may be used. Again, this antibody needs to be produced in a different host to the subject species. Some monoclonal secondary antibodies are manufactured with covalently attached conjugate enzymes and these can be detected directly by addition of a suitable substrate. More commonly, a monoclonal recombinant antibody specific to the serum antibody (i.e., antisheep IgG1 produced in goat) is used to detect the sheep IgG1 molecules, then an antisheep Ig antibody-conjugate produced in another species (i.e., mouse, anti-goat Ig) is used to detect all bound goat antibodies. Care should be taken when ordering these antibodies pairs to ensure that they are compatible and in ready supply. The need to change suppliers of a secondary or conjugate antibody may change the background of the ELISA signal, and alter the specific end point titers observed. Many commercial primary or secondary conjugated antibodies are provided at around 1.0 mg/mL and can be diluted to 0.1–0.5 μg/mL (1:200–01:10,000) or as manufacturer recommends. A general protocol for direct-bind ELISA can be found in Table 14.5 and indirect-bind ELISA in Table 14.6.

Due to the lack of internal references in end point-titer analysis, strict data interpretation and statistical analysis should be used to ensure robust and reproducible results. The key to end point ELISA data analysis is the robust determination of the background for the assay. Background is present in all immune assays and is the signal

TABLE 14.5 Immune Response Direct ELISA Protocol

Step	Protocol
1	Coat an ELISA plate with 50 µL/well of antigen diluted in PBS, incubate for 2 h at 37°C or for overnight at 4°C
2	Wash plate three times with PBST (0.05% Tween-20)
3	Block plate with 300 µL/well of blocking buffer, incubate at room temperature for 1 h
4	Wash three times with PBST
5	Add 50 µL of sample to top row of plate and serially dilute down the plate
6	Incubate at room temperature for 1 h
7	Wash three times with PBST
8	Add 50 µL/well of the primary antibody in blocking buffer, incubate at room temperature for 1 h
9	Wash three times with PBST
10	Add 50 µL/well of the secondary antibody–enzyme (HRP) conjugate in blocking buffer, incubate at room temperature for 1 h
11	Wash five times with PBST
12	Add 50 µL/well of substrate according to manufacturers instructions, develop for 30 min at room temperature
13	Add 50 µL/well of substrate stop solution
14	Read plate at 450 nm

TABLE 14.6 Immune Response Indirect ELISA Protocol

Step	Protocol
1	Coat a plate with 50 µL/well of capture antibody diluted in PBS, incubate at 37°C for 2 h or for overnight at 4°C
2	Wash three times with PBST
3	Block plate with 300 µL/well of blocking buffer, incubate at room temperature for 1 h
4	Wash plate three times with PBST
5	Add 50 µL per well of antigen diluted in PBS, incubate at room temperature for 1 h
6	Wash three times with PBST
7	Add 50 µL of serum samples to top row of plate, serially dilute down the plate, incubate at room temperature for 1 h
8	Wash three times with PBST
9	Add 50 µL/well of the primary antibody in blocking buffer, incubate at room temperature for 1 h
10	Wash three times with PBST
11	Add 50 µL/well of the secondary antibody enzyme (HRP) conjugate in blocking buffer, incubate at room temperature for 1 h
12	Wash plate five times with PBST
13	Add 50 µL/well of substrate to plate according to manufacturers instructions, develop for 30 min at room temperature
14	Add 50 µL/well of substrate stop solution
15	Read plate OD values specific to substrate at 450 nm

produced through non-specific binding of antibodies to the plate surface or non-target ligands. The literature on immunoassays differs on what is considered "above background" signal, and the two most common methods are an absolute absorbance above which will be considered as valid, or an adaptive method where the mean signal of multiple nonserum or seronegative control groups are used to determine the threshold of validity.

Recently, the ability of several data analysis software packages to plot individual data points in single columns while also providing a mean figure have become popular. These plots are ideal as they allow the viewer to interpret the raw absorbance data to visually gauge how high and variable the background is. It is important to note that if titer data is to be statistically analyzed for significance above background or control group levels, then raw dilution figures cannot be used to perform standard parametric statistical analysis. The data in its raw form (e.g., titer of 1/100) is not normally distributed, rather the twofold relationship between dilution series follows a geometric distribution. Most software packages used for data analysis have methods of manipulating geometric datasets, and it is important to ensure that valid statistical tests are performed only when this limitation has been considered.

Problems are usually encountered with ELISA or Western blot analysis if antibody concentrations have not been optimized or the antibodies are inappropriate. A dot blot protocol and ELISA plate design have been given to optimize antibody concentration for Western blot and ELISA. The goal of performing these optimization protocols is to maximize the sensitivity of antibody detection, while promoting a robust and reproducible assay. Also, always know the source of your antibodies and what they were raised against. This can save considerable angst, for example, if the detection/primary antibody was raised against the native protein, it may not recognize the linear conformation. For Western analysis, another thing to consider is the isoelectric point of target protein. Should the isoelectric point of target protein be reached close to pH 8.3 (the pH of the transfer buffer), the protein of interest will have no charge to enable transfer to the membrane. Should this be the case, do not equilibrate the gel but use the charge on the protein of interest given by SDS in the running gel.

14.5 OTHER TECHNOLOGIES AND TECHNIQUES

The aim of this chapter is to discuss immunoassay methodologies for the production, characterization, and *in vivo* testing of plant-made vaccines for veterinary purposes. We have limited the protocols and discussion to techniques that are achievable with general laboratory equipment present in an agricultural molecular biology laboratory, and have excluded techniques that require significant investment in specialized equipment and training. However, the end goal of any vaccination program is to characterize the intensity and type of immune response raised, and while we have discussed the antibody humoral response to vaccination, cell-culture and biochemical analysis tools are available for assessing induced cytokines. Cytokine production may be determined in a cell-free or cell-culture method. Most simply, ELISA can be used to determine the concentration of cytokines present in the periphery. However, due to the low concentrations and relative instability of cytokine proteins, these signals are often difficult to detect within the sensitivity of this assay. There are several cytokine detection kits on the market that have been validated to resolve these low cytokine concentrations, but the majority of these kits are not useful for veterinarian purposes, only for human or rodent targets.

Due to the low concentrations of some circulating cytokines, tissue culture techniques can be employed to enrich a culture for lymphocytes expressing these molecules. These cells can be isolated from the lymphatic tissues of the animal under sterile conditions, separated from other cells types, and grown in culture. The enrichment of lymphocytes may be of a general nature, for example, removal of animal spleens and the selective hypotonic lysis of the erythrocytes, or may be performed by cell sorting. Cell sorting is generally performed in automated flow cytometry systems where specific cell-surface receptors are tagged with fluorescent marker and sorted by their presence or lack of the fluorescent marker. These systems have the added benefit of being able to process a higher throughput of samples and can be used to rapidly sort cells based on single or multiple surface markers. Smaller scale sorting is also possible with paramagnetic bead systems which can be used to pull-down cells expressing specific markers, and there are several commercially available kits that can be used to prepare small-scale enrichments of cells using these systems. There are also several small-scale cell enrichment columns available, but unfortunately, at present these have only been verified with cells from model organisms.

REFERENCES

Bruyns, A.-M., De Neve, M., De Jaeger, G., De Wilde, C., RouzeÂ, P., and Depicker, A. Quantification of heterologous protein levels in transgenic plants by ELISA. In Cunningham, C.and Porter, A.J.R. (Eds.), *Recombinant Proteins from Plants*, Humana Press, 1998, pp. 251–269.

Dogan, B., Mason, H. S., Richter, L., Hunter, J. B., and Shuler, M. L. Process options in hepatitis B surface antigen extraction from transgenic potato. *Biotechnol. Prog.*, **2000**, *16*, 435–441.

Floss, D. M., Falkenburg, D., and Conrad, U. Production of vaccines and therapeutic antibodies for veterinary applications in transgenic plants: an overview. *Transgenic Res.*, **2007**, *16*, 315–332.

He, Z.M., Jiang, X. L., Qi, Y., et al. Assessment of the utility of the tomato fruit-specific E8 promoter for driving vaccine antigen expression. *Genetica*, **2008**, *133*, 207–214.

Kulkarni, R.R., Parreira, V. R., Sharif, S., and Prescott, J. F. Oral immunization of broiler chickens against necrotic enteritis with an attenuated *Salmonella* vaccine vector expressing *Clostridium perfringes* antigens. *Vaccine*, **2008**, *26*, 4194–4203.

Mason, H. S., Lam, D. M. K., and Arntzen, C. J. Expression of hepatitis B surface antigen in transgenic plants. *Proc. Natl. Acad. Sci. U.S.A.*, **1992**, *89*, 11745–11749.

Mihaliak, C.A., Webb, S., Miller, T., Fanton, M., Kirk, D., Cardineau, G., Mason, H., Walmsley, A., Arntzen, C., and Van Eck, J. Development of plant cell produced vaccines for animal health applications. Proc. U.S. Anim. Health Assoc., **2005**.

Nanda, C. L., Ternouth, J. H., and Kondos, A. C. An improved technique for plant protein extraction. *J. Sci. Food Agric.*, **1975**, *26*, 1917–1924.

Smith, M.L., Keegan, M. E., Mason, H. S., and Shuler, M. L. Factors important in the extraction, stability and *in vitro* assembly of the hepatitis B surface antigen derived from recombinant plant systems. *Biotechnol. Prog.*, **2002**, *18*, 538–550.

USDA APHIS. USDA issues license for plant-cell-produced Newcastle disease vaccine for chickens, available at http://www.aphis.usda.gov/newsroom/content/2006/01/ndvaccine.shtml, **2006**.

Walmsley, A.M. and Arntzen, C. J. Plants for delivery of edible vaccines. *Curr. Opin. Biotechnol.*, **2000**, *11*, 126–129.

Walmsley, A. M., Alvarez, M. L., Jin, Y., Kirk, D. D., Lee, S. M., Pinkhasov J., Rigano, M. M., Arntzen, C. J., and Mason, H. M. Expression of the B subunit of *Escherichia coli* heat-labile enterotoxin as a fusion protein in transgenic tomato. *Plant Cell Rep.*, **2003**, *10*, 1020–1026.

CHAPTER 15

IMMUNOASSAY AS A GM DETECTION METHOD IN INTERNATIONAL TRADE

Ray Shillito
Thomas Currier

Immunoassays in Agricultural Biotechnology, edited by Guomin Shan
Copyright © 2011 John Wiley & Sons, Inc.

15.1 INTRODUCTION

The main reasons for developing and using analytical methods to detect genetically engineered (or recombinant DNA) (GE or GM) traits in crops, grain, food, and animal feeds are for excluding unwanted materials from a consignment, for ensuring that the amount of such materials, if present, is below a threshold, and for ensuring purity of materials, particularly seed. All of these require that the methods are validated and fit for the assigned purpose. Such validation may be an internal process (where tests are only used internally in a single laboratory), a multi-laboratory test in a few laboratories, or a full international collaborative ring trial using many laboratories. Appropriate validation of methods is critical when applying them to materials that will be sold or traded to another party. If the test is not reliable and repeatable, the second party may obtain a different result from the first. This is particularly problematic where the levels of material being measured are low, in which case the largest source of error will be sampling, especially for particulate matter such as grains and seed.

Immunoassays are generally used for testing nonprocessed materials, such as grain and seed. They rely on the presence of a particular protein. If the protein is denatured due to processing, then an immunoassay method is not generally appropriate (is not fit for purpose). Thus, immunoassays tend to be used for seed and grain, rather than for processed products. They are also used in testing individual plants, although that is not the subject of this chapter.

In this chapter, we will refer to the solid-phase immunoassay as a lateral flow device (LFD), also known as lateral flow strip (LFS) or immunostrip.

15.2 USE OF IMMUNOASSAYS IN REGULATORY SYSTEMS

Immunoassays are not widely used as an aid to surveillance and enforcement of regulatory requirements. Exceptions to this include the use of immunoassay in the United States by the U.S. Federal Drug Administration (FDA) as a means of measuring the amount of a pesticidal protein (e.g., a *Bacillus thuringiensis* (Bt) protein) in plants, the use of lateral flow devices to measure the amount of certain commodities such as rice (e.g., by Canada), and the use of a LFD to test seed lots and/or plants in the field by some countries (notably, Argentina, Brazil, and Mexico).

In these cases, commercially available LFD kits have been adopted and validated as test methods. In the case of Brazil, LFD are now acceptable as a means of testing cottonseed lots for the presence of low levels of rDNA events. Brazilian National Biosafety Technical Commission (CTNBio) took the decision to set the limit of 1% for transgene presence in conventional cottonseed (on a seed count basis). Seed is also tested by government laboratories using real-time quantitative polymerase chain reaction (PCR).

15.3   IMPLICATIONS OF TESTING FOR TRADE

15.3.1   Use of LFD for Grain Testing to Meet Thresholds

Seed and grain are traded internationally. For example, maize (corn in the United States) is exported in large quantities from the major producing countries. In these exporting countries, primarily the United States, Brazil, and Argentina, agricultural GM technology is employed on a significant and growing percentage of acreage. New GM events may be approved in one country before being approved in others (asynchronous approval). It is industry practice not to commercialize a GM event until there is approval in major export markets. However, some markets do not yet have an efficient or fully functional system for reviewing applications for approval. As a result, some commercialization has occurred when certain export markets were not accepting of such material. In these cases, grains for export to those markets are typically tested in country to make sure they do not contain the unapproved events. In some cases, if a crop that is authorized in an exporting country is not authorized for use in the importing country (and is therefore illegal), then the presence of such material must be excluded (a policy termed zero tolerance). From a scientific perspective, it is not possible to test to zero unless every seed/grain is tested, and testing via immunochemical means is generally destructive, thus leaving no product to trade. Accordingly, zero tolerance stance on imports of feedstock containing unauthorized GM materials has already "practically stopped the import of maize gluten feed and corn distillers grain" into the European Union (EU) (AIC, 2008).

This situation makes the development of reliable, fit testing methods using appropriate technology critically important for trade. In addition, many large importing countries/trade blocks have imposed regulatory systems that require that GM products be labeled, especially for food use. In cases where consumer acceptance of the product is low, food processors and retailers have often avoided placing "GM-labeled" foods on the market, and on their store shelves. This attitude has recently been showing signs of changing, at least in some EU countries such as the United Kingdom (Waitrose, 2007; Tesco, 2009).

The reluctance to include GM materials in foods (and in some cases feed) conflicts with the fact that the major commodity sources of the maize and soybean raw ingredients contain significant amounts of, if not pure, GM-derived grain (>70% of the global soybean crop and 24% of the maize consist of GM varieties) (James, 2008). A significant identity-preserved market has therefore arisen to supply the needs of food processors and producers to source raw materials that have very low concentrations of GM materials. Such an identity-preserved market chain requires stringent segregation of grain, augmented by targeted testing to ensure that the segregation has achieved the desired goal of excluding the biotechnology products.

It is almost impossible, especially in the United States and other major exporters of biotechnology-derived crops to exclude all low-level presence of authorized biotechnology products from the product stream, but it is possible, at significant cost, to reduce inclusion rates to a level that avoids a regulated consumer label in the intended market. It must be recognized that identity-preserved systems incur significant costs in segregating the grain. This begins with certified planting seed that has

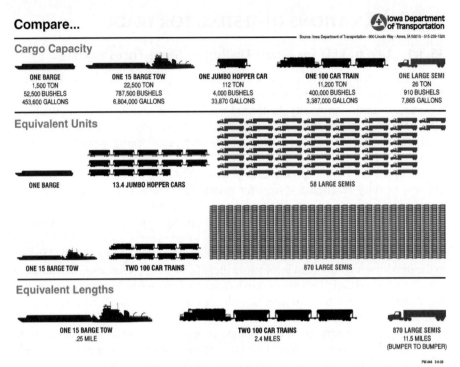

Figure 15.1 Comparison of the capacity of trucks, rail cars, and barges. (*Source*: Compare chart. © 2008, Iowa Department of Transportation.)

been shown to contain no more than very low levels of biotechnology products, through to testing that may be performed by the end user (e.g., food or feed producer).

The costs of testing are significant. For example, every truck delivering grain from the contract farmer to the elevator dealing in identity-preserved grain may be tested. The loads are then consolidated into trains or barges for transport to the ports (Figure 15.1; Compare, 2008). To load one barge (1500 tons) requires in the range of 50 large truckloads (termed semi's in Figure 15.1). A grain shipment may consist of up to 65,000 tons of grain (a large Panamax vessel); this requires about 43 barges, or more than five 100-car trains. This means that about 2166 large trucks full of grain are consolidated to create one shipload, and these must be handled through the grain channels via train and/or barge. Most individual shipments are smaller than this, although 50,000 ton shipments are common (~1666 trucks). In addition to testing each truckload, tests may also be performed on the barge loads and again at other stages of the transport chain.

Many importing countries or trading blocks, such as the EU, have set the standard for tolerance levels using the PCR as the criterion and use real-time quantitative PCR (RTQPCR) to enforce the regulations. A typical cost for such a test in the United States is in the range of $150–300 and this can be more expensive in other regions (e.g., Brazil where the cost was recently US$ 400). Thus, PCR-based testing is prohibitively expensive for this sort of application. In addition, multiple tests may be required to make sure that all the possible biotechnology events have been excluded from the lot, further increasing the costs. Such testing costs are passed to the end consumer who is requesting "GM-free" materials. Using PCR, the costs of the testing every truck used to fill one ship ($250,000–1,000,000) would significantly impact the final cost to the customer of the food and feed made from the grain. In addition to this obvious economic cost, the tests must be carried out quickly (within 10–15 min) in order to make a decision whether or not to accept and unload the truck or redirect it to another market. PCR-based testing is unable to meet these time constraints and therefore is not the most appropriate technology for such an application. In comparison, the cost of an immunoassay-based test (LFD or ELISA) is about 100-fold lower than that of a PCR test and much quicker in providing the needed results. For these reasons, PCR is neither a practicable nor a cost-efficient method for testing at point of delivery, and immunoassays have therefore become the *de facto* workhorse of the identity-preserved grain trade. They are used for quality control of the grain delivered by the farmer, and for some quality control downstream of the initial grain delivery.

15.3.2 Use of Immunoassays to Meet Seed Thresholds

Seeds are traded internationally, and from 1997 to 2007 international trade in seeds has increased sixfold (ISF, 2009; Figure 15.2) in U.S. dollar terms. Many of the major seed producers are also located in countries that grow transgenic crops (ISF, 2009; Table 15.1), resulting in an increased need for reliable, fast, and inexpensive testing methods to ensure seed purity.

Immunoassays meet these requirements most of the time and are extensively used in seed testing. Seed standards such as those administered by ISTA (www. seedtest.org) are measured in terms of number of seeds. Thresholds are common for conventional seed quality and purity, but have yet to be established on a global or national scale for GM seeds. In some cases of GM seeds, the amount of protein expressed per seed may be the same from seed to seed (repeatable), whereas in others it may vary somewhat depending on the individual seed or due to the conditions under which it was grown. These variations should be understood and the immunoassays validated using seed showing these variable levels of expression. Used properly, an

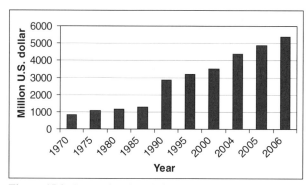

Figure 15.2 International trade in seeds, 1997–2006. (*Source*: ISF, 2009.)

TABLE 15.1 Top 20 Countries in Value (US$) of Seed Exported (2007)

Country	Agricultural seeds[a]	Vegetable seeds	Total
Netherlands	186	854	1040
United States[b]	650	369	1019
France[b]	698	216	914
Germany[b]	442	41	483
Canada[b]	265	82	347
Denmark	281	44	325
Chile[b]	124	80	204
Hungary	186	10	196
Italy	114	70	184
Mexico	162	9	171
Belgium	139	3	142
Argentina[b]	97	21	118
Austria	102	3	105
Japan	30	71	101
Spain[b]	54	35	89
China[b]	41	47	88
Australia[b]	66	13	79
United Kingdom	44	28	72
Israel	9	62	71
Brazil[b]	45	8	53

[a]Agricultural seeds are row crops and other crops grown on a large scale, such as maize, soybean, alfalfa, wheat, and rice.

[b]Countries with significant biotechnology plantings in 2007

immunoassay can give a reliable measure of the percentage of GM seeds in a sample, especially for low levels of material, and particularly if a subsampling approach is used in conjunction with a statistical model based on probability theory such as Seedcalc that is an Excel Spreadsheet using a binomial distribution function (Remund et al., 2001; Laffont et al., 2005). This is analogous to the use of such methods to determine the (low) percentage of an event in a grain sample. In addition, a large number of immunoassays are used in seed testing to determine seed purity; purity can be measured by testing a chosen number of either individual seeds or germinating seedlings from a seed lot.

15.3.3 Use to Determine Low-Level Presence of Seeds or Plants in a Sample

Immunoassays (predominantly LFD) are used extensively in testing seed lots for the presence of low levels of nondesired seeds in a lot. The LFD is suitable for this use (where validated) and is sold commercially. The application is very similar to that for grain and the results are interpreted using Seedcalc or other appropriate statistical approaches.

In addition, LFD and the enzyme-linked immunosorbent assay (ELISA) are used to test leaf samples for the presence of undesired traits, or to determine in the field which plants carry the desired trait, for example, during breeding. In the first case, leaves from individual plants can be bulked together and the samples sent to a laboratory where the ELISA or LFD test is carried out. In the second case, the individual plants are either tested using an LFD in the field, which takes only a few minutes, or the samples can be sent to a testing laboratory.

15.4 USE IN ADHERING TO FOOD AND FEED LABELING REGULATIONS/THRESHOLDS

15.4.1 Loss of Immunoreactivity During Processing

During processing, proteins present in the matrix become denatured, and thereby lose their tertiary (three-dimensional) structure. As a result, recognition of the proteins by antibodies (the basis of immunoassays) in most test kits cannot occur. This issue is discussed in more detail in Chapters 9 and 11. Therefore, highly processed foods are generally not suitable for assay by immunoassay. An exception is the immunoassay developed for Roundup Ready® Soy.[1] The ELISA and lateral flow immunoassays marketed by Strategic Diagnostics Inc. as GMOChek™ RUR Soya Meal Test Kit and TraitChek® RUR Toasted Meal Kit, respectively, are able to recognize the CP4 EPSPS protein in soybean toasted meal, concentrates and isolates, as the antibodies recognize both the native and denatured forms of the protein found in this material.

Due to the need for the tertiary structure, the sensitivity of immunoassays will often be affected by processing. The majority of immunoassays are designed to be sensitive to the native, nondenatured proteins and can be validly used to assay foodstuffs that are not highly processed. An example was the use of immunoassays to detect the Cry9C protein in some maize-derived foods that were not highly processed (Shillito et al., 2001). However, the more highly processed a food is, the less able immunoassays are to detect the target protein, and therefore the limit of detection of the assay for the protein will be increased as the food is processed more extensively. The measurement of the percentage of the transgenic material in food is therefore often not reliable (except for the RUR assay mentioned above), and immunoassays are unreliable for measurement of content against a threshold for highly processed foods. For similar reasons, immunoassays are infrequently used in the testing of processed feeds, although they are used for testing of incoming grains at elevators that then are used to make up the finished feed.

Thus, immunoassays are used primarily as a first screen of incoming grain, which may then be tested again by PCR before processing into products, and then further tested by PCR during the processing phases.

---

[1] Roundup Ready® is a trademark of Monsanto Company.

15.5   GLOBAL HARMONIZATION EFFORTS

A number of organizations are engaged in harmonization of immunoassay methods for specific analytes or matrices; however, their approaches, processes, and methods are not always consistent. However, communication between International Organization for Standardization (ISO), Codex Alimentarius, International Seed Testing Association (ISTA), and other organizations is ongoing in an attempt to improve consistency in the use of such assays along the seed, grain, feed, and food chain. Organizations such as CropLife International (CLI) are supportive of these harmonization activities.

15.5.1   Codex Alimentarius

Codex has been interested for several years in the harmonization of methods for the testing of food and food products for the presence of products of modern biotechnology (rDNA products). The Codex Ad Hoc Intergovernmental Task Force on Biotechnology requested input on criteria for validation of methods and agreed to a collaborative exchange with the Codex Committee on Methods and Sampling (CCMAS) in 2000. A working group of CCMAS concluded that existing detection methods had not been validated, and in 2004 initiated work on a document entitled, "Consideration of the methods for the detection and identification of foods derived from biotechnology: general approach and criteria for the methods." This draft document has been commented on by countries and updated several times over the intervening years (CX/MAS, 2009). In 2008, CCMAS requested that the document be accepted as new work and entered into the formal step process for development of Codex reference guidelines. The 31st Session of Codex Alimentarius Commission (ALINORM 08/31/ REP, 2009) subsequently approved this work, supporting the Project Document, "Proposal of new work for the development of guidelines on criteria for methods for the detection and identification of foods derived from biotechnology."

The purpose of this new work was to give guidance to Codex member countries and industries on the use of methods for the detection and identification of foods, including those derived from biotechnology. Existing Codex guidelines did not provide guidance on the use of immunoassays to detect large molecules such as proteins, and they were therefore included in this document. As of 2009, there existed a significant amount of scientific opinion that the draft guidelines needed to focus on DNA- and immunoassay-based methods of analysis, rather than specifically on products of biotechnology. In March 2010, CCMAS agreed on a final guidance document for detection methods for specific DNA sequences and specific proteins (ALINORM 10/33/23, 2010), and it was adopted by Codex Alimentarius Commission at their June 2010 meeting.

15.5.2   International Organization for Standardization

ISO developed a standard covering the use of immunoassays for the detection of GM products (termed GMOs within ISO) in plants. This standard, "ISO 21572—Foodstuffs—Methods for the detection of genetically modified organisms and derived products—Protein based methods," was developed by the European Standards Center (CEN) under the Vienna agreement (Vienna, 1990) and covers the use of ELISA methods such as sandwich assays. It does not include LFD-based methods. However, as of early 2009, the standard is being reviewed and updated and it is probable that LFD-based methods will be included in the next iteration of this standard.

In 2008, ISO formed a new subcommittee (SC 16) under Technical Committee 34 (TC 34—Food and food products). This subcommittee, ISO TC (2008), covers biomolecular testing and is charged with the development and review of standards covering testing for the presence of GM products in food, feed, and seed, and is chaired by the United States. The SC 16 scope is "Standardization of biomolecular testing methods applied to foods, feeds, and seeds and other propagules of food and feed crops." The scope includes methods that analyze nucleic acids (e.g., PCR, genotypic analysis, and sequencing), proteins (e.g., ELISA), and other suitable methods. The scope also includes variety identification and detection of plant pathogens. Because food microbiological methods are covered in another subcommittee (TC 34/SC 9), the scope of SC 16 does not include those methods.

15.5.3 International Seed Testing Association

ISTA published a position paper in 2001 on "Methods for the detection, identification and quantification of genetically modified seed in conventional seed lots" (ISTA, 2001). The strategy focused on the use of PCR although it did leave the option for ISTA to consider different methods such as bioassays and protein-based methods, as well as DNA-based methods. ISTA subsequently set up a "GMO Task Force" in 2002. The GMO Task Force was centrally involved in the elaboration of new information within "Chapter 8" of the ISTA rules, which, since 2006, covers biomolecular tests for varietal purity and testing for specific impurities (e.g., GMOs).

This ISTA GMO Task Force is also responsible for proficiency tests of laboratories for the ability to detect GM products in seed. Until recently, this focused only on low-level presence of such products, and most analyses were carried out using PCR. More recent proficiency rounds have included samples with high percentages of GM seeds to test proficiency in purity testing. Immunoassay methods can be used to test such samples both for the low-level presence of GM seeds and for purity testing. However, the focus of most participants of the ISTA proficiency rounds looking for low-level presence has been on using PCR. This may reflect the predominance of EU laboratories in these proficiency rounds. These proficiency tests are a critical part of the accreditation program for laboratories involved in the testing of seed of biotechnology products.

In addition, this ISTA Task Force promotes information exchange in the form of a webpage and workshops (ISTA, 2009). The ISTA Task Force is also interested in reference material distribution in the framework of the performance-based approach to GM seed testing, although no reference materials have been distributed by ISTA.

A major contribution of ISTA is in the development of the Seedcalc application for Microsoft Excel. This application can be used to develop testing plans, and assess producer and consumer (customer) risk (probability of Type I or Type II errors) when developing testing plans. The calculations that Seedcalc performs are critical to the application of subsampling regimes for determining the presence of low levels of GM materials in seeds and grain, and in devising purity testing schemes. In version 7.1, it includes modules for zero tolerance testing and for estimating the uncertainty of the results of multiple real-time quantitative PCR methods where multiple subsamples are analyzed. Tables for use in devising statistical sampling plans are also available from the U.S. Department of Agriculture Grain Inspection Packers and Stockyards Administration (USDA GIPSA, 2009).

15.5.4 AEIC

The Analytical Excellence through Industry Collaboration (AEIC, formerly the Analytical Environmental Immunochemical Consortium, www.aeicbiotech.org) meets twice per year in the United States to discuss issues surrounding harmonization of analytical methods. Its members include the producers of GM crops, the grain industry, testing laboratories, and test kit manufacturers. In addition, government laboratory scientists also attend. One of the goals of this consortium is to increase the quality of immunoassay-based test methods. They have published two papers on the proper use of immunoassays (Lipton et al., 2000; Grothaus et al., 2006), and members are actively involved in ISO, Codex, and other harmonization activities. The organization's Web site offers downloadable training documents and presentations that explain sampling and the appropriate use of immunoassays and PCR as detection methods.

15.5.5 AACC International

AACC International (AACC, formerly the American Association of Cereal Chemists) has a Biotechnology Committee that has been active in the performance of multiple site validations of immunoassay methods. Such international collaborative studies (Horwitz, 1988) are the standard for full international acceptance of methods. So far, three methods have been completed: one ELISA for Mon810 (AACC Official method 1110) and two for Cry9C ELISA methods (AACC 1120 and 1121) for the quantification of Starlink™. The latter are presently being reviewed and are expected to become AACC official methods in the near future. As of the time of writing, there are no plans to carry out further international collaborative studies of immunochemical methods for detection of GM crops products.

The AACC also administered a check-sample program (proficiency samples) using Roundup Ready® Soy, LibertyLink®,[2] StarLink, and Mon810 corn in 2001 and 2002. This fee-based program has now been discontinued due to the availability of the free GIPSA proficiency program.

The AACC Biotechnology Committee has been active in helping ISO with the formulation of positions on proposed standards and supports the U.S. Technical Advisory Group and activities of ISO/TC 34/SC 16.

---

[2] LibertyLink® is a registered trademark of Bayer.

In 2002, AACC organized training workshops on detection methods for the U.S. FDA and EPA, and has been active with International Life Sciences Institute (ILSI) in organizing workshops on sampling and detection methods for crops and foods containing GM products in many countries (see below). AACC played an important role in organizing a workshop for attendees at the 2004 Meeting of the CCMAS.

15.5.6 International Life Sciences Institute

Since 2004, the ILSI International Food Biotechnology Committee (IFBiC) has joined with AACC and other groups in organizing a series of sampling and detection methods workshops in Brazil, Argentina, India, Singapore, China, Chile, and Colombia. These educational workshops often involve a practical hands-on session that teaches the use of immunochemical methods to detect grain containing GM products. The workshops are organized in conjunction with the local ILSI branch organizations. They include participants from governments, academia, and industry, are often regional in nature and organized in conjunction with local regulatory authorities. IFBiC has also sponsored a number of lectures at international meetings that also covered the use of immunochemical methods to detect GM products.

In 2007, IFBiC sponsored a report on the state of harmonization of methods (Bridges, 2007) and held a workshop to facilitate discussions about harmonization between those government laboratories in Canada, the United States, and Mexico that carry out testing for GM products. The workshop participants specifically recognized the need for consistent test results for a product through the food supply chain, and that the testing marketplace for GM needs standards and standardization, that all methods are not created equal, and that proper sampling is critical.

15.5.7 CropLife International

CropLife International (www.croplife.org) is a global federation representing the plant science industry and a network of regional and national associations in 91 countries. CLI has a Detection Methods Project Team that seeks the implementation of harmonized and practical laws, regulations, or policies for the development, validation, and use of detection methods for plant biotechnology products including the production, distribution, and use of their reference materials by regulatory authorities globally.

15.5.8 Other Organizations

National standards institutes have not been particularly active themselves in developing standards for immunoassay-based testing, except that they have provided input to the ISO and Codex activities.

The EU has concentrated primarily on the use of PCR methodologies for detecting GM products. This supports Regulation EC 1829/2003 and related guidance documents that require the use of PCR as the regulatory control method. However, some European institutions (Van den Bulcke et al., 2007) have recently shown some interest in better understanding the possible applications of immunoassay-based methods as evidenced by a Co-Extra project on protein-based GM testing (Co-Extra, 2009).

In addition, some interest in harmonization of methods for detection of GM seeds, food, and feed has been expressed in the International Bureau of Weights and Measures (Bureau Internationale de Poids et Measures, BIPM). However, no action by BIPM is planned as of 2009 (R. Wielgos, personal communication).

15.6 PROFICIENCY TESTING

Validation of ELISA and LFD methods is routinely carried out by the manufacturers of such tests. In addition, users of the tests should carry out internal validations (e.g., as required for conformance to ISO 17025), before using them. The USDA GIPSA laboratory also certifies LFD methods for use in the United States.

As described above, AACC International carried out validation studies of three ELISA methods in 2000 and 2001. Such validations are time and resource intensive. At present, there is no other active program to validate immunochemical methods internationally.

Proficiency testing consists of an organizer (administrator) sending out samples of known nominal value (concentration) to multiple laboratories. The assay results are returned to the administrator and compared to the expected, and the average value reported by the laboratories taking part. Participating laboratories then receive summary information including the correct values and comparison of their performance to the other laboratories in the proficiency round. These results are used to improve the performance of laboratories and are an important part of ensuring harmonization of test results globally.

There are several active proficiency testing (check sample) programs available to laboratories that are carrying out GM analyses. The largest of these is administered by the USDA GIPSA laboratory. This program is offered two times a year and supplies samples of ground corn and soy to laboratories that request the samples. More than 170 laboratories have participated in this program, and between 40 and 70 typically take part in each cycle of testing (Ron Jenkins, GIPSA, personal communication). The majority of participants carry out the analyses using PCR. For example, in the October 2007 round of testing (USDA GIPSA, 2007), 42 laboratories submitted qualitative results based on PCR, and only 5 laboratories submitted immunoassay-based results. As the material sent was powder rather than grain, subsampling techniques could not be used. Therefore, many of the samples were of too low a concentration to be detected using available immunoassays. However, the report stated that "Detection by lateral flow strips displayed good overall accuracy." In the 2008 round (USDA GIPSA, 2008), only three laboratories submitted immunoassay-based results, and these were 100% correct.

As described above, ISTA also administers a proficiency program for seed, but is strongly focused on low-level presence and the samples are not generally assayed using immunoassays.

15.7 INTERNATIONAL CONSIDERATIONS

As most international shipments are governed by contract and are assessed using PCR methods, immunoassays play a minor role in current international once the product reaches the export location. The existence of different regulatory regimes worldwide is further complicated by differing interpretations of the results of the methods. Although immunoassays are not used to a major extent in international trade, the uncertainty of PCR results and the application of nonstandard sampling plans have the potential to lead to differing conclusions and consequently to different regulatory outcomes. Choice of analytical method and reporting of results (e.g., less than defined limit of detection, present but not quantifiable, etc.) is most critical for DNA-based testing methods, as these are generally used in regulatory situations.

15.8 CERTIFICATION OF PROTEIN-BASED METHODS

Once assays are developed, stable test kits are produced and marketed by test kit suppliers. As stated above, these test kits are extensively tested by the manufacturers both during development and before release of each lot. Commercial kits are also used and validated for specific applications in users' laboratories, and by authorities in certain countries such as the United States (USDA GIPSA, 2004).

15.9 COSTS OF TESTING

Estimates of the costs of testing vary widely. Immunochemical assays generally have lower costs than testing via PCR methods. Test kits are produced and marketed by test kit suppliers, although some large organizations may prepare their own. Commercially produced kits are generally more reliable and convenient. As stated above, the cost of an immunoassay may be lower than that for PCR when applied to grain and seed, although in high-throughput situations, PCR costs can be reduced so as to be comparable. Developments that reduce the labor and handling costs and increased automation will also affect the cost. Thus, costs are quite variable and depend on the organization that is applying them and on the number of tests carried out. Although highly variable, LFD and ELISA costs, when applied to a small number of samples, tend to run into the range of a few U.S. dollars per sample to less than one U.S. dollar.

15.10 CONCLUSIONS

Immunochemical test methods are of great value to the grain handlers and particularly valuable and suited to the receipt of grain at elevators. Immunochemical tests are also a critical and valuable tool for seed purity testing. They are fast, cost-effective, easy to use, and easily available. By reducing the costs of identity-preserved grain, they reduce the cost to consumers.

Thus, immunochemical-based testing is a critical and effective tool for the whole seed, grain, and food and feed chain, and an important method for the detection of GM products for those whose task is to make sure that materials conform to international regulations.

REFERENCES

AIC Feed Executive Chairman Tony Bell quoted by E. Watson in "Food Manufacture," 2008, available at http://www.foodmanufacture.co.uk/news/printpage.php/aid/7122/Europe_s_animal_feed_is_running_-dry,_warn_importers.html, (accessed on April 20, 2009).

ALINORM 08/31/REP. Report of the 31st Session of Codex Alimentarius Commission, available at http://www.codexalimentarius.net/download/report/698/al31REPe.pdf (accessed on April 4, 2009).

ALINORM 10/33/23. Report of the 31st Session of the Codex Committee on Methods of Analysis and Sampling, Budapest, Hungary, March 8–12, 2010. Appendix III: Proposed Draft Guidelines on performance criteria and validation of methods for detection, identification and quantification of specific DNA sequences and specific proteins in food, available at http://www.codexalimentarius.net/web/index_en.jsp. (accessed on August 10, 2010).

Bridges, A. Sampling and detection methods for products of modern agricultural biotechnology in NAFTA countries. Background Paper for a Workshop, International Life Sciences Institute, Washington, DC, 2007.

CCMAS. Report of the 30th Session of the Codex Committee on Methods of Analysis and Sampling, Balatonalmádi, Hungary, March 9–13, 2009 (AL32_23E), available at http://www.codexalimentarius.net/download/report/720/al32_23e.pdf (accessed on April 20, 2009).

Co-Extra project number 220 "Protein based GMO testing," available at http://www.coextra.eu/projects/project220.html (accessed on July 7, 2009).

Compare chart. © 2008, Iowa Department of Transportation, Ames, IA, USA, available at www.iowadot.gov/compare.pdf, 2008.

CX/MAS 09/30/8 Proposed Draft Guidelines on criteria for the methods for the detection and identification of foods derived from biotechnology, available at ftp://ftp.fao.org/codex/ccmas30/ma30_08e.pdf (accessed on April 20, 2009).

Grothaus, G. D., Bandla, M., Currier, T., Giroux, R., Jenkins, G. R., Lipp, M., Shan, G., Stave, J. W., and Pantella, V. Immunoassay as an analytical tool in agricultural biotechnology. *J. AOAC Int.*, **2006**, *89*, 913–928.

Horwitz, W. Protocol for the design, conduct and interpretation of method performance studies. *Pure Appl. Chem.*, **1988**, *60*, 855–864.

ISF. Growth in the International Seed Trade. International Seed Federation, Switzerland, available at www.worldseed.org (accessed on May 22, 2009).

ISO/TC 34/SC 16. Horizontal methods for molecular biomarker analysis, 2008.

ISTA's strategy regarding methods for the detection, identification and quantification of genetically modified seed in conventional seed lots, 2001, available at http://www.seedtest.org/upload/cms/user/42ISTAPositionPaperonGMOapproved14.11.2001.pdf (accessed on April 20, 2009).

ISTA. Information platform for GM seed: International Seed Technologists Association, available at http://www.seedtest.org/en/info_platform_for_gm_seed_content---1--1195.html (accessed on April 24, 2009).

James, C. *Global Status of Commercialized Biotech/GM Crops*. ISAAA Brief No. 39. ISAAA, Ithaca, NY, 2008.

Laffont J. L., Remund, K. M., Wright, D. L., Simpson, R. D., and Grégoire, S. Testing for adventitious presence of transgenic material in conventional seed or grain lots using quantitative laboratory methods: statistical procedures and their implementation. *Seed Sci. Res.*, **2005**, *15*, 197–204.

Lipton, C. R., Dautlick, X., Grothaus, G. D., Hunst, P. L., Magin, K. M., Mihaliak, C. A., Ubio, F. M., and Stave, J. W. Guidelines for the validation and use of immunoassays for determination of introduced proteins in biotechnology enhanced crops and derived food ingredients. *Food Agric. Immunol.*, **2000**, *12*, 153–164.

Remund, K. M., Dixon, D. A., Wright, D. L., and Holden, L. R. Perspectives in seed technology: statistical considerations in seed purity testing for transgenic traits. *Seed Sci. Res.*, **2001**, *11*, 101–120.

Shillito, R. D., MacIntosh, S., and Kowite, W. J.Detection of Cry9C protein in dry milled, wet milled and masa processed fractions and processed foods made from 100% StarLink™ grain. Aventis report CM00B01418 April, 18 April 2001, available at http://www.epa.gov/oppbppd1/biopesticides/pips/old/stlink/b003244_cm00B014-final.pdf (accessed on April 4, 2009).

Tesco CEO Sir Terry Leahy, Speaking at the Annual City Food Lecture, held in London, February 2009, available at http://greenbio.checkbiotech.org/news/leahy_says_supermarkets_too_hasty_judging_gm (accessed on April 20, 2009).

USDA GIPSA Program Directive 9181.2, Performance verification of qualitative mycotoxin and biotech rapid test kits, 2004, available at http://www.gipsa.usda.gov/GIPSA/documents/GIPSA_Documents/9181-2.pdf (accessed on April 24, 2009).

USDA GIPSA. Proficiency program: testing for the presence of biotechnology events in corn and soybeans. October 2007 sample distribution results, available at http://archive.gipsa.usda.gov/biotech/quarterly_reports/oct_2007_final_report.pdf (accessed on April 24, 2009).

USDA/GIPSA proficiency program: testing for the presence of biotechnology events in corn and soybeans. May 2008 sample distribution results, available at http://archive.gipsa.usda.gov/biotech/quarterly_reports/May_2008_final_report.pdf (accessed on April 24, 2009).

USDA GIPSA. Sampling for the detection of biotech grains, available at http://archive.gipsa.usda.gov/biotech/sample2.htm (accessed on April 24, 2009).

Van den Bulcke, M., De Schrijver, A., De Bernardi, D., Devos, Y., MbongoMbella, G., Leunda Casi, A., Moens, W., and Sneyers, M. Detection of genetically modified plant products by protein strip testing: an evaluation of real-life samples. *Eur. Food Res. Technol.*, **2007**, *225*, 49–57.

Vienna Agreement. Agreement on technical co-operation between ISO and CEN, 1990, available at http://publicaa.ansi.org/sites/apdl/Documents/Standards%20Activities/Background%20Papers/Supporting%20Documents/ISOCEN%20VA.pdf (accessed on April 20, 2009).

Waitrose Managing Director Mark Price, on BBC Hardtalk, 26 October 2007, available at http://www.bbc.co.uk/newsa/n5ctrl/progs/07/hardtalk/price_26oct.ram, (accessed on April 20, 2009).

CHAPTER *16*

FUTURE PERSPECTIVES AND CHALLENGES

Zi Lucy Liu
Ai-Guo Gao
Leslie A. Harrison
Kerrm Y. Yau
John Lawry
Guomin Shan

16.1 INTRODUCTION

Immunoassays have proven to be a great protein detection method, which have met numerous needs of the agricultural biotechnology field in the past 15 years. At present, genetically engineered (GE or GM) technology has entered a new era featuring more complex and sophisticated biotech products, more stringent regulatory requirements, and higher standards of the quality control. This has brought new challenges for the detection methods that are critical for new trait discovery, product development, seed production, and product stewardship. In general, there are four challenges a new analytical technology has to face: sensitivity, specificity, speed, and adaptability. First, the zero tolerance policy for any unintended GM traits in almost every country

Immunoassays in Agricultural Biotechnology, edited by Guomin Shan
Copyright © 2011 John Wiley & Sons, Inc.

requires a detection sensitivity of GM presence at 0.1% or lower, and this is currently predominated by the polymerase chain reaction (PCR) method. Can an immunoassay reach such level of sensitivity? Second, some GM products may only have a single or a few amino acids difference from plant endogenous protein such as GA21 protein (a modified maize EPSPS protein); a highly specific assay is a basic requirement for every phase of product development. It is not an easy task to develop such a specific immunoassay. Third, during the discovery and product development, from event sorting of trait development to seed production, a high-throughput assay is required. Recently, stacking of multiple traits in a single crop product has become a common strategy and practice in agricultural biotechnology. Some products may contain eight or more genes, which require an analytical technology to be able to detect multiple proteins in a single assay. With all the aforementioned challenges, an ideal new technology should be simple, rapid, sensitive, nondestructive, semiquantitative, and suitable for field use. In addition, some proteins such as membrane-bound proteins are extremely difficult to be extracted and an immunoassay may not be applicable. In the past few years, some new technologies have emerged to assist researchers tackling these challenges. In this chapter, a brief review of the development of new technologies such as immuno-PCR (I-PCR), biolayer interferometry (BLI), aptamer technology, and molecularly imprinted polymer (MIP). Immuno-PCR aims for ultrasensitive detection by taking the advantage of amplification from nucleic acid polymer chain reaction. Biolayer interferometry technology is simple, fast, and label-free, and enables real-time detection. Both aptamer and MIP technologies aim to produce specific antibody-like binding partners in animal-free systems. To effectively detect multiple proteins in the plant matrices, multiplexing immunoassay and LC/MS/MS are the leading technologies, which have been actively researched in recent years and have shown promise for use in agricultural biotechnology.

16.2 IMMUNO-PCR

First described by Sano et al. (1992), immuno-PCR is a technology for the ultrasensitive analysis of proteins, including biopharmaceutical compounds, therapeutic antibodies, and cytokines. It combines well-established ELISA methodology with the signal amplification power of polymerase chain reaction. Extensive reviews from Dr. Barletta (2006) and Dr. Niemeyer's group (Alder et al., 2008) in Germany have recently been published and provided detail information on this protein detection technology.

The basic concept of I-PCR is a conventional ELISA with a twist. The antigen recognition power of an antibody to specifically detect the target molecule remains the same, but a fragment of double-stranded DNA that can be amplified and quantified using quantitative PCR technology has replaced the enzyme part of the assay. The assay procedure can be divided into two phases: the initial phase is identical to a conventional ELISA, that is, target molecule immobilization and separation of unbound detector antibody. The binding of detector antibody to target molecule, instead of being detected by an enzyme as commonly used in ELISA, is registered by a DNA sequence conjugated to the detector antibody. This DNA-containing molecule has similar functions as in ELISA; that is, it recognizes specifically the first binding event and at the same time provides a reporter molecule, in this case DNA. This piece of DNA then enters the second phase of the assay, amplification by PCR to a detectable level for recording and quantifying the amount of DNA template. As a consequence, the amplified signal is correlated to the numbers of binding events, hence the quantity of the target antigen. The limit of detection (LOD) is claimed to be enhanced 100–10,000-fold (Alder et al., 2008).

Due to use in discovery, clinical, and environmental applications, there are many commercial technology providers. University development in this area tends to be less of a driver than commercial developers since the concepts are straightforward. Potential hurdles encountered in development include reagent creation, signal generation and detection, and background noise reduction to improve assay performance. Using commercially available reagents and/or services simplifies the development process. The performance of the assay, however, is highly dependent on the applications, for example, formats of the assay, available instruments, required sensitivity, sample matrix, and so on. Just like any other immuno-based assays, optimizations and validations are required before routine use.

16.3 BIOLAYER INTERFEROMETRY TO SURFACE PLASMON RESONANCE

The biolayer interferometry is a proprietary technology in the Octet system (offered by the FortéBio) used to measure the binding of both large and small molecules to small biosensor probes and to measure protein–protein interactions (Abdiche et al., 2008). Similar to the surface plasmon resonance (SPR) instruments in the market (e.g., Biacore 3000 or ProteOn XPR36) that provide label-free real-time analysis, BLI uses a different detection system. Both technologies use optic systems that measure the change in optical properties of light during the binding events on the sensor surface. SPR measures the change in total internal reflection angle (see Chapter 2), while BLI measures the change in interference on reflected light. In addition, while the SPR system uses microfluidics to handle samples and buffer, the BLI system brings the sensor probe to sample. In that sense, the dip-and-read format in BLI is designed with multiple probes (from 8 to 12 probes at a time) to facilitate high-throughput analysis, versus the single sample analysis in SPR. Unlike the SPR, however, the probes in BLI are used only once for a single measurement, and then discarded. SPR, on the other hand, allows for the regeneration of the chip after a measurement so that it can be used repeatedly, in many cases up to 20–30 times. SPR can attain higher sensitivity and is better suited for kinetic analysis (Abdiche et al., 2008).

Binding measurements using BLI are very fast, able to determine concentrations of protein samples in as little as 3 min (data not shown). This is in contrast to an ELISA assay that may require many hours of incubation and washing procedures, and to SPR that requires about 2 h for eight measurements. BLI assays can be rapidly developed, as opposed to ELISA assays that require extensive development time. Although not as sensitive as the Biacore, it is most likely sufficiently sensitive for screening expression experiments and similar applications. In addition, BLI takes much less engineering time compared to lateral flow devices and is therefore more appropriate for discovery screening campaigns.

16.4 APTAMER TECHNOLOGY

When the detection system requires a biomolecular recognition event, antibody-based detection methodologies are still considered the standard in environmental, food, and clinical assays. These assays are well established, and they have demonstrated sufficient sensitivity and selectivity. However, the use of antibodies in multianalyte detection methods and in the analysis of very complex sample matrices may encounter limitations deriving from the nature of these protein receptors. In order to circumvent some of these drawbacks, other recognition molecules are being explored as alternatives.

The knowledge that nucleic acids, RNAs in particular, can assume stable tertiary structures and can be easily synthesized and functionalized has led to the idea of selecting new nucleic acids ligands called aptamers. Aptamers are single-stranded DNA or RNA ligands that can be selected for different targets starting from a huge library of molecules containing randomly created sequences (Tombelli et al., 2005). The selection process is called systematic evolution of ligands by exponential enrichment (SELEX), first reported in 1990 (Ellington and Szostak, 1990; Tuerk and Gold, 1990). The SELEX process involves iterative cycles of selection and amplification starting from a large library of oligonucleotides with different sequences (generally 10^{15} different structures). After incubation with immobilized specific targets and partitioning of the bound from the unbound, the oligonucleotides selected are amplified to create a new mixture enriched in those nucleic acid sequences with a higher affinity for the target. After several cycles of the selection process, the pool is enriched in the high-affinity sequences at the expense of the low-affinity binders. Unlimited amounts of the aptamer can easily be achieved by chemical synthesis and modified to introduce enhanced stability, affinity, and specificity.

An IgE-specific aptamer with sensitivity of 0.5 nM, comparable to an antibody-based biosensor, exhibits a 10-fold higher dynamic range (Liss et al., 2002). An aptamer piezoelectric biosensor for HIV Tat protein demonstrated better sensitivity than antibody-based detection (Minunni et al., 2004). Reports have been published for peptides (Baskerville et al., 1999), proteins (Klug et al., 1999; Wen et al., 2001), and also whole cells (Herr et al., 2006) or bacteria (Homann and Göringer, 1999). For the first time, an aptamer has been recently approved by the U.S. Food and Drug Administration for the clinical treatment of age-related ocular vascular disease (Ng et al., 2006). In bioanalytical chemistry, aptamers can be used for macromolecule purification through affinity chromatography (Clark and Remcho, 2002; Michaud et al., 2004).

Using aptamer and immuno-PCR technology, the sensitivity of protein detection and quantification can be greatly enhanced. For example, Fischer et al. (2008) identified and modified an aptamer that can bind to the protein thrombin with high affinity ($K_d \sim 75$ nM). More importantly, with the addition of a linker and primer binding sites, the oligonucleotide can be amplified and measured using quantitative PCR and/or rolling circle amplification (RCA). Similar to immuno-PCR (see the above section), the aptamer in this case serves as both the recognition and the reporting unit, thereby omitting the tedious conjugation steps required for protein–DNA complex synthesis. Because of the amplification of PCR, the sensitivity is also radically increased in detecting the thrombin molecules, with the detection dynamic range from 1 μM down to low pM concentrations.

16.5   MOLECULARLY IMPRINTED POLYMER (MIP)

Naturally occurring protein binders are proteinaceous molecules, for example, receptors, antibodies, or nucleic acids that are also involved in certain functions, such as catalysis and/or signal transduction. In an effort to mimic these binding properties, organic chemists have been able to produce polymer networks with specific recognition and binding sites that are complementary to a template, which may be either a biomolecule or a synthetic compound. This technique is referred to as molecular imprinting. The concept is similar to making a rubber mold to tightly fit the template. The product of molecular imprinting is called a molecularly imprinted polymer. The major difference between a mold and an MIP is that the latter allows a macromolecular template to penetrate the polymer network to access the binding pockets. MIPs are also known as synthetic polymeric receptors or robust artificial antibodies and therefore have been termed "plastibodies." MIPs have attracted attention owing to their inherent simplicity, robust polymer network, and cost-effectiveness. Recent reviews have extensive descriptions of the design of polymers for the synthesis of macromolecular recognition structures (Bossi et al., 2001, 2007; Ge and Turner, 2008). Compared to the affinity matrices prepared using cell receptors or mono- or polyclonal antibodies, the man-made protein imprinted polymers offer a highly stable, labor-extensive, cost-effective, and time-saving alternative to the existing techniques (Mallik et al., 1994). In order to use MIPs in diagnostics for biomolecules, several hurdles have to be overcome. The increased structural complexity and large size of a macromolecule lead to more nonspecific and heterogeneous binding sites, and as a result, the MIPs display poor recognition behavior. Owing to the complex properties of some biomacromolecules, such as proteins and cells, the conditions of biomacromolecular imprinting need to be close to their natural environment to ensure that their conformational integrities and binding activities are maintained. Different proteins, however, have been successfully imprinted using various technologies, albeit for proof of concept purposes, for example, hemoglobin (Guo et al., 2004), lysozyme (Ou et al., 2004), and bovine serum albumin (Pang et al., 2006; Hua et al., 2008). In the latter case, Hua et al. (2008) synthesized a BSA imprinted hydrogel in aqueous solution. The resultant polymer showed notable selectivity and binding capacity toward the target BSA molecules. It is important to note that the synthetic step of the hydrogel polymer, which is in aqueous solution, is compatible with protein and therefore more feasible for biological macromolecules. More importantly, the authors also demonstrated the use of the hydrogel as affinity matrix to successfully extract BSA from crude bovine serum without contamination of other serum proteins such as immunoglobulins or plasma proteins.

16.6 MULTIPLEXING IMMUNOASSAY

In pharmaceutical drug discovery, simultaneously and effectively detecting many proteins in a single sample is essential to provide and track the metabolic profiles of disease-associated changes, which may lead to the discovery of disease-associated biomarkers, thus assisting in diagnosis, prognosis, disease progress predicting, patient response monitoring, and possibly identifying drug candidates for therapeutic purposes (Joos et al., 2000; Liu et al., 2006; Kopf and Zharhary, 2007). In agricultural biotechnology, multiplexing assays are needed for cost-efficiency, gene stacking product development, and possible biomarker discovery. While some multiplexing technologies/platforms have been around for years in pharmaceutical industry, they are still relatively new in the agbiotech field. In this section, the current and emerging multiplexing technologies, their applications, and the remaining challenges in agricultural sciences will be briefly reviewed and discussed.

16.6.1 Protein Microarrays and Multiplexed Sandwich Immunoassays

16.6.1.1 Protein Arrays In the late 1980s, Roger Ekins described the basic principles of protein microarray technology (Ekins, 1989). In the past decade, with the further development of nucleic acid arrays for gene expression profiling of cells at the mRNA level, the interest in microarray-based proteomics has increased enormously (Albala and Humphrey-Smith, 1999; Templin et al., 2004). Primarily, protein microarrays can be divided into two classes: proteome array and reverse array (Nielsen and Geierstanger, 2004). In proteome arrays, a large number of purified proteins are immobilized onto the array, while in reverse arrays, antisera or lysate samples are immobilized onto the array. In agricultural biotech applications, arrays immobilized with antigens may be used for profiling the reactivity of antisera while reverse arrays containing immobilized antibodies can be very useful for the detection of multiple transgenic proteins. Two major types of antibody microarrays, planar microarrays or bead-based arrays, can be employed for multiplexed immunoassay development.

16.6.1.2 Planar Microarrays Theoretically, planar microarrays can be generated with hundreds or thousands of different capture spots (Templin et al., 2004). Making an antibody array requires a proper printing platform support, which assures that the antibody retains its functionality and stability after spotting. The most commonly used supports for antibody printing include chemically activated glass slides, glass slides coated with polymers (such as hydrogels or nitrocellulose polymers), or nitrocellulose membranes (Turtinen et al., 2004; Liu et al., 2006; Kopf and Zharhary, 2007). Detection of planar arrays can be performed by chemiluminescence, radioactivity, mass spectrometry, or fluorescence (Templin et al., 2004;

TABLE 16.1   Comparison of Multiplexing ELISA Platform

Technology	Microarray printer	Sensitivity	Dynamic range	Multiplexing
ELISA	No	~10 pg/mL	2–3 logs	None
Protein chips	Yes	Low pg/mL	3–4 logs	15–30
MSD	Yes	Low pg/mL	5 logs	15–20
Luminex	No	20–50 pg/mL	3 logs	Up to 10
SearchLight	Yes	Low pg/mL	3–4 logs	16

Kopf and Zharhary, 2007). The current technologies available for antibody printing are based on planar two-dimensional or three-dimensional supports onto which the antibodies are immobilized. The use of antibody arrays is mainly for the initial screening of large number of proteins to identify drug candidates or to discover biomarkers for disease/cancers. Antibody arrays may be used for monitoring changes in protein abundance, posttranslational modifications (such as phosphorylation, acetylation, glycosylation, or others), and protein–protein interactions in a complex environment. More recent developments with planar antibody microarrays in cancer research can be found in a review article by Kopf and Zharhary (2007). There are a number of companies with commercially available array technologies, kits, and/or services for comparative protein profiling, reverse screening, and cytokine profiling. MSD and SearchLight multiplexed ELISA platform are used as examples to discuss the technology principles below.

MSD Platform   Meso Scale Discovery® (MSD) is a multiarray-based technology with a proprietary combination of electrochemiluminescence (ECL) detection and patterned arrays. Electrodes are built into the bottom of the plate and are energized within the instrument. The electrochemical reaction occurs within the plate and light is measured through a CCD camera. The unique ECL detection system allows this technology to achieve the high sensitivity and broad dynamic range, which is crucial for detecting multiple proteins with different expression levels in the same assay. Up to 10 analytes can be multiplexed (Table 16.1). MSD can be easily adapted for automation and high-throughput analysis.

Pierce SearchLight Platform   SearchLight™ proteome arrays are quantitative, plate-based antibody multiplexed arrays derived from traditional ELISA. Up to 16 analytes per well can be measured simultaneously. SuperSignal ELISA Femto Maximum Sensitivity Substrate is added and the enzyme–substrate (HRP-Super-Signal) reaction produces a luminescent signal that is detected with the SearchLight Plus CCD imaging system (Table 16.1).

16.6.1.3 Bead-Based Arrays   In addition to the planar microarray-based systems, which are best suited for screening a large number of target proteins in drug discovery, bead-based assays are a promising alternative technology (Moody et al., 2001; Kellar and Douglass, 2003; Templin et al., 2004; Elshal and McCoy,

2006). Multiplexing bead array-based assays were first reported in 1977 (Horan and Wheeless, 1977). Recently, multiplexed bead array assays have been used for protein expression profiling such as for cytokines and cell signaling molecules (Moody et al., 2001; Templin et al., 2004). The multiplexing capability in bead-based arrays depends on the number of distinguishable beads. The separation of beads is conducted via color and/or size coding of capture beads. Fluorescence is the preferred detection method for bound analytes in bead-based microarray assays. So far, most of the commercially available bead-based protein microarrays are based on the Luminex 100™ or 200™ xMAP® platform.

Luminex® xMAP® is a fluorescent, flow cytometry-based microbead platform that allows a single sample to be tested simultaneously for multiple analytes. The technology utilizes antibodies conjugated to 5.6 μm fluorescent dye coded polystyrene spheres. The dual laser system includes the red laser, which excites the microsphere's internal dyes (two classification fluorophores), thus identifying and characterizing each sphere, and the green laser, which excites the reporter dye conjugated to the detection antibody for quantification. Theoretically, high number of targets can be simultaneously quantified (up to 500 with FlexMap3D that utilizes three different dyes to identify the beads). It is more cost-effective compared to many other multiplexing technologies requiring specialized services and/or reagents for detecting multiple proteins. One limitation of Luminex technology is the low throughput of the plate reading step. It takes about 30–45 min per plate for the Luminex 200™. If a large number of samples are being analyzed, the stability of resulting sandwich complex may become an issue in the course of reading many plates in one run.

16.6.1.4 Sandwich Microarrays: Multiplexed ELISAs

In addition to Luminex, SearchLight, and MSD, several other multiplexing technologies/platforms including QuanSys, GenTel, Oxonica, Illumina's MPL, and Panomics are also available or under development. Although there are different technology platforms, sandwich microarrays for multiplexing share at least one thing in common: they are based almost exclusively on the antibody–antigen reaction. The quality, sensitivity, and specificity of the assay depend mainly on the performance of the antibodies (Nielsen and Geierstanger, 2004). Since several second antibodies are applied simultaneously, cross-reactivity among these antibodies or from nonspecific proteins in the mixture may lead to false-positive readings or impact the dynamic range. Luminex for cytokine analysis has shown that anti-cytokine antibodies cross-react with other cytokines (Kellar and Douglass, 2003), cross-species antibodies (Earley et al., 2002; Kellar and Douglass, 2003), and other interfering substances (Pang et al., 2005).

Due to the cross-reactivity issues, the number of proteins being analyzed simultaneously in a multiplexing immunoassay may be limited. Literature reports that up to 40–50 sandwich assays can be run in a single microarray (MacBeath, 2002; Schweitzer and Kingsmore, 2002). However, small levels of cross-reactivity will reduce or severely restrict the dynamic range even after careful assay optimization (Horan and Wheeless, 1977; Ekins et al., 1990). Spotting different sets of capture antibodies into physically separated subarrays may avoid some of the cross-reactivities.

Compared to the conventional single sandwich ELISAs where a wide range of polyclonal antibodies, monoclonal antibodies, and recombinant antibodies can be used, the multiplexing immunoassay requires antibodies with high specificity and affinity (Mendoza et al., 1999; Moody et al., 2001).

16.6.2 Challenges of Multiplexed ELISA for Plant Transgenic Protein Analysis

With the increased number of stack crops containing multiple genes being developed, multiplexed ELISAs can play a significant role for plant transgenic protein analysis during product development and commercialization. A number of multiplexing assays or kits have been developed for pharmaceutical diagnostic and therapeutic applications, but few have been developed or are available for transgenic protein analysis in plant tissues. One application of multiplexing array in agricultural biotech is by applying a sandwich ELISA assay in 96- and 384-well format with standard laboratory equipment (such as liquid handling robots and automatic plates washer). In such an array platform, antibodies are printed on the solid surface (SearchLight and MSD) or attached to beads (Luminex) in a solution. After incubation with the sample, a labeled secondary antibody is applied and detected either by fluorescence or chemiluminescence.

A quantitative multiplexed immunoassay needs to demonstrate acceptable accuracy, precision, sensitivity, broad dynamic range, and suitability for rapid analysis as well as amenability to automation. The major challenge of multiplexed ELISA for plant transgenic protein analysis is the cross-reactivities between antibodies, from nontarget proteins in the assay mixture, or from other endogenous plant proteins with partial homology to the target proteins. Similar to regular sandwich ELISA, matrix effects can be problematic to multiplexing assays. Matrix effects are commonly caused by macromolecules such as protein, enzyme, lipid, starch, or other compounds (e.g., complex carbohydrates and phenol compounds) in the plant sample extracts. In addition, to achieve satisfactory extraction efficiency from a single extraction, identifying a common extraction buffer and method suitable for all target proteins is critical. Finally, widely varying expression levels of transgenic proteins in *plant* is another unique challenge for multiplexing assay. Some transgenic proteins may be expressed at levels of hundreds of *ppm* in plant tissue while others may only be expressed at *ppb* levels. Thus, a multiplexing assay with broad dynamic range is needed to minimize multiple sample dilutions.

16.6.3 Selection and Evaluation of Multiplexed ELISA Formats

Commercially available multiplexing technologies and platforms differ from each other in several aspects such as cost, multiplexing capability, and the amenability to automation and high throughput analysis (Ekins and Chu, 1991; MacBeath and Schreiber, 2000). While evaluating and implementing a multiplexing technology platform for plant transgenic protein analysis, other technical features of the platform need to be considered including sensitivity, accuracy, dynamic range, precision, specificity, and ease of use.

The standard sandwich ELISA format is the basis for the evaluation of multiplexing assays. A newly developed multiplexing assay needs to be validated and bridged with plate format ELISA methods. In developing a multiplexing assay for cytokines, variations between the two technologies were found likely to be the results of the different capture and detection antibodies used (Moody et al., 2001; Haab et al., 2001). Therefore, when developing a new multiplexing application, identical antibody pairs need to be considered for the initial feasibility testing to minimize this source of variation.

In summary, multiplexed protein analysis is a rapidly growing area that has promising potential to assist gene-stacked product development. Multiplexed immunoassays can not only improve throughput by greatly increasing data point delivery obtained from a single experiment, but also decrease expenditures by using reducing amounts of assay reagents and plant extracts.

16.7 MASS SPECTROMETRY FOR PROTEIN DETECTION

Basic bioresearch and clinical laboratories are utilizing mass spectrometers for research with increasing frequency, often complementing traditional immunoassay-based methodologies. The beginning of mass spectrometry use in today's biological research can, in large part, be traced to technological developments of the 1970s, which enabled nondestructive ionization of macromolecules such as proteins (Watson and Sparkman, 2007). Prior to these developments, proteins were excluded from mass spectrometry analysis because the sample ionization procedures required of all mass spectrometers at that time were incompatible with thermally labile proteins. The potential for mass spectrometry in biology was unleashed as the ability to "softly" ionize proteins became more refined. These advancements moved mass spectrometry analysis into the biological realm, and the subsequent evolution of mass spectrometry in general is largely related to its use in bioresearch applications.

A mass spectrometer operates by selecting individual ions according to a ratio known as the m/z ratio. The two parts of this ratio are the mass of an ion (m) and the charge state (z) the ion has during its travel through the instrument. There are different types of mass spectrometers allowing for different types of analytical questions to be answered, but all mass spectrometers utilize the m/z ratio of any given ion to execute a specified analysis. Two of the most commonly used mass spectrometer types in bioresearch are the quadrapole mass analyzer and the time-of-flight (TOF) mass analyzer. In general, quadrapole mass analyzers are commonly used for accurate quantitation of analytes while TOF mass analyzers are more capable of obtaining an accurate mass value for identification, especially for large protein analytes. This is a generalization, as both types of mass analyzers can perform either function when applied in the correct manner. A tandem mass spectrometer is a mass spectrometer containing two separate mass analyzers. These mass analyzers may be of the same type as in a "triple quad" that uses only quadrapole type mass analyzers, or of different type, as in a QTOF instrument that utilizes one quadrapole mass analyzer followed by a time-of-flight mass analyzer. A tandem format permits two separate rounds of mass analysis (MS^2 analysis) to occur on a single analyte. The first mass analyzer selects an ion (precursor ion) of a specific m/z value and directs it into a collision cell whereby the precursor ion fragments into product ions. The second mass analyzer then determines the m/z values of the resultant product ions. If the m/z values of the resultant product ions match the predicted product ion m/z values of the known precursor ion, then analyte confirmation is achieved. This selection/fragmentation/detection scheme greatly increases method sensitivity and specificity. Also worthy of note here is the use of ion trap mass spectrometers, which use a single physical device, the ion trap, to collect and fragment ions. This type instrument is capable of carrying out multiple rounds of analysis resulting in MS^n experiments, which may be useful to examine posttranslational modifications in biological samples.

Selection reaction monitoring (SRM) is a mass spectrometry acquisition method well suited for analytical situations traditionally employing ELISA analysis. This MS^2 acquisition method is often referred to as signature peptide assay or multiple reaction monitoring (MRM), which is used hereafter. The MRM assay employs a triple quad mass spectrometer to profile a targeted list of m/z values previously determined to be unique (signatory) for a protein of interest. The target list is organized in a pairwise fashion of precursor/product ion m/z values (transition pairs). The first m/z value in a transition pair is specific to a targeted precursor ion, and the second m/z value is specific for a predicted product ion produced upon fragmentation of the targeted precursor ion. The quadrapole mass analyzers work in tandem to continually profile through the transition pair list, only allowing ions capable of both correct precursor and product ion m/z value to reach the detector. A response to the increasing demand for rapid and inexpensive analysis of complex samples is a procedural adaptation using a hybrid mass spectrometer (possessing both triple quadrupole and ion trap functionality) to perform MS^3 acquisition capable of quantitation (Cesari et al., 2010). Using a third round of analysis (selection) could improve signal detection in crude matrix samples and reduce the costs of sample preparation by enabling a more "dilute and shoot" approach.

The ability to quantitate multiple proteins from the same sample in a multiplex format is readily achieved by monitoring transition pairs specific for different proteins of interest. Identification of single amino acid differences and posttranslational modifications can also be determined in this manner (Kitterman et al., 2009; Cui and Thomas, 2009).

The scouting efforts to determine suitable signature peptides require only a few micrograms of target protein standard. Although having a protein standard of very high purity is desired if antibodies are to be raised against it, a known protein sample at reduced purity will serve to identify the signature peptides to be used for MRM analysis. Once the signature peptides are determined, they can be commercially synthesized to known amount and purity in approximately 3–4 weeks. Peptide standards can be labeled with stable isotope (^{13}C and/or ^{15}N) containing amino acids to serve as internal standards for absolute quantitation (Kirkpatrick et al., 2005).

A stably labeled peptide is physiochemically equivalent to the endogenous target peptide in regard to chromatographic elution, ionization energy, and fragmentation ratios. But incorporating a stably labeled amino acid creates a known mass shift relative to the endogenous target peptide, and this mass shift is exploited by the mass spectrometer to filter between the endogenous target peptide and the labeled peptide standard. The signal detected from a known concentration of stably labeled peptide fortified into an unknown sample will directly correlate to the concentration of peptide derived from the target protein in the original sample. The use of synthetic peptides as standards in place of purified whole protein can greatly reduce the time and effort required to begin a targeted analysis.

Since mass spectrometry does not require immunoreactive recognition, concern for maintaining target protein confirmation is not required. The choices of extraction reagents are not restricted to mild conditions, offering increased flexibility in sample handling and protein extraction. However, the ionization process required for mass spectrometry requires the final sample buffer to be volatile. Having the final sample solubilized in an aqueous solvent (water, methanol, acetonitrile) and free of non-volatile agents such as salts (KH_2PO_4, Tris, NaCl) and detergents (Tween, SDS, Triton) can avoid ionization suppression and loss of signal. Compatible volatile buffers frequently used are ammonium acetate and ammonium bicarbonate at concentrations of 20 mM. Post extraction sample preparation begins by reducing and denaturing the protein(s) and then using a protease of known cleavage site specificity to completely digest all sample proteins into peptide fragments. By knowing the cleavage specificity of the protease and the amino acid sequence of a targeted protein, the mass values of the individual target peptide fragments are predictable. In a complete protease digestion, the target peptides exist in equal molar amounts to their respective whole protein and can serve as surrogates for whole protein quantitation.

Matrix effects need to be considered due to other compounds that may be present upon sample extraction and certain control samples are required (Taylor, 2005). A "null" matrix sample void of all independent variables must be tested to account for potential interference from any isobaric ions present in the matrix. A null matrix sample fortified with all independent variables to be measured provides critical parameters including the peptide chromatographic profile prior to samples introduction into the mass spectrometer and the relative fragmentation ratios in the presence of matrix for each targeted peptide. Instrument scanning functions also have the switching ability to measure all ions present (full scan mode) at the particular moment when an MRM transition pair is being detected to further ensure the proper precursor ion was selected. Adjustments to sample preparation protocols may be required if it is determined that the matrix interferes with peptide quantitation.

The application of mass spectrometry to analytical situations traditionally handled via immunoassay has demonstrated great potential and created a growing awareness, helping to accelerate the acceptance of this technology within the immunoassay community at large. As demand for this technology continues to grow, it will encourage more technology development addressing the specific challenges present in basic biological research, as well as clinical laboratories (Vogeser and Seger, 2008; Feng et al., 2008).

REFERENCES

Abdiche, Y., Malashock, D., Pinkerton, A., and Pons, J. Determining kinetics and affinities of protein interactions using a parallel real-time label-free biosensor, the Octet. *Anal. Biochem.*, **2008**, *377*, 209–217.

Albala, J. S. and Humphrey-Smith, I. Array-based proteomics: high-throughput expression and purification of IMAGE consortium cDNA clones. *Curr. Opin. Mol. Ther.*, **1999**, *1*, 680–684.

Alder, M., Wacker, R., and Niemeyer, C. M. Sensitivity by combination: immuno-PCR and related technologies. *Analyst*, **2008**, *133*, 702–718.

Barletta, J. Applications of real-time immuno-polymerase chain reaction (rt-IPCR) for the rapid diagnoses of viral antigens and pathologic proteins. *Mol. Aspects Med.*, **2006**, *27*, 224–253.

Baskerville, S., Zapp, M., and Ellington, A. D. Anti-Rex aptamers as mimics of the Rex-binding element. *J. Virol.* **1999**, *73*, 4962–4971.

Bossi, A., Piletsky, S. A., Piletska, E. V., Righetti, P. G., and Turner, A. P. F. Surface-grafted molecularly imprinted polymers for protein recognition. *Anal. Chem.*, **2001**, *73*, 5281–5286.

Bossi, A., Bonini, F., Turner, A. P. F., and Piletsky, S. A. Molecularly imprinted polymers for the recognition of proteins: the state of the art. *Biosen. Bioelectron.*, **2007**, *22*, 1131–1137.

Cesari, N., Fontana, S., Montanari, D., and Braggio, S. Development and validation of a high-throughput method for the quantitative analysis of D-amphetamine in rat blood using liquid chromatography/MS^3 on a hybrid triple quadrapole-linear ion trap mass spectrometer and its application to a pharmacokinetic study. *J. Chromatogr. B*, **2010**, *878*, 21–28.

Clark, S. L. and Remcho, V. T. Aptamers as analytical reagents. *Electrophoresis*, **2002**, *23*, 1335–1340.

Cui, Z. and Thomas, M. J. Phospholipid profiling by tandem mass spectrometry. *J. Chromatogr. B*, **2009**, *877*, 2709–2715.

Earley, M. C., Vogt, R. F., Shapiro, H. M., Mandy, F. F., Kellar, K. L., Bellisario, R., Pass, K. A., Marti, G. E., Stewart, C. C., and Hannon, W. H. Report from a workshop on multianalyte microsphere arrays. *Cytometry*, **2002**, *50*, 239–242.

Ekins, R. P. Multi-analyte immunoassay. *J. Pharm. Biomed. Anal.*, **1989**, *7*, 155–168.

Ekins, R., Chu, F., and Biggart, E. Multispot, multianalyte, immunoassay. *Ann. Biol. Clin. (Paris)*, **1990**, *48*, 655–666.

Ekins, R. P. and Chu, F. W. Multianalyte microspot immunoassay—microanalytical "compact disk" of the future. *Chin. Chem.*, **1991**, *37*, 1955–1967.

Ellington, A. D. and Szostak, J. W. *In vitro* selection of RNA molecules that bind specific ligands. *Nature*, **1990**, *346*, 818–822.

Elshal, M. F. and McCoy, J. P. Multiplex bead array assays: performance evaluation and comparison of sensitivity to ELISA. *Methods*, **2006**, *38*, 317–323.

Feng, X., Liu, X., Luo, Q., and Liu, B.F. Mass spectrometry in systems biology: an overview. *Mass Spectrom. Rev.*, **2008**, *27*, 635–660.

Fischer, N. O., Tarasow, T. M., and Tok, J. B. Protein detection via direct enzymatic amplification of short DNA aptamers. *Anal. Biochem.*, **2008**, *373*, 121–128.

Ge, Y. and Turner, A. P. F. Too large to fit? Recent developments in macromolecular imprinting. *Trends Biotechnol.*, **2008**, *26*, 218–224.

Guo, T., Xia, Y., Hao, G., Song, M., and Zhang, B. Adsorptive separation of hemoglobin by molecularly imprinted chitosan beads. *Biomaterials*, **2004**, *25*, 5905–5912.

Haab, B. B., Dunham, M. J., and Brown, P. O. Protein microarrays for highly parallel detection and quantitation of specific proteins and antibodies in complex solutions. *Genome Biol.*, **2001**, *2*, RESEARCH0004.

Herr, J. K., Smith, J. E., Medley, C. D., Shangguan, D., and Tan, W. Aptamer-conjugated nanoparticles for selective collection and detection of cancer cells. *Anal. Chem.*, **2006**, *78*, 2918–2924.

Homann, M. and Göringer, H. U. Combinatorial selection of high affinity RNA ligands to live African trypanosomes. *Nucl. Acids Res.*, **1999**, *27*, 2006–2014.

Horan, P. K. and Wheeless, L. L. *Quantitative single cell analysis and sorting*, **1977**, *198*, 149–157.

Hua, Z., Chen, Z., Li, Y., and Zhao, M. Thermosensitive and salt-sensitive molecularly imprinted hydrogel for bovine serum albumin. *Langmuir*, **2008**, *24*, 5773–5780.

Joos, T. O., Schrenk, M., Hopfl, P. L., Kroger, K., Chowdhury, U., Stoll, D., Schorner, D., Durr, M., Herick, K., Rupp, S., Sohn, K., and Hammerle, H. A microarray enzyme-linked immunosorbent assay for autoimmune diagnostics. *Electrophoresis*, **2000**, *21*, 2641–2650.

Kellar, K. L. and Douglass, J. P. Multiplexed microsphere-based flow cytometric immunoassays for human cytokines. *J. Immunol. Methods*, **2003**, *279*, 277–285.

Kirkpatrick, D. S., Gerber, S. A., and Gygi, S. P. The absolute quantification strategy: a general procedure for the quantitation of proteins and post-translational modifications. *Methods*, **2005**, *38*, 265–273.

Kitterman, N. R., Jenkins, R. E., Laney, C. S., Elliot, V. L., and Park, B. K. Multiple reaction monitoring for quantitative biomarker analysis in proteomics and metabolomics. *J. Chromatogr. B*, **2009**, *877*, 1229–1239.

Klug, S. J., Huttenhofer, A., and Famulok, M. *In vitro* selection of RNA aptamers that bind special elongation factor SelB, a protein with multiple RNA-binding sites, reveals one major interaction domain at the carboxyl terminus. *RNA*, **1999**, *5*, 1180–1190.

Kopf, E. and Zharhary, D. Antibody arrays—an emerging tool in cancer proteomics. *Int. J. Biochem. Cell Biol.*, **2007**, *39*, 1305–1317.

Liss, M., Petersen, B., Wolf, H., and Prohaska, E. An aptamer-based quartz crystal protein biosensor. *Anal. Chem.*, **2002**, *74*, 4488–4495.

Liu, B. C., Zhang, L., Lv, L. L., Wang, Y. L., Liu, D. G., and Zhang, X. L. Applications of antibody array technology in the analysis of urinary cytokine profiles in patients with chronic kidney disease. *Am. J. Nephrol.*, **2006**, *26*, 483–490.

MacBeath, G. and Schreiber, S. L. Printing proteins as microarrays for high-throughput function determination. *Science*, **2000**, *289*, 1760–1763.

MacBeath, G. Protein microarrays and proteomics. *Nat. Genet.*, **2002**, *32*, 526–532.

Mallik, S., Johnson, R. D., and Arnold, F. H. Synthetic bis-metal ion receptors for bis-imidazole "protein analogs". *J. Am. Chem. Soc.*, **1994**, *116*, 8902–8911.

Mendoza, L. G., McQuary, P., Mongan, A., Gangadharan, R., Brignac, S., and Eggers, M. High-throughput microarray-based enzyme-linked immunosorbent assay (ELISA). *BioTechniques*, **1999**, *27*, 778.

Michaud, M., Jourdan, E., Ravelet, C., Villet, A., Ravel, A., Grosset, C., and Peyrin, E. Immobilized DNA aptamers as target-specific chiral stationary phases for resolution of nucleoside and amino acid derivative enantiomers. *Anal. Chem.*, **2004**, *76*, 1015–1020.

Minunni, M., Tombelli, S., Gullotto, A., Luzi, E., and Mascini, M. Development of biosensors with aptamers as bio-recognition element: the case of HIV-1 Tat protein. *Biosens. Bioelectron.*, **2004**, *20*, 1149–1156.

Moody, M. D., Van Arsdell, S. W., Murphy, K. P., Orencole, S. F., and Burns, C. Array-based ELISAs for high-throughput analysis of human cytokines. *BioTechniques*, **2001**, *31*, 186.

Ng, E. W., Shima, D. T., Calias, P., Cunningham, E. T. Jr., Guyer, D. R., and Adamis, A. P. Pegaptanib, a targeted anti-VEGF aptamer for ocular vascular disease. *Nat. Rev. Drug Discov.*, **2006**, *5*, 123–132.

Nielsen, U. B. and Geierstanger, B. H. Multiplexed sandwich assays in microarray format. *J. Immunol. Methods*, **2004**, *290*, 107–120.

Ou, S.H., Wu, M. C., Chou, T. C., and Liu, C. C. Polyacrylamide gels with electrostatic functional groups for the molecular imprinting of lysozyme. *Anal. Chim. Acta*, **2004**, *504*, 163–166.

Pang, S., Smith, J., Onley, D., Reeve, J., Walker, M., and Foy, C. A comparability study of the emerging protein array platforms with established ELISA procedure. *J. Immunol. Methods*, **2005**, *302*, 1–12.

Pang, X., Cheng, G., Lu, S., and Tang, E. Synthesis of polyacrylamide gel beads with electrostatic functional groups for the molecular imprinting of bovine serum albumin. *Anal. Bioanal. Chem.*, **2006**, *384*, 225–230.

Sano, T., Smith, C. L., and Cantor, C. R. Immuno-PCR: very sensitive antigen detection by means of specific antibody–DNA conjugates. *Science*, **1992**, *258*, 120–122.

Schweitzer, B. and Kingsmore, S. F. Measuring proteins on microarrays. *Curr. Opin. Biotechnol.*, **2002**, *13*, 14–19.

Taylor, P. J. Matrix effects: the Achilles heel of quantitative high performance liquid chromatography–electrospray-tandem mass spectrometry. *Clin. Biochem.*, **2005**, *38*, 328–334.

Templin, M. F., Stoll, D., Bachmann, J., and Joos, T. O. Protein microarrays and multiplexed sandwich immunoassay: what beats the beads? *Comb. Chem. HTP Screen.*, **2004**, *7*, 223–229.

Tombelli, S., Minunni, M., and Mascini, M. Analytical application of aptamers. *Biosens. Bioelectron.*, **2005**, *20*, 2424–2434.

Tuerk, C. and Gold, L. Systematic evolution of ligands by exponential enrichment: RNA ligands to bacteriophage T4 DNA polymerase. *Science*, **1990**, *249*, 505–510.

Turtinen, L. W., Prall, D. N., Brener, L. A., Nauss, R. E., and Hartsel, S. C. Antibody array-generated profiles of cytokine release from THP-1 leukemic monocytes exposed to different amphotericin B formulations. *Antimicrob. Agents Chemother.* **2004**, *48*, 396–403.

Vogeser, M. and Seger, C. A decade of HPLC–MS/MS in the routine clinical laboratory: goals for further developments. *Clin. Biochem.*, **2008**, *41*, 649–662.

Watson, J. T. and Sparkman, O.D. *Introduction to Mass Spectrometry*, Wiley, West Sussex, England, 2007, pp. 691–692.

Wen, J. D., Gray, C. W., and Gray, D. M. SELEX selection of high affinity oligonucleotides for bacteriophage Ff gene 5 protein. *Biochemistry*, **2001**, *40*, 9300–9310.

INDEX